AF361879

Applications of Information Theory to Epidemiology

Applications of Information Theory to Epidemiology

Editor

Gareth Hughes

MDPI • Basel • Beijing • Wuhan • Barcelona • Belgrade • Manchester • Tokyo • Cluj • Tianjin

Editor
Gareth Hughes
Scotland's Rural College
UK

Editorial Office
MDPI
St. Alban-Anlage 66
4052 Basel, Switzerland

This is a reprint of articles from the Special Issue published online in the open access journal *Entropy* (ISSN 1099-4300) (available at: https://www.mdpi.com/journal/entropy/special_issues/epidemic).

For citation purposes, cite each article independently as indicated on the article page online and as indicated below:

LastName, A.A.; LastName, B.B.; LastName, C.C. Article Title. *Journal Name* **Year**, *Volume Number*, Page Range.

ISBN 978-3-0365-0316-5 (Hbk)
ISBN 978-3-0365-0317-2 (PDF)

Contents

About the Editor

Gareth Hughes is Emeritus Professor of Plant Disease Epidemiology at Scotland's Rural College (SRUC), UK. Before joining SRUC in 2010, he had held faculty positions at the University of the West Indies from 1977 to 1981 and the University of Edinburgh from 1981 to 2010. In 2000 he received the Lee M. Hutchins Award of the American Phytopathological Society and in 2003 he was made a Fellow of the Institute of Mathematics and its Applications. Professor Hughes' work includes *The Study of Plant Disease Epidemics* (co-authored with Laurence Madden and Frank van den Bosch, 2007) and *Applications of Information Theory to Epidemiology* (2012), both books published by APS Press. Analysis of the crop protection decision-making problem is at the center of Professor Hughes' research interests.

Preface to "Applications of Information Theory to Epidemiology"

Applications of Information Theory to Epidemiology collects together a new review article written by William Benish with ten original research articles covering aspects of the analysis of diagnostic decision making and epidemic dynamics. Overall, there is a balance of theory and applications, presented from both clinical medicine and plant pathology perspectives. Previously, epidemiological applications of information theory have tended to be widely scattered through the literature, featured in specialist medical, phytopathological and statistical journals, for example. While this diversity will no doubt continue, the current collection now provides a focal point from which new developments can in future emerge and ramify.

Gareth Hughes
Editor

Editorial

Applications of Information Theory to Epidemiology

Gareth Hughes

SRUC, Scotland's Rural College, The King's Buildings, Edinburgh EH9 3JG, UK; gareth.hughes@sruc.ac.uk

Received: 23 November 2020; Accepted: 4 December 2020; Published: 9 December 2020

This Special Issue of *Entropy* represents the first wide-ranging overview of epidemiological applications since the 2012 publication of *Applications of Information Theory to Epidemiology* [1]. The Special Issue comprises an outstanding review article by William Benish [2], together with 10 research papers, five of which have been contributed by authors whose primary interests are in phytopathological epidemiology, and five by authors primarily interested in clinical epidemiology. Ideally, all readers will study Benish's review—it is just as relevant for phytopathologists as it is for clinicians—and then clinicians and phytopathologists will take advantage of the opportunity to read about each other's current approaches to epidemiological applications of information theory.

This opportunity arises especially where there turns out to be an overlap of interests between the two main groups of contributors. For example, Benish's review provides detailed insight into the analysis of diagnostic information via pre-test probabilities and the corresponding post-test probabilities (predictive values). This theme is then pursued further by means of the predictive receiver operating characteristic (PROC) curve, a graphical plot of positive predictive value (PPV) against one minus negative predictive value (1−NPV) [3–5]. Although this format recalls the familiar receiver operating characteristic (ROC) curve, the dependence of the PROC curve on pre-test probability has made it more difficult to characterize and deploy. The articles presented here contribute to an improved understanding of the way that ROC and PROC curves can jointly contribute to the analysis of diagnostic information. An alternative approach to the diagrammatic analysis of diagnostic information via pre-test and post-test probabilities is presented in [6] and then taken up for practical application in [7].

Four articles in the Special Issue apply information-theoretic methods to analyze various aspects of epidemic dynamics [8–11]. Here, the balance is tipped towards contributions from clinical epidemiology, but information-theoretic applications of time series analysis are presented from both clinical and phytopathological perspectives. Epidemic analyses of observational studies of course depend on the availability of appropriate sample data. In this context, Dalton et al. [12] address the limitations of statistics used to assess balance in observational samples and present an application of the Jensen–Shannon divergence to quantify lack of balance.

Together, the authors whose contributions are presented in this Special Issue have provided a range of novel information-theoretic applications of interest to epidemiologists and diagnosticians in both medicine and plant pathology. While these articles represent the current state of the art, this Special Issue represents only a beginning in terms of what is possible.

Acknowledgments: On behalf of the authors whose work is presented in this Special Issue of the journal *Entropy*, I should like to thank all the anonymous peer-reviewers who have read and critiqued the submissions. As Academic Editor, I offer my personal thanks to all the MDPI editorial staff who have worked behind the scenes to make the Special Issue a success.

Conflicts of Interest: The author declares no conflict of interest.

References

1. Hughes, G. *Applications of Information Theory to Epidemiology*; APS Press: St. Paul, MN, USA, 2012.
2. Benish, W.A. A review of the application of information theory to clinical diagnostic testing. *Entropy* **2020**, *22*, 97. [CrossRef] [PubMed]
3. Hughes, G. On the binormal predictive receiver operating characteristic curve for the joint assessment of positive and negative predictive values. *Entropy* **2020**, *22*, 593. [CrossRef] [PubMed]
4. Oehr, P.; Ecke, T. Establishment and characterization of an empirical biomarker SS/PV-ROC plot using results of the UBC® *Rapid* Test in bladder cancer. *Entropy* **2020**, *22*, 729. [CrossRef] [PubMed]
5. Hughes, G.; Kopetzky, J.; McRoberts, N. Mutual information as a performance measure for binary predictors characterized by both ROC curve and PROC curve analysis. *Entropy* **2020**, *22*, 938. [CrossRef] [PubMed]
6. Hughes, G.; Reed, J.; McRoberts, N. Information graphs incorporating predictive values of disease forecasts. *Entropy* **2020**, *22*, 361. [CrossRef] [PubMed]
7. Gottwald, T.; Poole, G.; Taylor, E.; Luo, W.; Posny, D.; Adkins, S.; Schneider, W.; McRoberts, N. Canine olfactory detection of a non-systemic phytobacterial citrus pathogen of international quarantine significance. *Entropy* **2020**, *22*, 1269. [CrossRef] [PubMed]
8. Muhammad Altaf, K.; Atangana, A. Dynamics of Ebola disease in the framework of different fractional derivatives. *Entropy* **2019**, *21*, 303. [CrossRef] [PubMed]
9. De la Sen, M.; Nistal, R.; Ibeas, A.; Garrido, A.J. On the use of entropy issues to evaluate and control the transients in some epidemic models. *Entropy* **2020**, *22*, 534. [CrossRef] [PubMed]
10. Sun, S.; Li, Z.; Zhang, H.; Jiang, H.; Hu, X. Analysis of HIV/AIDS epidemic and socioeconomic factors in Sub-Saharan Africa. *Entropy* **2020**, *22*, 1230. [CrossRef] [PubMed]
11. Choudhury, R.A.; McRoberts, N. Characterization of Pathogen Airborne Inoculum Density by Information Theoretic Analysis of Spore Trap Time Series Data. *Entropy* **2020**, *22*, 1343. [CrossRef] [PubMed]
12. Dalton, J.E.; Benish, W.A.; Krieger, N.I. An information-theoretic measure for balance assessment in comparative clinical studies. *Entropy* **2020**, *22*, 218. [CrossRef] [PubMed]

Publisher's Note: MDPI stays neutral with regard to jurisdictional claims in published maps and institutional affiliations.

Review

A Review of the Application of Information Theory to Clinical Diagnostic Testing

William A. Benish

Department of Internal Medicine, Case Western Reserve University, Cleveland, OH 44106, USA; wab4@cwru.edu

Received: 12 October 2019; Accepted: 9 January 2020; Published: 14 January 2020

Abstract: The fundamental information theory functions of entropy, relative entropy, and mutual information are directly applicable to clinical diagnostic testing. This is a consequence of the fact that an individual's disease state and diagnostic test result are random variables. In this paper, we review the application of information theory to the quantification of diagnostic uncertainty, diagnostic information, and diagnostic test performance. An advantage of information theory functions over more established test performance measures is that they can be used when multiple disease states are under consideration as well as when the diagnostic test can yield multiple or continuous results. Since more than one diagnostic test is often required to help determine a patient's disease state, we also discuss the application of the theory to situations in which more than one diagnostic test is used. The total diagnostic information provided by two or more tests can be partitioned into meaningful components.

Keywords: entropy; information theory; multiple diagnostic tests; mutual information; relative entropy

---

1. Introduction

Information theory was developed during the first half of the twentieth century to quantify aspects of communication. The pioneering work of Ralph Hartley and, subsequently, Claude Shannon was primarily motivated by problems associated with electronic communication systems [1,2]. Information theory was probably first used to quantify clinical diagnostic information by Good and Card in 1971 [3]. Subsequent papers helped to clarify the ability of information theory to quantify diagnostic uncertainty, diagnostic information, and diagnostic test performance, e.g., [4–9]. Although applications of information theory can be highly technical, fundamental concepts of information theory are not difficult to understand. Moreover, they are profound in the sense that they apply to situations in which "communication" is broadly defined.

Fundamental information theory functions are defined on random variables. The ubiquity of random processes accounts for the wide range of applications of the theory. Examples of areas of application include meteorology [10], molecular biology [11], quantum mechanics [12], psychology [13], plant pathology [14], and music [15]. The random variables of interest to the present discussion are an individual's disease state (D) and diagnostic test result (R). We require that the possible disease states be mutually exclusive and that, for each diagnostic test performed, one result is obtained. Hence, it is meaningful to talk about the probability that an individual randomly selected from a population is in a certain disease state and has a certain test result.

The primary purpose of this review is to understand the answers that information theory gives to the following three questions:

(1)   How do we quantify our uncertainty about the disease state of a given individual?

(2)   After a diagnostic test is performed and a specific test result is obtained, how do we quantify the information we have received about the tested individual's disease state?

(3)   Prior to performing a diagnostic test, how do we quantify the amount of information that we expect to receive about the disease state of the tested individual?

The answers that information theory gives to these questions are calculated using pretest and posttest probabilities. Whenever the pre-test and post-test probabilities differ, the test has provided diagnostic information [16]. The functions are applicable to situations in which any number of disease states are under consideration and in which the diagnostic test can yield any number of results (or continuous results) [17]. Moreover, a given test result can alter the probabilities of multiple possible disease states.

Since information theory functions depend only upon the probabilities of states, the information content of an observation does not take into consideration the meaning or value of the states [18] (p. 8). For example, the statement that a patient died who had been given a 50-50 chance of survival contains the same amount of information, from an information theory perspective, as the statement that a tossed coin turned up heads.

More than one diagnostic test is often required to help clarify a patient's disease state. Hence, an additional goal of this review is to answer questions 2 and 3, above, for the case in which two or more diagnostic tests are performed. We find that it is possible to quantify both the information that we have received from each of two or more diagnostic tests as well as the information that we expect to receive by performing two or more diagnostic tests.

The foundational theorem of information theory is the statement proved by Shannon that the entropy function, discussed below, is the only function that satisfies certain criteria that we require of a measure of the uncertainty about the outcome of a random variable [2]. As an alternative to this axiomatic approach to deriving information theory functions, we employ the concept of the surprisal, with the goal of achieving a more intuitive understanding of these functions. The surprisal function is explained in the following section. It is then used in Section 3 to answer the above three questions and, in doing so, derive expressions for three fundamental information theory functions: the entropy function (Section 3.1), the relative entropy function (Section 3.2), and the mutual information function (Section 3.3). The application of information theory functions to situations in which more than one diagnostic test is performed is considered in Section 4. Section 5 provides a brief review of the history of the application of information theory to clinical diagnostic testing. Examples which offer insight into what information theory can teach us about clinical diagnostic testing are presented in Section 6. The paper concludes by briefly summarizing and clarifying important concepts.

2. The Surprisal Function

The surprisal function, μ, quantifies the unlikelihood of an event [19,20]. It is a function of the probability (p) of the event. As its name suggests, it can be thought of as a measure of the amount we are surprised when an event occurs. Hence, this function assigns larger values to less likely events. Another reasonable requirement of the surprisal function is that, for independent events a_1 and a_2, the surprisal associated with the occurrence of both events should equal the sum of the surprisals associated with each event. Since a_1 and a_2 are independent, $p(a_1,a_2) = p(a_1)p(a_2)$. We therefore require that $\mu[p(a_1)p(a_2)] = \mu[p(a_1)] + \mu[p(a_2)]$. The only non-negative function that meets these requirements is of the form:

$$\mu(p) = -\log(p) \tag{1}$$

Ref. [21] (pp. 2–5). The choice of the base of the logarithm is arbitrary in the sense that conversion from one base to another is accomplished by multiplication by a constant. Two is often selected as the base of the logarithm, giving measurements in units of bits (binary digits). Some authors use the natural logarithm (giving measurements in units of nats) or log base 10 (giving measurements in units of hartleys) [22]. Using log base two, the surprise when a fair coin turns up heads is quantified as one bit, since $-\log_2(1/2) = 1$. Figure 1 plots the surprisal function (in units of bits) over the range of probabilities. Observe that the surprisal associated with the occurrence of an event that is certain to occur is zero, and that there is no number large enough to quantify the surprise associated with the occurrence of an impossible event.

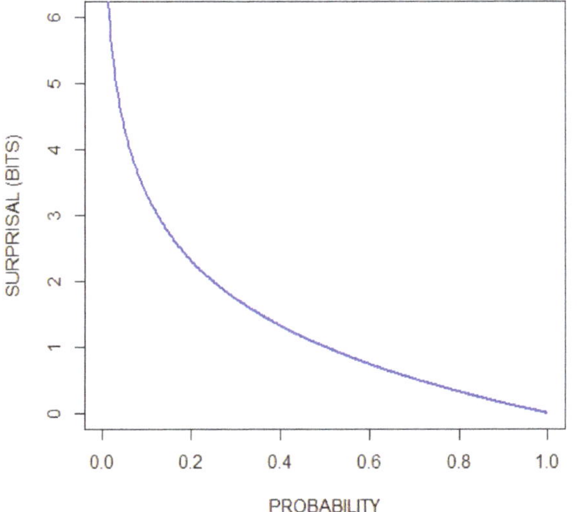

Figure 1. Surprisal (in bits) as a function of probability.

3. Answers to the Questions Asked in the Introduction

3.1. Entropy Quantifies the Uncertainty about the Disease State

Suppose that the possible causes of a patient's condition consist of four disease states, $d_1, \ldots, d_4$, with respective probabilities 1/8, 1/2, 1/8, and 1/4. How uncertain are we about the disease state? The more certain we are about the disease state the less surprised we will be, on average, when the disease state becomes known. This suggests that diagnostic uncertainty be quantified as the expected value of the surprisal. For the current example, the surprisals corresponding to the four probabilities are 3 bits, 1 bit, 3 bits, and 2 bits, respectively. To calculate the expected value of the surprisal we multiply each surprisal by its probability and then sum the four terms:

$$\left(\frac{1}{8}\right)(3 \text{ bits}) + \left(\frac{1}{2}\right)(1 \text{ bit}) + \left(\frac{1}{8}\right)(3 \text{ bits}) + \left(\frac{1}{4}\right)(2 \text{ bits}) = 1.75 \text{ bits}.$$

This procedure yields Shannon's entropy (H) of D, where D is the random variable associated with the four disease states. For the general case in which there are n possible disease states [2,23]:

$$H(D) = -\sum_{i=1}^{n} p(d_i) \log_2 p(d_i). \tag{2}$$

We saw above that the surprisal associated with a tossed coin turning up heads is 1 bit. Consequently, the uncertainty associated with the two possible outcomes of a coin toss is $(1/2)(1 \text{ bit}) + (1/2)(1 \text{ bit}) = 1$ bit. The uncertainty about the outcome of equally likely events increases as the number of possible events increases; for example, the uncertainty associated with three, four, and five equally likely events is 1.59 bits, 2 bits, and 2.32 bits, respectively.

Another way to think about the meaning of entropy is in terms of the average number of yes/no questions required to learn the outcome of the random variable. This works for cases like the current example, in which, before asking each question the remaining events can be partitioned into two groups of equal probability. For the current example, we first ask if the individual is in state d_2, and then, if necessary, ask if the individual is in state d_4, and finally, if necessary, ask if the individual is in state d_1 (or state d_3). We find that, on average, we will ask 1.75 questions.

In Shannon's axiomatic approach to the definition of the entropy function, a key requirement relates to the way in which an entropy calculation can be partitioned [18] (p. 49). As applied to the current problem, Shannon required, for example, that

$$H\left(\frac{1}{8}, \frac{1}{2}, \frac{1}{8}, \frac{1}{4}\right) = H\left(\frac{1}{8}, \frac{7}{8}\right) + \frac{7}{8}H\left(\frac{4}{7}, \frac{1}{7}, \frac{2}{7}\right).$$

This corresponds to first determining the entropy associated with whether the individual is in state d_1 and, if not, determining the entropy of the remaining three options. This latter entropy is weighted by $7/8$, the probability that the individual is not in state d_1.

Some authors refer to entropy as self-information [23] (p. 12). In this review, we restrict the use of the term information (diagnostic information) to measures of the magnitude of changes in the probabilities of states (disease states) that result from observations (diagnostic test results).

3.2. Relative Entropy Quantifies the Diagnostic Information Provided by a Specific Test Result

Table 1 presents hypothetical data showing characteristics of a population of 96 individuals, each of whom is in one of four disease states and who, when tested, will yield one of three possible results. The probabilities that an individual randomly selected from this population will be in the four disease states is identical to the probabilities in the above example: $1/8$, $1/2$, $1/8$, and $1/4$, respectively. If the diagnostic test is performed and result r_3 is obtained, the respective probabilities become $1/8$, $1/4$, $1/2$, and $1/8$. Because the post-test probabilities are the same as the pretest probabilities, even though the order has changed, the uncertainty about the disease state remains 1.75 bits. Has this test provided us with diagnostic information and, if so, how much?

Table 1. Hypothetical data showing the number of individuals in a given disease state (d_1, d_2, d_3, or d_4) and with a given test result (r_1, r_2, or r_3).

	d_1	d_2	d_3	d_4	
r_1	8	24	4	2	38
r_2	2	20	0	20	42
r_3	2	4	8	2	16
	12	48	12	24	96

The test result, r_3 identifies the patient as belonging to a subset within the larger population. It provides us with diagnostic information because the probabilities of the disease states are different within this subset than they are within the larger population. We quantify diagnostic information as the expected value of the reduction in the surprisal that results from testing. To calculate the amount of information obtained from this test result, we first note that the probabilities change from

$$\left[\frac{1}{8}, \frac{1}{2}, \frac{1}{8}, \frac{1}{4}\right] \text{ to } \left[\frac{1}{8}, \frac{1}{4}, \frac{1}{2}, \frac{1}{8}\right],$$

respectively; the surprisals (in units of bits) change from

$$[3, 1, 3, 2] \text{ to } [3, 2, 1, 3],$$

respectively; and the reductions in the surprisals (in units of bits) are

$$[0, -1,\ 2,\ -1],$$

respectively. To calculate the expected value of the reduction in the surprisal, we use the updated probabilities obtained by testing:

$$\left(\frac{1}{8}\right)(0 \text{ bits}) + \left(\frac{1}{4}\right)(-1 \text{ bit}) + \left(\frac{1}{2}\right)(2 \text{ bits}) + \left(\frac{1}{8}\right)(-1 \text{ bit}) = \frac{5}{8}\text{bits}.$$

Hence, test result r_3 provides 5/8 bits of information about the disease state.

For the general case with pretest probabilities: $p(d_1)$, $p(d_2),\ldots,p(d_n)$ and posttest probabilities after receiving result r_j: $p(d_1|r_j)$, $p(d_2|r_j)$, $\ldots,p(d_n|r_j)$, the reduction in the surprisal for the i-th disease state is

$$\left[-\log_2 p(d_i)\right] - \left[-\log_2 p(d_i|r_j)\right] = \log_2 \frac{p(d_i|r_j)}{p(d_i)},$$

with the expected value calculated in terms of the post-test distribution giving

$$D(post\|pre) = \sum_{i=1}^{n} p\left(d_i|r_j\right)\log_2 \frac{p\left(d_i|r_j\right)}{p(d_i)}. \tag{3}$$

$D(post\|pre)$ is called the relative entropy (or the Kullback-Leibler divergence) from pre (the pretest probability distribution) to post (the posttest probability distribution) [23,24]. Its value is always nonnegative [23]. Relative entropy is sometimes thought of as a measure of distance from one probability distribution (pre) to another probability distribution (post). Since it is an asymmetric function, i.e., $D(post\|pre)$ and $D(pre\|post)$ are not necessarily equal, and because it does not satisfy the triangle inequality, it does not qualify as a true distance metric [23] (p.18). As illustrated by the above example, the expected value of the reduction in the surprisal (5/8 bits) is different than the reduction in the expected values of the surprisal (0 bits), i.e., the diagnostic information, in this case, is not simply pretest entropy minus posttest entropy.

3.3. Mutual Information Quantifies the Diagnostic Information That We Expect to Receive by Testing

Using the same data set (Table 1) we consider the question of how much information we expect to receive if we randomly select and test an individual from this population. Hence, the question we are now asking is from the pretest perspective, in contrast to the posttest perspective of the preceding subsection. Once again, we quantify diagnostic information as the expected value of the reduction in the surprisal that results from testing. We found above that if the test result is r_3, then we obtain 5/8 = 0.625 bits of information. Using the relative entropy function (Equation (3)), we can also calculate that r_1 provides 0.227 bits of information and r_2 provides 0.343 bits of information. The probabilities of obtaining each of the three possible test results are 0.396, 0.438, and 0.167, respectively. Therefore, the amount of diagnostic information, on average, that we will receive by performing this test is

$$(0.396)(0.227 \text{ bits}) + (0.438)(0.343 \text{ bits}) + (0.167)(0.625 \text{ bits}) = 0.345 \text{ bits}.$$

The expected value of the amount of diagnostic information to be obtained by testing is the expected value of the relative entropy. For the general case, this is

$$I(D; R) = \sum_{j=1}^{m} p\left(r_j\right) \sum_{i=1}^{n} p(d_i|r_j)\log_2 \frac{p(d_i|r_j)}{p(d_i)} = \sum_{i=1}^{n} \sum_{j=1}^{m} p\left(d_i, r_j\right)\log_2 \frac{p\left(d_i, r_j\right)}{p(d_i)p\left(r_j\right)}, \tag{4}$$

where $p(d_i)$ is the probability that a patient randomly selected from the population is in disease state d_i, $p(r_j)$ is the probability that a patient randomly selected from the population has test result r_j, and $p(d_i, r_j)$ is the probability that a patient randomly selected from the population is both in disease state d_i and has test result r_j. $I(D; R)$ is known as the mutual information between the disease state and the test result [23]. It is called the mutual information between D and R because knowing value of D provides the same information about the value of R, on average, as knowing the value of R provides about the value of D, on average, i.e., $I(D; R) = I(R; D)$.

Established consequences of the definitions of entropy and mutual information are that, for random variables X and Y,

$$I(X; Y) = H(X) + H(Y) - H(X, Y), \tag{5}$$

and

$$H(X|Y) = H(X, Y) - H(Y), \tag{6}$$

where $H(X, Y)$ is the entropy of the random variable defined by the joint occurrence of the events defining X and Y and $H(X|Y)$ is the entropy of the random variable defined by the events defining X conditional upon the events defining Y [23].

A consequence of Equations (5) and (6) is:

$$H(D|R) = H(D) - I(D; R), \tag{7}$$

i.e., performing a diagnostic test decreases the uncertainty about the disease state, on average, by the mutual information between D and R. Recall that, for the current example, $H(D) = 1.75$ bits and $I(D; R) = 0.345$ bits. Hence, the remaining uncertainty after performing this test is, on average, 1.405 bits. A perfect test would provide 1.75 bits of information.

In the preceding subsection we noted that relative entropy is not generally equal to pretest entropy minus posttest entropy. Here, however, where we are calculating the expected value of the amount of information that a test will provide, it is equal to pretest entropy minus posttest entropy: rearranging Equation (7) gives $I(D; R) = H(D) - H(D|R)$.

The mutual information provided by a diagnostic test is a single parameter measure of the performance of the test. It is dependent upon the pretest probabilities of disease. What is known as the channel capacity is the maximum possible value of the mutual information across all possible distributions of pretest probabilities [23].

4. Quantifying the Diagnostic Information Provided by Two or More Tests

More than one diagnostic test is often required to characterize a patient's disease state. In this section we extend the theory to situations in which more than one diagnostic test is performed.

4.1. Relative Entropy Applied to the Case of Multiple Diagnostic Tests

Let $p_0(d_i)$ be the pretest probability of the i-th disease state. Let $p_a(d_i)$ be the probability of the i-th disease state after performing test A and obtaining result r_a. Let $p_b(d_i)$ be the probability of the i-th disease state after performing test B and obtaining result r_b. Finally, let $p_{ab}(d_i)$ be the probability of the i-th disease state after performing both tests A and B and obtaining results r_a and r_b. The amount of information provided by test A for the subgroup of patients with result r_a, as we saw in Section 3.2, is $D(p_a \| p_0)$. Similarly, the amount of information provided by test B for the subgroup of patients with result r_b is $D(p_b \| p_0)$, and the amount of information provided by both tests for the subgroup of patients with results r_a and r_b is $D(p_{ab} \| p_0)$.

Now consider a patient belonging to the subset of patients with both result r_a and result r_b. How much diagnostic information is obtained if only test A is performed? The reduction in the surprisal for the i-th disease state is

$$\left[-\log_2 p_0(d_i)\right] - \left[-\log_2 p_a(d_i)\right] = \log_2 \frac{p_a(d_i)}{p_0(d_i)}.$$

To quantify the diagnostic information, we calculate the expected value of the reduction in the surprisal. Since the patient belongs to the subset of patients with results r_a and r_b the expectation is calculated using the $p_{ab}(d)$ distribution. This gives

$$\sum_i p_{ab}(d_i) \log_2 \frac{p_a(d_i)}{p_0(d_i)}. \tag{8}$$

We will call this the modified relative entropy (I. J. Good called this trientropy [25]). We can think of it as the distance from the $p_0(d)$ probability distribution to the $p_a(d)$ probability distribution when the true probability distribution is $p_{ab}(d)$. Expression (8) can yield negative diagnostic information values. This occurs when the pretest probability distribution is a better estimate of the true probability distribution than the posttest probability distribution. In Appendix A, we show that the modified relative entropy satisfies the triangle inequality but still fails to meet the criteria for a distance metric.

As an example of the application of Expression (8) to a case in which two diagnostic tests are performed, consider a situation in which a person is being evaluated for possible cancer. Assume two disease states, cancer and not cancer, and that a screening test increases the probability of this individual having cancer from 0.05 to 0.3, but that a subsequent, more definitive test, decreases the probability of cancer to 0.01. We can imagine that this person belongs to a theoretical population (A) in which 5% of its members have cancer. Screening identifies this patient as belonging to a subset (B) of A in which 30% of its members have cancer. Finally, the second test identifies the patient as belonging to a subset (C) of B in which 1% of its members have cancer. Using Expression (8) we calculate that the screening test provided -0.410 bits of information (*from* a probability of cancer of 0.05 *to* a probability of cancer of 0.3 given that the probability of cancer is actually 0.01) and that the second test provided 0.446 bits of information (*from* a probability of cancer of 0.3 *to* a probability of cancer of 0.01 given that the probability of cancer is actually 0.01). The two tests together provided 0.036 bits of information. We obtain this final value either by summing the information provided by each of the two tests or by calculating the relative entropy (Equation (3)) given the pretest probability of cancer of 0.05 and the posttest probability of cancer of 0.01. Although the screening test shifted the probability of cancer in the wrong direction for this specific individual, there is no reason to conclude that the result of the screening test was a mistake. The screening test did its job by properly identifying the individual as a member of subset B.

4.2. Mutual Information Applied to the Case of Multiple Diagnostic Tests

The mutual information common to random variables X, Y and Z is defined as

$$I(X;Y;Z) = I(X;Y) - I((X;Y)|Z) \tag{9}$$

where $I((X;Y)|Z) = I((X|Z);(Y|Z))$ is the mutual information between and Y conditional upon Z [23] (p. 45). Hence, from Equations (5), (6), and (9):

$$I(X;Y;Z) = H(X) + H(Y) + H(Z) - H(X,Y) - H(X,Z) - H(Y,Z) + H(X,Y,Z). \tag{10}$$

Although the mutual information between two random variables is always nonnegative, the mutual information among three random variables can be positive, negative, or zero [23] (p. 45).

The expected value of the amount of information that two diagnostic tests, A and B, will provide about the disease state is $I(D; (R_A, R_B))$. This can be expressed in terms of entropies (per Equation (5)) as

$$I(D; (R_A, R_B)) = H(D) + H(R_A, R_B) - H(D, R_A, R_B), \tag{11}$$

and it can be partitioned:

$$I(D; (R_A, R_B)) = I(D; R_A) + I(D; R_B) - I(D; R_A; R_B). \tag{12}$$

Equation (12) can be proved by using Equations (5), (10), and (11) to replace the four mutual information terms with their entropy equivalents. Hence, the expected value of the information that tests A and B provide about disease state D is equal to the sum of the expected values of the information provided by each test minus $I(D; R_A; R_B)$, a term that quantifies the interaction among D, R_A, and R_B. Since $I(D; R_A; R_B)$ can be positive, negative, or zero, $I(D; (R_A, R_B))$ can be less than, greater than, or equal to the sum of $I(D; (R_A))$ and $I(D; (R_B))$, respectively.

Alternatively, we can use Equations (9) and (12) to partition $I(D; (R_A, R_B))$ as follows:

$$I(D; (R_A, R_B)) = I(D; R_A) + I((D; R_B)|R_A), \tag{13}$$

where $I((D; R_B)|R_A) = I(D|R_A; R_B|R_A)$ is the average incremental information provided by test B after performing test A. $I((D; R_B)|R_A)$ can be expressed in terms of entropies using Equations (5) and (6):

$$I((D; R_B)|R_A) = H(D, R_A) + H(R_A, R_B) - H(R_A) - H(D, R_A, R_B). \tag{14}$$

Although the expressions become more complicated as the number of diagnostic tests increase, the mutual information between the disease state and the results of multiple diagnostic tests can be partitioned in fashions analogous to Equations (12) and (13). For the case in which there are three diagnostic tests:

$$\begin{aligned}
I(D; (R_A, R_B, R_C)) &= I(D; R_A) + I(D; R_B) + I(D; R_C) - I(D; R_A; R_B) - I(D; R_A; R_C) \\
&\quad - I(D; R_B; R_C) + I(D; R_A; R_B; R_C) \\
&= I(D; R_A) + I((D; R_B)|R_A) + I((D; R_C)|(R_A, R_B)).
\end{aligned} \tag{15}$$

These equations can be proven, once again, by replacing the mutual information terms with their entropy equivalents, recognizing that

$$I(D; R_A; R_B; R_C) = I(D; R_A; R_B) - I((D; R_A; R_B)|R_C) \tag{16}$$

The entropy function, expressed as Equation (2), is not defined for continuous random variables. Nevertheless, the mutual information between or among continuous random variables, which is defined, can be approximated numerically as the sum and differences of entropies using Equations (5), (10), (11), and (14) [23] (pp. 231–232).

5. Historical Background

With this understanding of basic information theory functions, we can briefly consider the development of information theory and the evolution of its application to a clinical diagnostic testing. The concept of entropy is probably most familiar within the context of thermodynamics, where it is a measure of the "degree of randomness" of a physical system [18] (p. 12). Although an understanding of the basic principles of thermodynamics preceded the development of information theory, the entropy of thermodynamics can be understood to be an application of the concept of entropy stated by Equation (2). The difference between the two functions is that in thermodynamics Equation (2) is multiplied by the Boltzmann constant to provide the appropriate physical dimensions (joules per kelvin) [26] (p. 30).

As mentioned in Section 1, Hartley and Shannon were early developers of information theory. Hartley published a paper in 1928 concerning the relationship between the quantity of information transmitted over a system and the width of the frequency range of the transmission [1]. He defined entropy (which he called information) for situations in which the possible states are equally likely. The more general concept of entropy, stated by Equation (2), was defined in 1948 by Shannon in "A Mathematical Theory of Communication" [2]. This foundational paper also defined mutual information and channel capacity. Relative entropy was introduced by Kullback and Leibler in 1951 [24].

The applicability of information theory to a clinical diagnostic testing was not immediately recognized, has been slow in its development, and remains an area of research. As noted in the introduction, Good and Card probably published the first paper on the subject in 1971 [3]. Their contribution was not recognized by many subsequent authors interested in this subject. To a large extent, the history of the application of information theory to clinical diagnostic testing is the history of the discovery of concepts previously understood by Good and Card. They recognized that mutual information (what they called mean information transfer) quantifies the expected value of the amount of information provided by a test and that this function can be used regardless of the number of disease states and test results. Implicit in their report is the use of relative entropy and, what we have called, modified relative entropy (in their language, dinegentropy and trientropy, respectively) to quantify the information provided by specific test results. They also quantified the information gained by sequential testing [3,27].

The "weight of evidence in favor of a hypothesis" is a central concept in the Good and Card paper [3]. The concept was developed independently by C.S. Peirce [28] and A.M. Turing (possibly in collaboration with I.J. Good) [29]. The weight of evidence in favor of disease state d_i given result r_j, as opposed to the other disease states, $\bar{d}_i$ can be expressed as

$$\log\left[\frac{p(r_j|d_i)}{p(r_j|\bar{d}_i)}\right].$$

This is equal to

$$\log\frac{p(d_i|r_j)}{p(d_i)} - \log\frac{p(\bar{d}_i|r_j)}{p(\bar{d}_i)}$$
$$= [-\log p(d_i) - -\log p(d_i|r_j)] - [-\log p(\bar{d}_i) - -\log p(\bar{d}_i|r_j)].$$

As pointed out by Good and Card, we find by looking at each of the above two expressions in brackets (which are reductions in surprisals) that weight of evidence can be interpreted in terms of quantities of information; in this case, as the amount of information that r_j provides about d_i minus the amount of information that r_j provides about $\bar{d}_i$. A second important observation about the weight of evidence is that it is equal to the logarithm of a likelihood ratio. This point has been used to advantage by Van den Ende et al. to provide clinicians with an accessible approach to interpreting diagnostic tests, including the fact that the logarithm of the pretest odds plus the weight of evidence equals the logarithm of the posttest odds [30]. Since weight of evidence can be interpreted in terms of quantities of information, the logarithm of a likelihood ratio is an information quantity and so has information units. When working in log base 10 (as in [30]) the appropriate unit is the hartley. To convert from hartleys to bits, divide by $\log_{10} 2 = 0.301$.

Most papers on the application of information theory to clinical diagnostic testing are founded upon a report published by Metz, Goodenough, and Rossmann in 1973 [4]. They derived the expression for the information content (mutual information) of a diagnostic test as a function of the pretest probability of disease and the test's true positive rate (probability of a positive result given disease) and false positive rate (probability of a positive result given no disease), i.e., they used these parameters to calculate the posttest probability distribution and then used the pretest and posttest distributions to calculate mutual information. They applied the theory to the evaluation of radiographic systems and noted that this statistic can be used to compare points on the same or different receiver operating characteristic (ROC) curves (defined below in Section 6.1) [31]. The area under the ROC curve (AUC) is a popular measure of diagnostic test performance [32]. Relationships between the AUC and mutual information are discussed in the example presented in Section 6.1. Metz et al. also suggested that the performance of a diagnostic test be quantified as the maximum of the set of information contents associated with the points on a test's ROC curve (I_{max}). Subsequent authors suggested that I_{max} can be used in the selection of the point that partitions test results into normal results and abnormal results [5,7,33], i.e., the diagnostic cutoff. The use of a diagnostic cutoff, however, can result in some loss of diagnostic information [6,34,35]. This is illustrated by examples presented in Sections 6.2 and 6.3.

Diamond and colleagues applied information theory in 1981 to the quantification of the performance of the exercise electrocardiogram (ECG) in the diagnosis of coronary heart disease (CHD) [6]. This paper is discussed in Section 6.2. The primary theoretical contribution of their paper is the recognition that it is not necessary to select a single diagnostic cutoff in order to calculate the information content (mutual information) provided by a diagnostic test. This concept is implicit in the work of Good and Card [3].

The relative entropy function was applied to clinical diagnostic testing in 1999 by Lee [36] and, independently, by Benish [8]. Lee used the relative entropy between the distributions of test results for diseased subjects and disease-free subjects to characterize the potential of a diagnostic test to rule in (confirm) and rule out (exclude) disease. A different approach to characterizing the potential of a diagnostic test to rule in or rule out disease states is illustrated by examples presented in Sections 6.2 and 6.4. Benish recognized that the relative entropy function allows for calculation of the information provided by a specific test result. Once again, this observation is implicit in the paper by Good and Card [3]. Use of the relative entropy function for this purpose is discussed above in Sections 3.2 and 4.1 and is demonstrated in Sections 6.2, 6.4 and 6.5. Benish also discussed the channel capacity of a medical diagnostic test [37]. Hughes, writing from the perspective of a plant disease epidemiologist, published the only book on the application of information theory to diagnostic testing in 2012 [9].

Section 4, above, develops concepts found in the work by Good and Card regarding the quantification of information provided by multiple diagnostic tests [3,27]. These functions are illustrated in the examples presented in Sections 6.4 and 6.5.

6. Examples

R code for the calculations and figures in the following examples are available in the Supplementary Materials.

6.1. The Relationship between I(D;R) and the AUC

ROC curves are often used to describe the performance of a diagnostic test when the test results lie on a continuum or are otherwise ordered [31]. This methodology is applicable when two disease states are under consideration, e.g., disease present and disease absent. A ROC curve plots the tradeoff between the true positive rate (test sensitivity) and the false positive rate (1 − test specificity) as the cutoff point for defining normal and abnormal test results is moved along the ordered set of results. As noted above, the AUC is a popular measure of diagnostic test performance [32]. Both the AUC and $I(D;R)$ are single-parameter measures of diagnostic test performance. It is helpful to understand some of their differences.

A classic approach to explaining ROC curves is to assume that test results are normally distributed for both healthy (d−) and diseased (d+) individuals. This is illustrated by the Figure 2 insert. The ROC curve is then constructed, as noted above, by plotting test sensitivity as a function of 1−test specificity for all possible diagnostic cutoffs. As the distance between the means of the two distributions increases, the ROC curve shifts upward and to the left, increasing the AUC from a value of 0.5 toward its maximal value of one. This is illustrated in Figure 2, which includes a plot of the AUC as a function of the separation between the means, for the case in which the standard deviations of both distributions are one. $I(D; R)$, but not the AUC, is a function of the pretest probability of disease. This is illustrated in the figure by plots of $I(D; R)$ as a function of the distance between the means of the same two distributions for three pretest probabilities of disease: 0.1, 0.2, and 0.5. The figure also plots a transformation of the AUC, AUC*, which is equal to $2(\mathrm{AUC}) - 1$. This transformation of the AUC changes its range from [0.5,1] to [0,1]. Collectively, these plots demonstrate that the AUC and $I(D; R)$ are qualitatively different statistics.

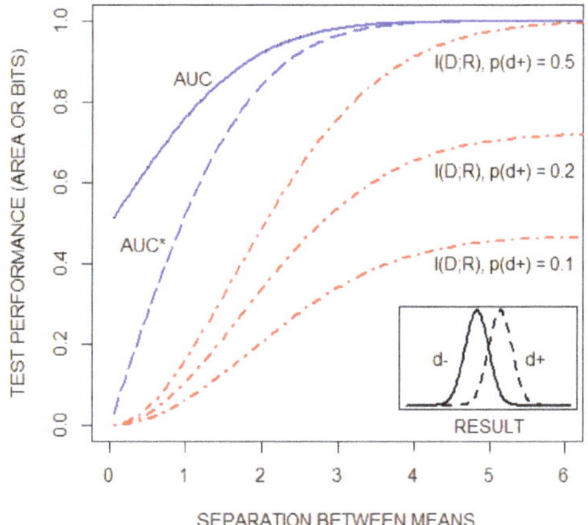

Figure 2. The area under the receiver operating characteristic curve (AUC), a transformation of the AUC (AUC*), and the mutual information between the disease state and the test result ($I(D; R)$ in units of bits), as a function of the distance between the means of the distributions of test results for both healthy (d−) and diseased (d+) individuals (see text). $I(D; R)$ is plotted for three pretest probabilities of disease: 0.1, 0.2, and 0.5.

6.2. Diagnostic Information from the Exercise Electrocardiogram (ECG)

As noted in the preceding section, Diamond et al. used information theory to evaluate the performance of the exercise ECG in the diagnosis of CHD [6]. Depression of the ST segment (a portion of the ECG tracing) during exercise is an indicator of coronary artery disease. The data in Table 2 shows their estimates of the probability of ST segment depression falling into six different categories as a function of whether the patient has significant CHD.

They first analyzed the data by selecting a criterion to dichotomize the results into positive and negative categories. For example, if a positive test is defined as ST depression ≥ 1 mm, then, as seen from the table, $p(r+|d+)$ becomes $0.233 + 0.088 + 0.133 + 0.195 = 0.649$ and $p(r+|d-)$ becomes $0.110 + 0.021 + 0.012 + 0.005 = 0.148$. Recognizing that $p(d_i, r_j) = p(r_j|d_i)p(d_i)$ and $p(r_j) = p(d+, r_j) + p(d-, r_j)$, Equation (4) can then be used to calculate the information content (mutual information) for the test for this cutoff as a function of the pretest probability of disease.

Table 2. Data from Diamond et al. [6] showing the probabilities of various categories of ST segment depression (the result, *r*) during an exercise electrocardiogram as a function of the presence (d+) and absence (d−) of significant CHD.

ST Depression (mm)	$p(r|d+)$	$p(r|d-)$
$0 \leq ST < 0.5$	0.143	0.625
$0.5 \leq ST < 1.0$	0.208	0.227
$1.0 \leq ST < 1.5$	0.233	0.110
$1.5 \leq ST < 2.0$	0.088	0.021
$2.0 \leq ST < 2.5$	0.133	0.012
$2.5 \leq ST < \infty$	0.195	0.005

They contrast this with a calculation of the information content (mutual information) if the results are not dichotomized, but rather left partitioned into six categories. If the ST segment is depressed by 2.2 mm for example, it makes sense to calculate the posttest probability using the more accurate test operating characteristics that apply to the narrower interval of [2, 2.5) than the operating characteristics that apply to the larger interval of [1, ∞). Equation (4) is again used to make the calculation, but in this case, there are six possible test results rather than two.

Figure 3 (reconstructed from their report with permission) compares mutual information as a function of pretest probability of significant CHD for the dichotomized and non-dichotomized approaches. The curve labeled IDEAL is the pretest diagnostic uncertainty as a function of pretest probability. It indicates the average amount of information that an ideal test would provide, i.e., the average amount of information needed to reduce the diagnostic uncertainty to zero (by yielding a posttest probability of either zero or one). We observe that, for most of the range of pre-test probabilities, approximately one third of the diagnostic information is lost by dichotomizing the results with a diagnostic cutoff of 1 mm. The issue of information lost as a consequence of dichotomizing test results is considered again in the following subsection.

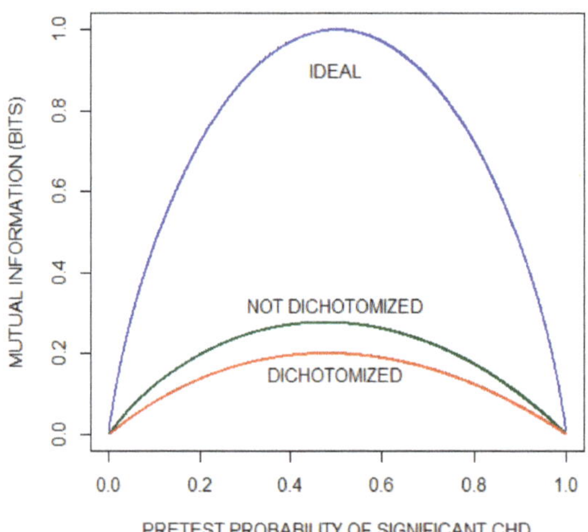

Figure 3. Mutual information as a function of pretest probability of significant coronary heart disease (CHD) for the exercise electrocardiogram. The plot compares the performance of a theoretical ideal test with the actual performance when the results are either (1) dichotomized using the criterion of ST segment depression of ≥ 1 mm or (2) not dichotomized. This plot has been reconstructed with permission from the paper by Diamond et al. [6].

Although, on average, the exercise ECG does not provide much information about whether a patient has significant CHD, the possibility remains that specific test results are informative. To illustrate this, we consider the two results that lie on opposite ends of the test result spectrum: ST depression < 0.5 mm and ST depression ≥ 2.5 mm. Recall that relative entropy (Equation (3)) quantifies the amount of diagnostic information provided by a given test result. Figure 4 plots relative entropy as a function of the pretest probability of significant CHD for these two test results. For comparison, the figure includes relative entropy plots for a theoretical ideal test when significant CHD is present (d+) and when significant CHD is absent (d−). Inspecting these curves, we conclude that an ST depression of <0.5 mm is not helpful in ruling out significant CHD. On the other hand, when significant CHD is present and as the pre-test probability increases, the information provided by an ST depression of ≥2.5 mm approaches the information provided by the ideal test.

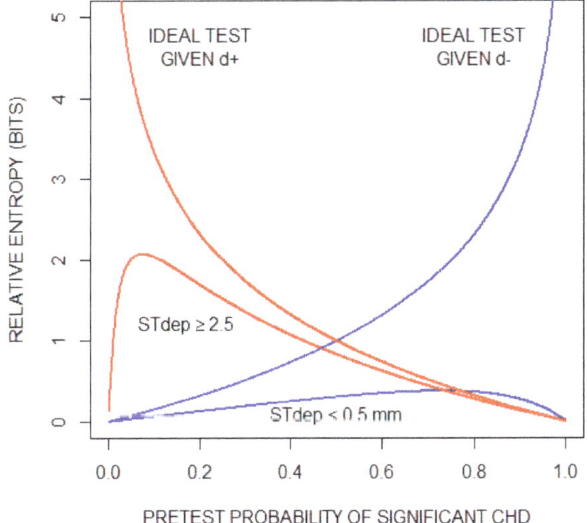

Figure 4. Diagnostic information (relative entropy) provided by the findings of an ST segment depression (STdep) < 0.5 mm and an ST depression (STdep) ≥ 2.5 mm as a function of pretest probability of significant coronary heart disease (CHD). Also shown are relative entropy plots for a theoretical ideal test when significant CHD is present (d+) and when significant CHD is absent (d−). For the theoretical ideal test when significant CHD is present, relative entropy increases indefinitely as pretest probability of significant CHD approaches zero; and for the theoretical ideal test when significant CHD is absent, relative entropy increases indefinitely as pretest probability of significant CHD approaches one.

6.3. Diagnostic Information Lost by Selecting a Diagnostic Cutoff

Diagnostic test results are often continuous or lie along a continuum, e.g., body temperature, serum glucose, histologic grade. As observed in the preceding example, dichotomizing test results by selecting a diagnostic cutoff can result in some loss of diagnostic information. As an additional illustration of this, consider again the classic example in which the probability densities of test results are normally distributed for both the healthy (d−) and diseased (d+) populations (as illustrated by the Figure 2 insert). Let us assume that the pretest probability of disease is 0.2, that the standard deviations of the two distributions are equal to one, and that the means of the two distributions are separated by one standard deviation.

The cutoff that maximizes the mutual information provided by this test lies between the means of the two distributions, approximately 0.66 standard deviations from the mean of the healthy population (determined by inspection; see the Supplementary Materials). This results in a test sensitivity, $p(r+|d+)$, of 0.63 and a test specificity, $p(r-|d-)$, of 0.75. Recalling that $p(d_i, r_j) = p(r_j|d_i)p(d_i)$ and $p(r_j) = p(d+, r_j) + p(d-, r_j)$, Equation (4) can be used to calculate that the average amount of information gained by performing the test with this cutoff is 0.071 bits.

Alternatively, we can calculate the posttest probabilities directly from the obtained results. Since the test results are continuous, we modify Equation (4) to calculate $I(D; R)$ as follows:

$$\sum_{i=1}^{2} \int_{-\infty}^{\infty} \rho(d_i, r) \log_2 \frac{\rho(d_i, r)}{p(d_i)\rho(r)} dr.$$

Because r is a continuous variable, we have replaced the summation over index j with an integral and, for terms involving r, replaced the probabilities (indicated by p) with probability densities (indicated by ρ). Note that $\rho(d_i, r) = \rho(r|d_i)p(d_i)$ and $\rho(r) = \rho(d+, r) + \rho(d-, r)$. This calculation gives a mutual information of 0.106 bits. Hence, in this example, approximately one third of the expected value of the information provided by the test is discarded by selecting a cutoff to dichotomize the results.

6.4. Diagnostic Information Provided by Two Tests with Discrete Results

A study that investigated the value of combining two diagnostic tests in the diagnosis of deep vein thrombosis (DVT) [38] provides a convenient data set to illustrate information theory functions that apply when more than one test is used (see Section 4). A DVT is a blood clot of the deep veins, typically in the lower extremities, that can be fatal if it detaches and travels to the lungs. One of the tests is a clinical index, based on the patient's medical history and physical exam findings, that classified the patient as being at low, moderate, or high risk for a DVT. The other test is a blood test that detects a protein, the d-dimer, that is often elevated in the presence of a DVT. The d-dimer was reported as positive or negative. The number of patients found to be in each of the 3×2 test result categories as a function of whether they were ultimately diagnosed with a DVT is shown in Table 3.

Table 3. Data from Anderson et al. [38]. The number of patients with and without a DVT as a function of the clinical index and the d-dimer test.

Clinical Index	d-Dimer	DVT+	DVT−
Low risk	−	3	313
	+	17	113
Moderate risk	−	15	243
	+	61	93
High risk	−	15	50
	+	79	55

The study included 1057 patients, 190 of whom were diagnosed as having a DVT. Therefore, the probability of being diagnosed with a DVT in this population is $190/1057 = 0.180$. The uncertainty about whether a patient randomly selected from this population was diagnosed with a DVT is calculated using the entropy function (Equation (2)). We find that $H(D) = 0.680$ bits. Given that only two disease states are under consideration, the range of possible entropy values is 0–1 bits.

If the clinical index (test A) is applied as a single test within this population, the diagnostic uncertainty will decrease, on average, by $I(D; R_A) = 0.111$ bits (calculated using Equation (5)). Similarly, the d-dimer test (test B) will decrease the diagnostic uncertainty, on average, by $I(D; R_B) = 0.125$ bits. Using the information provided by both tests will decrease the diagnostic uncertainty, on average, by $I(D; (R_A, R_B)) = 0.197$ bits (calculated using Equation (11)), which is less than the sum of the information provided by each test separately by a value of $I(D; R_A; R_B) = 0.039$ bits (see Equation (12);

this value was calculated using Equation (10)). The residual uncertainty after performing both tests is substantial: $H(D) - I(D; (R_A, R_B)) = 0.483$ bits (an uncertainty reduction of 29%). A perfect test or combination of tests would reduce the uncertainty to zero.

If the clinical index is found to be high and no additional testing is performed, the posttest probability of a DVT is 0.472. Using the relative entropy function (Equation (3)), we calculate that the test has provided 0.323 bits of information. An isolated negative d-dimer yields a posttest probability of a DVT of 0.052 and 0.106 bits of information. The range of possible relative entropy values is bounded by the relative entropies associated with reducing the diagnostic uncertainty to zero. For this example, 2.476 bits of information are required to rule in a DVT (going from the pretest probability of 0.180 to a posttest probability of one) and 0.286 bits of information are required to rule out a DVT (going from the pretest probability of 0.180 to a post-test probability of zero). Using the pre-test probabilities and these boundary information values, we calculate that the expected value of the amount of information provided by a theoretical perfect test is $(0.180)(2.476 \text{ bits}) + (0.820)(0.286 \text{ bits}) = 0.680 \text{ bits} = H(D)$.

Next, consider a patient who belongs to the subset of patients with both a high clinical index and a negative d-dimer. The probability of a DVT given both results is 0.231 and, per the relative entropy function (Equation (3)), this combination of results provides 0.012 bits of information. Because the two test results are discordant, their net effect is to provide very little diagnostic information. Furthermore, by applying Expression (8) we find that for this subpopulation (patients with a high clinical index and a negative d-dimer) the information provided by each of the two tests performed separately is negative (-0.168 bits for the finding of a high clinical index and -0.254 bits for the finding of a negative d-dimer). The negative results indicate that the baseline probability of a DVT is a more accurate estimate than the posttest probability when only one test is performed.

Although these two tests do not, on average, provide much information about whether a patient has a DVT, they have the potential to help rule out a DVT. We saw above that an isolated negative d-dimer decreases the probability of a DVT from 0.180 to 0.052 and provides 0.106 bits of information. An isolated low clinical index decreases the probability to 0.045 and provides 0.120 bits of information. The combination of both findings decreases the probability to 0.009 and provides 0.229 bits of information (out of the 0.286 bits of information required to decrease the probability of a DVT to zero).

6.5. Diagnostic Information Provided by Two Tests with Continuous Results

Finally, we consider a hypothetical example in which the results of two diagnostic tests (A and B) are normally distributed for both healthy and diseased patients. We define the means, variances, and covariance for the binormal distribution of the test results for the healthy population as:

$$\mu_A = 0, \ \mu_B = 0, \ \sigma_A^2 = 1, \ \sigma_B^2 = 2, \ \sigma_{AB} = 1,$$

and the parameters for the binormal distribution of the test results for the diseased population as:

$$\mu_A = 2, \ \mu_B = 1, \ \sigma_A^2 = 2, \ \sigma_B^2 = 1, \ \sigma_{AB} = -1.$$

Figure 5 shows probability density contour plots for the two binormal distributions, with test results expressed as standard deviations from the means of the distribution for healthy patients $(\mu_A = \mu_B = 0)$.

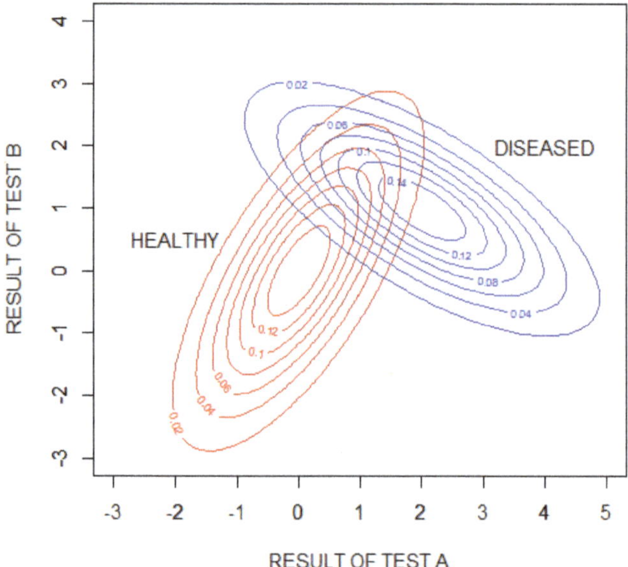

Figure 5. Contour plots showing the probability densities of the results of tests A and B for the healthy and diseased populations. The test results are expressed as standard deviations from the means of the distribution for healthy patients ($\mu_A = \mu_B = 0$).

We will assume that the pretest probability of disease is 0.2. The pretest uncertainty of the disease state, $H(D)$, is then found to be 0.722 bits (calculated using Equation (2)). Test A applied as a single test within this hypothetical population will decrease the diagnostic uncertainty, on average, by $I(D; R_A) = 0.281$ bits (calculated numerically using Equation (5)). Test B applied as a single test within this population will decrease the diagnostic uncertainty, on average, by $I(D; R_B) = 0.083$ bits. Test A is, on average, more informative than test B because the separation between the means of the healthy and diseased populations is larger for test A than for test B. Using the information provided by both tests will decrease the diagnostic uncertainty, on average, by $I(D; (R_A, R_B)) = 0.424$ bits (calculated numerically using Equation (11)), which is more than the sum of the information provided by each test separately by a value of $-I(D; (R_A, R_B)) = 0.060$ bits (see Equation (12); this value was calculated numerically using Equation (10)). The residual uncertainty after performing both tests is: $H(D) - I(D; (R_A, R_B)) = 0.298$ bits (an uncertainty reduction of 59%).

Figure 6 is a contour plot showing the amount of diagnostic information (relative entropy in units of bits) provided by possible combinations of the results of the two tests, with test results expressed as standard deviations from the means of the distribution for healthy patients ($\mu_A = \mu_B = 0$). The information value is bounded in two opposing quadrants by the amount of information required to rule in disease, $\log_2(1/0.2) \approx 2.322$ bits, and in the other two opposing quadrants by the amount of information required to rule out disease, $\log_2(1/0.8) \approx 0.322$ bits. More information is required to rule in disease than to rule out disease because the pretest probability of disease is less than 0.5. Given that the performance of the tests is characterized by binormal distributions, it is theoretically impossible to rule in or rule out disease with certainty. In the case of ruling in disease, for example, as the results of test A increase and the results of test B decrease, the probability of disease approaches (but never equals) one and the amount of information obtained approaches (but never equals) $\log_2(1/0.2)$ bits.

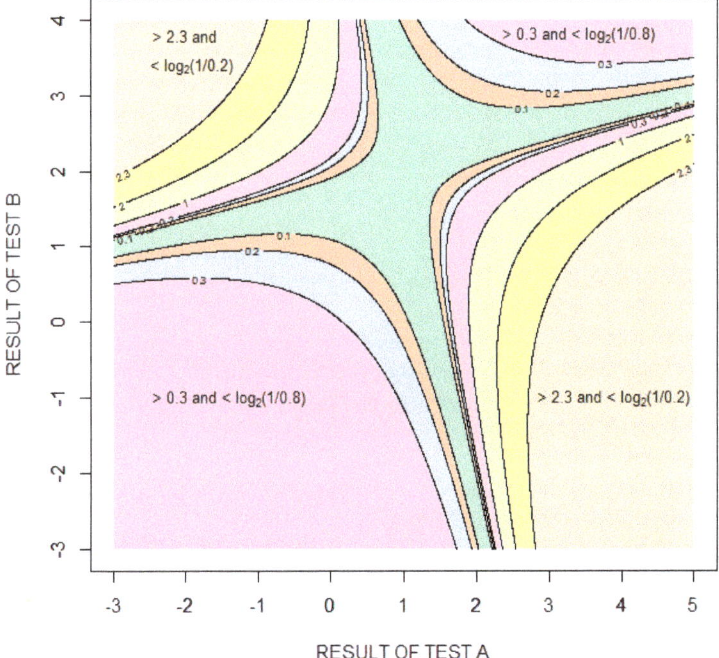

Figure 6. Contour plot showing the information (relative entropy in units of bits) provided by specific combinations of results of tests A and B. The test results are expressed as standard deviations from the means of the distribution for healthy patients ($\mu_A = \mu_B = 0$).

7. Discussion

7.1. Summary

Information theory functions are defined on random variables. The recognition that an individual's disease state and diagnostic test result are random variables allows for the direct application of fundamental information theory functions to clinical diagnostic testing. The concept of the surprisal (discussed in Section 2) provides an intuitive understanding of these functions. The uncertainty about a patient's disease state is quantified by the entropy function (discussed in Section 3.1). The amount of diagnostic information provided by a specific test result is quantified by the relative entropy function (discussed in Section 3.2). Prior to performing a diagnostic test, the expected value of the amount of information that the test will provide is quantified by the mutual information function (discussed in Section 3.3). The mutual information associated with a diagnostic test is a single parameter measure of diagnostic test performance that is dependent upon the pretest probabilities of the disease states. The information theory functions are applicable given any number of disease states and any number of test results (or continuous test results). Information theory functions can also be used to evaluate the conjoint performance of multiple diagnostic tests: the information provided by each of several test results can be calculated (discussed in Section 4.1), and the mutual information between the disease state and the results of multiple diagnostic tests can be partitioned into components corresponding to the contributions of the individual tests and interactions among the disease states and test results (discussed in Section 4.2).

7.2. Points of Emphasis and Clarification

- Diagnostic information does not necessarily decrease diagnostic uncertainty. A screening test, for example, might change the probability of disease from 0.005 to 0.5. The test has provided 2.826 bits of information (quantified by the relative entropy function), while the uncertainty has increased from 0.045 bits to one bit (quantified by the entropy function). This is consistent with the common experience of becoming more perplexed about an issue as more is learned about it.
- The three questions posed in the Introduction all include the word "we", implying a general agreement upon the probabilities used to calculate diagnostic uncertainty and diagnostic information. This is often not the case. Probability estimates usually include a subjective component. Consequently, two individuals can obtain different amounts of information from the same test result or observation.
- Mutual information, but not relative entropy, is equal to pretest uncertainty minus posttest uncertainty. Both functions are equal to the expected value of the reduction in the surprisal.
- The random variables of interest in this review are an individual's disease state and test result. The same theory applies when the random variables are defined as any type of state and any type of observation.
- The goal of clinical diagnostic testing is not to make the diagnosis but, rather, to assign accurate probabilities to the possible disease states.
- Some diagnostic information is typically lost by dichotomizing continuous or ordered test results.

7.3. Conclusions

Information statistics have a useful role to play in the evaluation and comparison of diagnostic tests. In some cases, information measures may complement useful concepts such as test sensitivity, test specificity, and predictive values. In other situations, information measures may replace more limited statistics. Mutual information, for example, may be better suited as a single parameter index of diagnostic test performance than alternative statistics. Furthermore, information theory has the potential to help us learn about and teach about the diagnostic process. Examples include concepts illustrated above, including the importance of pretest probability as a determinant of diagnostic information, the amount of information lost by dichotomizing test results, the limited potential of some diagnostic tests to reduce diagnostic uncertainty, and the ways in which diagnostic tests can interact to provide diagnostic information. These are concepts that can all be effectively communicated graphically.

It is hoped that this review will help to motivate new applications of information theory to clinical diagnostic testing, especially as data from newer diagnostic technologies becomes available. The challenge will be to develop systems that accurately diagnosis and treat patients by integrating increasingly large amounts of personalized data [39,40]. A potential role for information theory functions in this process is suggested by their applicability to multidimensional data.

Supplementary Materials: R code for calculations and figures are available online at http://www.mdpi.com/1099-4300/22/1/97/s1.

Funding: This research received no external funding.

Acknowledgments: The author is very grateful for multiple helpful suggestions by reviewers and the academic editor.

Conflicts of Interest: The author declares no conflict of interest.

Appendix A Modified Relative Entropy Is not a Distance Metric

We first show that the modified relative entropy function (Expression (8)) satisfies the triangle equality. Let the modified relative entropy from probability distribution b to probability distribution a when the true probability distribution is c be expressed as

$$d_c(a,b) = \sum_i c_i \log \frac{a_i}{b_i}.$$

Then,

$$d_c(x,y) + d_c(y,z) = \sum_i c_i \log\frac{x_i}{y_i} + \sum_i c_i \log\frac{y_i}{z_i} = \sum_i c_i \log \frac{x_i}{z_i} = d_c(x,z).$$

Despite satisfying the triangle inequality, $d_c(a,b)$ does not meet the criteria for a distance metric [41] (p. 117) because it can be negative. If we try to circumvent this problem by defining the measure as the absolute value of $d_c(a,b)$, then the triangle inequality is still satisfied:

$$\left|d_c(x,y)\right| + \left|d_c(y,z)\right| \geq \left|d_c(x,y) + d_c(y,z)\right| = \left|d_c(x,z)\right|.$$

Nevertheless, $\left|d_c(a,b)\right|$ still fails to qualify as a distance metric because it is not necessarily the case that $\left|d_c(a,b)\right| = 0$ implies that $a = b$. For example, if $a_1 = b_2 = 1/4$, $a_2 = b_1 = 3/4$, and $c_1 = c_2 = 1/2$, then $\left|d_c(a,b)\right| = 0$ but $a \neq b$.

References

1. Hartley, R.V.L. Transmission of information. *Bell Syst. Tech. J.* **1928**, *7*, 535–563. [CrossRef]
2. Shannon, C.E. A mathematical theory of communication. *Bell Syst. Tech. J.* **1948**, *27*, 379–423. [CrossRef]
3. Good, I.J.; Card, W.I. The diagnostic process with special reference to errors. *Methods Inf. Med.* **1971**, *10*, 176–188. [PubMed]
4. Metz, C.E.; Goodenough, D.J.; Rossmann, K. Evaluation of receiver operating characteristic curve data in terms of information theory, with applications in radiography. *Radiology* **1973**, *109*, 297–303. [CrossRef] [PubMed]
5. McNeil, B.J.; Keeler, E.; Adelstein, S.J. Primer on certain elements of medical decision making. *NEJM* **1975**, *293*, 211–215. [CrossRef] [PubMed]
6. Diamond, G.A.; Hirsch, M.; Forrester, J.S.; Staniloff, H.M.; Vas, R.; Halpern, S.W.; Swan, H.J. Application of information theory to clinical diagnostic testing. The electrocardiographic stress test. *Circulation* **1981**, *63*, 915–921. [CrossRef]
7. Somoza, E.; Mossman, D. Comparing and optimizing diagnostic tests: An information-theoretical approach. *Med. Decis. Mak.* **1992**, *12*, 179–188. [CrossRef]
8. Benish, W.A. Relative entropy as a measure of diagnostic information. *Med. Decis. Mak.* **1999**, *19*, 202–206. [CrossRef]
9. Hughes, G. *Applications of Information Theory to Epidemiology*; APS Press: St. Paul, MN, USA, 2012.
10. Roulston, M.S.; Smith, L.A. Evaluating probabilistic forecasts using information theory. *Mon. Weather Rev.* **2002**, *130*, 1653–1660. [CrossRef]
11. Schneider, T.D.; Stormo, G.D.; Gold, L.; Ehrenfeucht, A. Information content of binding sites on nucleotide sequences. *J. Mol. Biol.* **1986**, *188*, 415–431. [CrossRef]
12. Schumacher, B. Quantum coding. *Phys. Rev. A* **1995**, *51*, 2738–2747. [CrossRef] [PubMed]
13. Coombs, C.H.; Dawes, R.M.; Tversky, A. *Mathematical Psychology: An Elementary Introduction*; Prentice-Hall: Oxford, UK, 1970; pp. 307–350.
14. Hughes, G.; McRoberts, N.; Burnett, F.J. Information graphs for binary predictors. *Phytopathology* **2015**, *105*, 9–17. [CrossRef] [PubMed]
15. Cohen, J.E. Information theory and music. *Behav. Sci.* **1962**, *7*, 137–163. [CrossRef]
16. Tribus, M.; McIrvine, E.C. Energy and information. *Sci. Am.* **1971**, *224*, 178–184. [CrossRef]

17. Benish, W.A. Mutual information as an index of diagnostic test performance. *Methods Inf. Med.* **2003**, *42*, 260–264. [CrossRef] [PubMed]

18. Shannon, C.E.; Weaver, W. *The Mathematical Theory of Communication*; University of Illinois Press: Urbana/Chicago, IL, USA, 1963.

19. Tribus, M. *Thermostatics and Thermodynamics*; D. van Nostrand: Princeton, NJ, USA, 1961.

20. Benish, W.A. Intuitive and axiomatic arguments for quantifying diagnostic test performance in units of information. *Methods Inf. Med.* **2009**, *48*, 552–557. [CrossRef] [PubMed]

21. Aczél, J.; Doróczy, Z. *On Measures of Information and Their Characterizations*; Academic Press: New York, NY, USA, 1975.

22. Harremoës, P. Entropy-New Editor-in-Chief and Outlook. *Entropy* **2009**, *11*, 1–3. [CrossRef]

23. Cover, T.M.; Thomas, J.A. *Elements of Information Theory*; John Wiley and Sons: New York, NY, USA, 1991.

24. Kullback, S.; Leibler, R.A. On information and sufficiency. *Ann. Math. Stat.* **1951**, *22*, 79–86. [CrossRef]

25. Good, I.J. What is the use of a distribution? In *Multivariate Analysis*; Krishnaiah, P.R., Ed.; Academic Press: New York, NY, USA, 1969; Volume 2, pp. 183–203.

26. Kittel, C.; Kroemer, H. *Thermal Physics*, 2nd ed.; W.H. Freeman and Company: New York, NY, USA, 1980.

27. Card, W.I.; Good, I.J. A logical analysis of medicine. In *A Companion to Medical Studies*; Passmore, R., Robson, J.S., Eds.; Blackwell: Oxford, UK, 1974; Volume 3, pp. 60.1–60.23.

28. Peirce, C.S. The probability of induction. In *The World of Mathematics*; Newman, J.R., Ed.; Simon and Schuster: New York, NY, USA, 1956; Volume 2, pp. 1341–1354.

29. Good, I.J. Studies in the history of probability and statistics. XXXVII. A. M. Turing's statistical work in World War II. *Biometrika* **1979**, *66*, 393–396. [CrossRef]

30. Ende, J.V.D.; Bisoffi, Z.; Van Puymbroek, H.; Van Der Stuyft, P.; Van Gompel, A.; Derese, A.; Lynen, L.; Moreira, J.; Janssen, P.A.J.; Lynen, L. Bridging the gap between clinical practice and diagnostic clinical epidemiology: Pilot experiences with a didactic model based on a logarithmic scale. *J. Evaluation Clin. Pr.* **2007**, *13*, 374–380. [CrossRef]

31. Green, D.M.; Swets, J.A. *Signal Detection Theory and Psychophysics*; John Wiley and Sons Inc.: New York, NY, USA, 1966.

32. Hanley, J.A.; McNeil, B.J. The meaning and use of the area under a receiver operating characteristic (ROC) curve. *Radiology* **1982**, *143*, 29–36. [CrossRef] [PubMed]

33. Somoza, E.; Soutullo-Esperon, L.; Mossman, D. Evaluation and optimization of diagnostic tests using receiver operating characteristic analysis and information theory. *Int. J. Biomed. Comput.* **1989**, *24*, 153–189. [CrossRef]

34. Rifkin, R.D.; Hood, W.B., Jr. Bayesian analysis of electrocardiographic exercise stress testing. *N. Engl. J. Med.* **1977**, *297*, 681–686. [CrossRef] [PubMed]

35. Rifkin, R.D. Maximum Shannon information content of diagnostic medical testing: Including application to multiple non-independent tests. *Med. Decis. Mak.* **1985**, *5*, 179–190. [CrossRef]

36. Lee, W.C. Selecting diagnostic tests for ruling out or ruling in disease: The use of the Kullback-Leibler distance. *Int. J. Epidemiol.* **1999**, *28*, 521–525. [CrossRef]

37. Benish, W.A. The channel capacity of a diagnostic test as a function of test sensitivity and test specificity. *Stat. Methods Med. Res.* **2015**, *24*, 1044–1052. [CrossRef]

38. Anderson, D.R.; Kovacs, M.J.; Kovacs, G.; Stiell, I.; Mitchell, M.; Khoury, V.; Dryer, J.; Ward, J.; Wells, P.S. Combined use of clinical assessment and d-dimer to improve the management of patients presenting to the emergency department with suspected deep vein thrombosis (the EDITED Study). *J. Thromb. Haemost.* **2003**, *1*, 645–651. [CrossRef]

39. Jameson, J.L.; Longo, D.L. Precision medicine—Personalized, problematic, and promising. *NEJM* **2015**, *372*, 2229–2234. [CrossRef]

40. Duffy, D.J. Problems, challenges and promises: Perspectives on precision medicine. *Brief Bioinform.* **2016**, *17*, 494–504. [CrossRef]

41. Munkres, J.R. *Topology: A First Course*; Prentice-Hall: Englewood Cliffs, NJ, USA, 1975.

Article

On the Binormal Predictive Receiver Operating Characteristic Curve for the Joint Assessment of Positive and Negative Predictive Values

Gareth Hughes

SRUC, Scotland's Rural College, The King's Buildings, Edinburgh EH9 3JG, UK; gareth.hughes@sruc.ac.uk

Received: 23 March 2020; Accepted: 5 May 2020; Published: 26 May 2020

Abstract: The predictive receiver operating characteristic (PROC) curve is a diagrammatic format with application in the statistical evaluation of probabilistic disease forecasts. The PROC curve differs from the more well-known receiver operating characteristic (ROC) curve in that it provides a basis for evaluation using metrics defined conditionally on the outcome of the forecast rather than metrics defined conditionally on the actual disease status. Starting from the binormal ROC curve formulation, an overview of some previously published binormal PROC curves is presented in order to place the PROC curve in the context of other methods used in statistical evaluation of probabilistic disease forecasts based on the analysis of predictive values; in particular, the index of separation (PSEP) and the leaf plot. An information theoretic perspective on evaluation is also outlined. Five straightforward recommendations are made with a view to aiding understanding and interpretation of the sometimes-complex patterns generated by PROC curve analysis. The PROC curve and related analyses augment the perspective provided by traditional ROC curve analysis. Here, the binormal ROC model provides the exemplar for investigation of the PROC curve, but potential application extends to analysis based on other distributional models as well as to empirical analysis.

Keywords: diagnostic test; evaluation; ROC curve; PROC curve; binormal; prevalence; positive predictive value; negative predictive value; Bayes' rule; leaf plot; expected mutual information

---

1. Introduction

The predictive receiver operating characteristic (PROC) curve is a diagrammatic format introduced by Shiu and Gatsonis [1] in the context of the statistical evaluation of probabilistic disease forecasts. Such forecasts are often evaluated by calculation of metrics defined conditionally on the actual disease status. Metrics defined conditionally on the outcome of the forecast—predictive values—are also important, although less frequently reported; this motivates the introduction of the PROC curve. Although this approach is potentially useful, as yet it has not been commonly applied [2]. One possible reason for this is the apparent complexity of patterns generated by PROC curve analysis [1,3]. Thus Shiu and Gatsonis note that "It is therefore essential to study and attempt to characterize the geometric properties of PROC curves before undertaking an investigation of how the curves can be used to evaluate the performance of a diagnostic test".

This article is intended as a contribution towards furthering an understanding of some of the properties of PROC curves as described by Shiu and Gatsonis [1]; and thus, hopefully, increasing applications of their analysis. The approach taken is to place the PROC curve in the context of some other methods for the statistical evaluation of probabilistic disease forecasts. In particular, we discuss the receiver operating characteristic (ROC) curve [4,5], the index of separation PSEP [6], and the leaf plot [7,8], in terms of their relationship to the PROC curve. The article is set out as follows. Section 2 provides background to the methods discussed as context for PROC curve analysis. Section 3 presents an analysis of some particular PROC curves, and the perspective provided by analyses based on corresponding contextual methods. Section 4 is a concluding general discussion.

2. Methods

The preliminary steps leading towards the calculation and analysis of a PROC curve largely follow the route mapped by Sackett et al. [4], particularly Chunk #2 and Chunk #3 of Chapter 4 on the interpretation of diagnostic data. The obvious difference is that the impetus of Sackett et al. is data-driven, whereas here the required observations are represented by normal distributions. As in ROC analysis, it is not necessary for test data to follow normal distributions. Here, the normality assumption is helpful in the investigation of the theoretical properties of the curve and the exploration of scenarios that lead to different shapes. To begin, we consider two groups of subjects for which the known actual ('gold standard') status is denominated case *'c'* or non-case *'nc'*. For each subject, a second observation is available, referred to generically as a risk score. The risk score may be useful as a proxy variable in diagnosis when obtaining the gold standard at the outset may be considered too time-consuming, difficult, or expensive; or when an early estimate of risk may facilitate preventative treatment. By convention, the risk score is calibrated so that *c* subjects tend to have higher scores than *nc* subjects, although typically there is overlap between the ranges of scores for the two groups.

Now consider a threshold on the risk score scale, such that a score above the threshold (designated '+') is taken as indicative of likely *c* status, while a score at or below the threshold (designated '−') is taken as indicative of likely *nc* status. The resulting two-way classification of subjects provides the basis for a 2×2 prediction-realization table, which may be based on numerical data as in Table 4-3 of Sackett et al. [4] or on probabilities as in the present analysis (Table 1). From Table 1, with $i = +, -$ (for the predictions) and $j = c, nc$ (for the realizations), we write $p_{i \cap j} = p_{j \cap i} = p_{i|j} \cdot p_j = p_{j|i} \cdot p_i$ via Bayes' rule. The p_j are taken as the Bayesian prior probabilities of case ($j = c$, 'prevalence') or non-case ($j = nc$) status, such that $p_{nc} = 1 - p_c$. We can write $p_i = p_{i|c} \cdot p_c + p_{i|nc} \cdot p_{nc}$ ($i = +$ or $-$) via the Law of Total Probability.

Table 1. The prediction-realization table for a test with two categories of realized (actual) status (c, nc) and two categories of prediction ($+$, $-$). In the body of the table are the joint probabilities.

Prediction (i)	Realization (j)		
	c	nc	Row sums
$+$	$p_{+ \cap c}$	$p_{+ \cap nc}$	p_+
$-$	$p_{- \cap c}$	$p_{- \cap nc}$	p_-
Column sums	p_c	p_{nc}	1

The conditional probability $p_{+|c}$ is referred to as the true positive proportion (TPP, sometimes *sensitivity*). The conditional probability $p_{-|nc}$ is referred to as the true negative proportion (TNP, sometimes *specificity*). We refer to the conditional probability $p_{+|nc}$ that is the complement of specificity as the false positive proportion (FPP = $1 - $ TNP). TPP and TNP are metrics often used in the evaluation of tests based on 2×2 tables. TPP characterizes the proportion of *c* subjects that had + test outcomes, while TNP characterizes the proportion of *nc* subjects that had − test outcomes. TPP and TNP are metrics defined conditionally on actual disease status, independent of prevalence.

Returning to the matter of the threshold on the risk score scale, consider now the problem of placement of this threshold. The effect of different threshold placements can be investigated by generating a set of 2×2 tables derived from a sequence of thresholds on the risk score scale, calculating the TPP and TNP values from each table, and then plotting the graph of TPP against FPP (= $1 - $ TNP). This graph is the receiver operating characteristic (ROC) curve, see [4,5]. Generally, risk score threshold values increase along an ROC curve from lower values in the top right-hand corner to higher values in the bottom left-hand corner. When both the case and non-case distributions of risk scores are modeled as normal distributions, we have a binormal ROC curve, a format that has been extensively studied, see for example Section 4.4 of [5].

The conditional probability $p_{c|+}$ is referred to as the positive predictive value (PPV). This characterizes the posterior probability of c status given a $+$ test outcome. The conditional probability $p_{nc|-}$ is referred to as the negative predictive value (NPV). This characterizes the posterior probability of nc status given a $-$ test outcome. The metrics PPV and NPV are also applicable to the evaluation of tests based on 2×2 tables, but they are less frequently reported than TPP and TNP. At least in part, this is likely because PPV and NPV vary with prevalence. The effect of prevalence on predictive values was illustrated diagrammatically by Sackett et al., see Figures 4–9 and 4–10 of [4], and more recently by Coulthard [7] and by Coulthard and Coulthard [8] as the leaf plot. This diagram shows how the posterior probabilities PPV and $1-$NPV vary with prevalence, given the TPP and TNP values of the test in question. In effect, the diagram provides a nomogram for calculating probability revisions resulting from use of a test, via application of Bayes' rule.

3. Results

3.1. Binormal ROC Curves

Section 3 of [1] is devoted to an exploration of properties of the theoretical PROC curve, and in particular to a detailed investigation of the shape properties of the PROC curve arising from the binormal ROC model. The results presented here begin with identification of characteristics of the binormal ROC curves corresponding to some qualitatively different PROC curves presented in [1] (Table 2). As in [1], the binormal ROC model is written so that the non-case distribution is standard normal (see Table 2).

From Table 2, we write $a = (\mu_c - \mu_{nc})/\sigma_c$ and $b = \sigma_{nc}/\sigma_c$, then the binormal ROC curve is TPP $= f(\text{FPP}) = \Phi(a+b\cdot\Phi^{-1}(\text{FPP}))$ [5] in which Φ denotes the standard normal cumulative distribution function and Φ^{-1} its inverse (Figure 1). Visually, the resulting ROC curves appear to be either symmetric about the negative diagonal of the graphical plot (Figure 1A), or skewed towards the upper axis (referred to as TNP-asymmetry, Figure 1B) or the left-hand axis (referred to as TPP-asymmetry, Figure 1C) of the plot. More formally, the shape (symmetry) properties of ROC curves have been characterized in terms of the Kullback-Leibler divergences $D(f_c\|f_{nc})$ and $D(f_{nc}\|f_c)$ between the case and non-case distributions [9–11] (with $D(f_c\|f_{nc})$ and $D(f_{nc}\|f_c) \geq 0$, and equality only if the case and non-case distributions are identical). In particular, binormal ROC curves may be symmetric, TNP-asymmetric or TPP-asymmetric [10]; for symmetric binormal ROC curves, $D(f_c\|f_{nc}) = D(f_{nc}\|f_c)$; while for TNP-asymmetric binormal ROC curves, $D(f_c\|f_{nc}) < D(f_{nc}\|f_c)$, and for TPP-asymmetric binormal ROC curves, $D(f_c\|f_{nc}) > D(f_{nc}\|f_c)$. For the binormal ROC curve in particular, these conditions reduce to symmetry when $\sigma_c = \sigma_{nc}$, TNP-asymmetry when $\sigma_c < \sigma_{nc}$, and TPP-asymmetry when $\sigma_c > \sigma_{nc}$ [10] (Table 2, Figure 1).

Table 2. Example data and binormal ROC curve terminology.

Example	Source	Case (f_c)		Non-case (f_{nc})		ROC Curve Symmetry [iii]	ROC Curve Proper or Improper [iv]	ROC Curve Crosses Diagonal [v]
		μ_c [i]	σ_c [ii]	μ_{nc} [i]	σ_{nc} [ii]			
1.	[1] Figure 1b [vi]	0.8	1	0	1	Symmetric	Proper	N/A
2.	[1] Figure 1b [vii]	0.8	1	0	1	Symmetric	Proper	N/A
3.	[1] Figure 2a	1	0.5	0	1	TNP-asymmetric	Improper	From below
4.	[1] Figure 2b	2	2	0	1	TPP-asymmetric	Improper	From above
5.	[1] Figure 6	1.134	1.704	0	1	TPP-asymmetric	Improper	From above

[i] Mean, [ii] Standard deviation, [iii] Terminology of [10]; [iv] Terminology of [12]; [v] As FPP increases; [vi] Prevalence = 0.7; [vii] Prevalence = 0.3.

Note that the main diagonal of an ROC plot, where TPP = FPP, is indicative of a situation where the distributions of risk scores for cases and non-cases are identical. A risk score threshold s anywhere on this line is thus characteristic of a test that provides no discrimination between cases and controls (a situation which, of course, is undesirable). Looking again at Figure 1, the ROC curves depicted appear either to fall entirely above the main diagonal of the plot (Figure 1A), or to cross it (Figure 1B,C). ROC curves of the former type are referred to as 'proper', and the latter type as 'improper' [12] (where the part of the curve under the diagonal is sometimes referred to as a 'hook'). More formally, the above-noted symmetry conditions also incorporate the conditions for which binormal ROC curves are either proper or improper. Symmetric curves are proper ($\sigma_c = \sigma_{nc}$, $b = 1$), TNP-asymmetric curves are improper ($\sigma_c < \sigma_{nc}$, $b > 1$) and cross the main diagonal of the plot from below with increasing FPP (Figure 1B), and TPP-asymmetric curves are improper ($\sigma_c > \sigma_{nc}$, $b < 1$) and cross the main diagonal of the plot from above with increasing FPP (Figure 1C). From [12], for improper curves we can calculate $\rho = (\mu_c - \mu_{nc})/(\sigma_c - \sigma_{nc})$, and then $t^* = \Phi(\rho)$ is the value of FPP where the ROC curve crosses the main diagonal and $s^* = -\rho$ is the risk score threshold on the curve at the point where it crosses the diagonal (Figure 1).

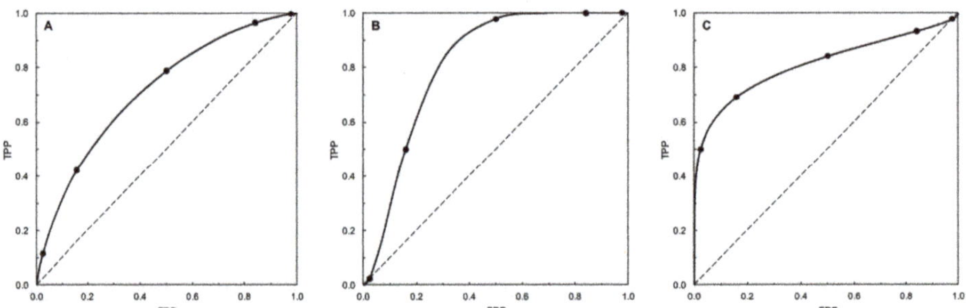

Figure 1. ROC curves based on data from Table 2. The curves are calibrated at intervals of the risk score threshold; starting in the top right-hand corner, $-2, -1, 0, 1, 2$. (**A**) Examples 1 and 2 (they have identical ROC curves), Symmetric. (**B**) Example 3, TNP-asymmetric, $t^* = 0.02275$, $s^* = 2$. (**C**) Example 4, TPP-asymmetric, $t^* = 0.97725$, $s^* = -2$.

3.2. The Corresponding PROC Curves

Figure 2 shows the PROC curves corresponding to the binormal ROC curves shown in Figure 1. These graphs appear in Figures 1 and 2 of [1] (see Table 2), minus the main diagonal and the calibration. The correspondence is as follows. Take a point on an ROC curve (as characterized by a particular risk score threshold value) in Figure 1 and note the matching TPP and FPP values. Now, given a value for the prevalence, we can calculate the corresponding PPV and $1-$NPV values via Bayes' rule. These values then define a point on the corresponding PROC curve in Figure 2 that is characterized by the same risk score threshold value as the point on the ROC curve from which we started. Thus we can denote the risk score threshold on a PROC curve at the point where it crosses the diagonal using the same notation (s^*) as for an ROC curve.

On the main diagonal of the ROC plot, where TPP = FPP, we find via Bayes' rule $PPV = 1 - NPV = p_c$ (the prior probability, estimated by prevalence). In words: if a test provides no discrimination between cases and non-cases, its application leaves the posterior probabilities unchanged from the prior. Typically, test evaluation on the basis of an ROC curve is concentrated on regions where the curve is above the main diagonal (i.e., TPP > FPP), because via Bayes' rule this implies $PPV > p_c$ and $NPV > p_{nc}$.

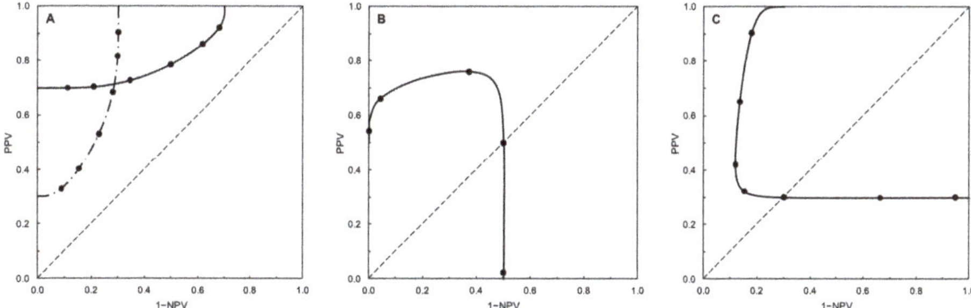

Figure 2. PROC curves based on data from Table 2. The curves are calibrated at intervals of the risk score threshold. (**A**) Example 1 (solid line, prevalence = 0.7; calibration points starting from the left-hand vertical axis are −3, −2, −1, 0, 1, 2) and Example 2 (dot-dash line, prevalence = 0.3; calibration points starting from the left-hand vertical axis are −1, 0, 1, 2, 3, 4). These examples have identical ROC curves (see Figure 1) but different PROC curves because the prevalence differs. (**B**) Example 3, prevalence = 0.5, calibration points starting from the left-hand vertical axis are −1, 0, 1, s^* = 2, 3. (**C**) Example 4, prevalence = 0.3, calibration points starting from the right-hand vertical upright are −4, −3, s^* = −2, −1, 0, 1, 2.

Figure 2A shows two PROC curves based on the same ROC curve (Figure 1A), the difference resulting from different prevalence values. The ROC curve is symmetric (proper, b = 1); the corresponding PROC curves are monotone [1]. In each case, PPV and 1−NPV increase as the risk score threshold increases (i.e., there is a trade-off between PPV and NPV along the PROC curve), and the curves do not cross the main diagonal of the plot.

Figure 2B,C show PROC curves based on improper ($b \neq 1$) ROC curves (Figure 1B,C, respectively). The PROC curves are non-monotone [1] and cross the main diagonal of the plot. In Figure 2B the PROC curve is based on a TNP-asymmetric ROC curve crossing the diagonal from below as FPP increases, the crossover point (s^* = 2) being in the bottom left-hand corner of the ROC plot. In Figure 2C the PROC curve is based on a TPP-asymmetric ROC curve crossing the main diagonal of the plot from above as FPP increases, the crossover point (s^* = −2) being in the top right-hand corner of the ROC plot. Although the PROC curves' shapes are very different to the shapes of the corresponding ROC curves, interpretation is aided if we note that in each case the same range of risk scores is of interest. In Figures 1B and 2B, risk scores < 2 fall above the main diagonal of the plots. In Figures 1C and 2C risk scores > −2 fall above the diagonals. In each case, these are the ranges of risk scores that would typically be considered as possible test thresholds. The part of the PROC curves in Figure 2B,C below the main diagonal of the plots, corresponding to the improper ROC 'hooks' in Figure 1B,C, appears more pronounced, but covers the same range of risk score thresholds in each case.

With these interpretations of PROC curves in relation to their corresponding ROC curves, we may investigate the properties of a putative test by first setting a risk score threshold based on designated TPP and FPP values, as is typical current practice. If we then locate this threshold value as a point on the corresponding PROC curve, we can trace from this point to the horizontal and vertical axes of the PROC plot to establish the x,y coordinates of the point, and so obtain the corresponding predictive values at that threshold. Thus ROC curves and PROC curves in combination may contribute to test evaluation.

3.3. Measures of Predictive Performance

Figure 3 shows the PROC curve corresponding to the Example 5 from Table 2. This example provides a context for Shiu and Gatsonis [1] to address the question of how to evaluate predictive performance based on the PROC curve. Here, we approach this problem by replotting the data on which Figure 3 is based in an alternative format. Figure 4 shows predictive values PPV and 1−NPV on the vertical axis, with the risk score threshold on the horizontal axis. Only risk score thresholds that fall above the main diagonal of the PROC plot shown in Figure 3 are included in Figure 4.

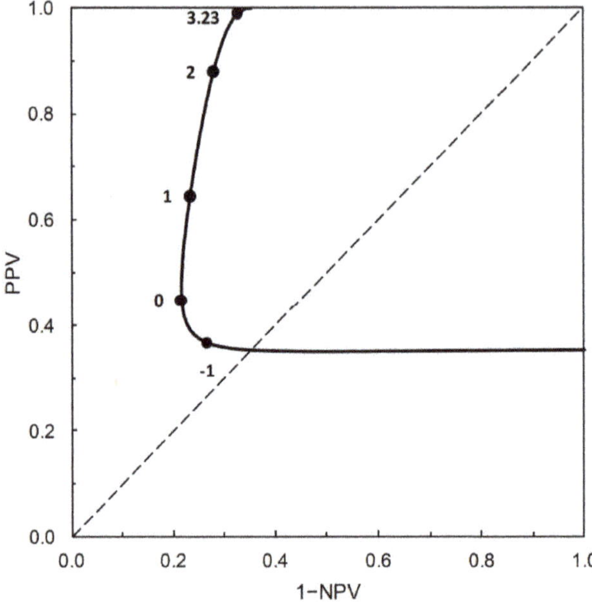

Figure 3. The PROC curve based on Example 5 (Table 2), with prevalence = 0.35 [13]. The curve is calibrated at intervals of the risk score threshold as it increases above the main diagonal at points $-1, 0, 1, 2$, and $s_{opt} = 3.23$ (at which PPV = 0.990, 1 − NPV = 0.326).

Figure 4 embodies two potential measures of predictive performance, the index of separation PSEP [6] and the distance to perfect prediction r [1]. PSEP provides a measure of 'prognostic information' in the situation where a test is applied to a validation data set, PSEP = PPV − (1−NPV). Note that PSEP is a probability measure [6], and not based on information in the Shannonian sense (see [14] for further discussion). In Figure 4, PSEP is the distance between the PPV and 1−NPV traces, varying with risk threshold score. Now suppose we start from the concept of a notionally perfect predictions, for which PPV = NPV = 1. Shiu and Gatsonis [1] define the distance from a given test to notional perfection as $r = (1-\text{PPV}) + (1-\text{NPV})$, varying with risk score threshold. In Figure 4, r is the sum of the distances from the PPV trace to 1 and from the 1−NPV trace to 0. Thus we have $r = 1-\text{PSEP}$.

Shiu and Gatsonis [1] interpret the minimum value of r (denoted here r_{opt}) as the notional best achievable test performance from a given PROC curve, and s_{opt} as the corresponding optimal risk score threshold. However, as Shiu and Gatsonis [1] point out, there may be practical restrictions that militate against the operational use of this optimal threshold. The variation in predictive values in the neighborhood of the threshold identified as optimal is thus likely to be of interest in the process of selecting a value for operational use.

Let us compare Figure 4 with the corresponding leaf plot [7,8]. To do so, we require values of TPP and TNP for the test characterized by the risk score threshold value of $s_{opt} = 3.23$ on the PROC curve for Example 5 (Table 2). Using the data for Example 5 as given in Table 2, we can calculate the ROC curve (not shown). We may then refer to the risk score threshold value of 3.23 on this ROC curve and so obtain the corresponding values of TPP and FPP as 0.109 and 0.001 respectively (so TNP = 1 − FPP = 0.999). These are the data required for completion of Table 3. The table is calculated as follows: $p_{+\cap c} = \text{TPP} \cdot p_c$, $p_{+\cap nc} = \text{FPP} \cdot (1 - p_c)$, $p_{-\cap c} = (1 - \text{TPP}) \cdot p_c$, $p_{-\cap nc} = (1 - \text{FPP}) \cdot (1 - p_c)$, results shown to 2dp. The resulting leaf plot is then shown in Figure 5. From the leaf plot, we see that PPV and 1–NPV increase as prevalence increases. The leaf plot (Figure 5) embodies the predictive performance measures PSEP and r in the same way as Figure 4. The vertical distance between the leaf margins at any given prevalence is a PSEP value, and $r = 1 - $ PSEP.

Table 3. The numerical prediction-realization table for a test with two categories of realized (actual) status (c, nc) and two categories of prediction (+, −) based on Example 5 (Table 2).

Prediction (i)	Realization (j)		Row sums
	c	nc	
+	$p_{+ \cap c} = 0.04$	$p_{+ \cap nc} = 0.00$	$p_{+} = 0.04$
−	$p_{- \cap c} = 0.31$	$p_{- \cap nc} = 0.65$	$p_{-} = 0.96$
Column sums	$p_{c} = 0.35$	$p_{nc} = 0.65$	1

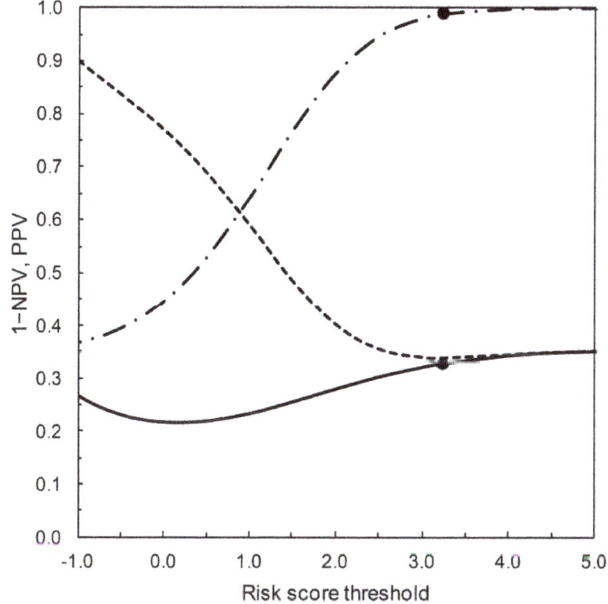

Figure 4. PPV and 1−NPV vary with the risk score threshold as characterized by the PROC curve in Figure 3. Only risk score thresholds that correspond to the part of the PROC curve above the main diagonal of the plot (as calibrated in Figure 3) are shown here. The PPV (dot-dash line) and 1−NPV (solid line) traces characterize the predictive performance measures PSEP = PPV − (1 − NPV) and $r = (1 - $ PPV$) + (1 - $ NPV$)$. The r trace (dashed line) reaches a minimum value of $r_{opt} = (1 - 0.990) + (1 - 0.674) = 0.336$ at risk score threshold $s_{opt} = 3.23$ [1]. The markers (•) on the predictive value traces at risk score threshold = 3.23 (PPV = 0.990, 1 − NPV = 0.326) indicate where this graphical plot coincides with Table 3.

On the vertical (probability scale) axis of both Figures 4 and 5 are PPV and 1−NPV. The two diagrams differ in terms of what drives variation in these predictive values. In Figure 4, the prevalence is constant and PPV and 1−NPV vary as TPP and FPP (and thus the risk score threshold) vary. In Figure 5 (the leaf plot), TPP and FPP (and thus the risk score threshold) are constant and PPV and 1−NPV vary as the prevalence varies. The numerical version of the 2 × 2 prediction-realization table (Table 3) for the test characterized by the risk score threshold value of $s_{opt} = 3.23$ on the PROC curve (Figure 3) describes the point at which Figures 4 and 5 coincide.

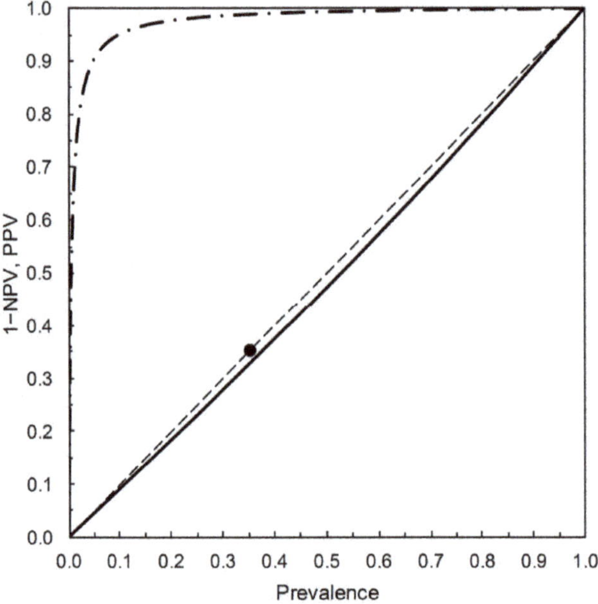

Figure 5. The leaf plot based on Example 5 (Table 2). The PPV trace (dot-dash line) shows that a + test outcome at almost any prevalence provides a useful indication of case status. However, the 1−NPV trace (solid line) shows that a − test outcome is of little value in ruling out non-case status. The marker (•) on the prevalence trace (dashed line) at 0.35 (where PPV = 0.990, 1 − NPV = 0.326) indicates where this graphical plot coincides with Table 3.

3.4. An Information Theoretic Perspective on Predictive Performance

The predictive performance metrics r and PSEP for PROC curves are measured on a probability scale; each provides a description of separation between prior and posterior probabilities which varies with prevalence. The corresponding approach to describing the performance of tests by means of ROC curves has been discussed in detail by Pepe, see Section 4.3 of [5]. In essence, an ROC curve provides a description of the separation between the distributions of risk scores for cases and non-cases. Indices that summarize ROC curves thus provide a summary of the separation between these two distributions. Separation between the distributions of risk scores is independent of prevalence, so we require this also of our summary indices. The most commonly-used such summary index is the area under the ROC curve.

If we wish instead to evaluate performance by measuring distances on an information theoretic scale, we must similarly distinguish between separation between prior and posterior probabilities (which depends on prevalence) and separation between (summaries of) distributions of risk scores (which does not). In both cases we may calculate relative entropies (Kullback-Leibler distances) in order to characterize separation. On the one hand, relative entropy calculations may be used to describe distances between distributions of risk scores [15,16], on the other, to describe distances between prior and posterior probabilities [17]; the details differ in each case as outlined in [18].

As applied in the present context, relative entropies are metrics that quantify diagnostic information from + and from − test outcomes (as expectations over both actual states). That is to say, diagnostic information is quantified in terms of prior and posterior probabilities. Then expected relative entropy, with the expectation calculated over both test outcomes, is equal to expected mutual information. From Table 3, we can directly calculate expected mutual information, I, via:

$$I = \sum_{i=+,-} \sum_{j=c,nc} p_{i \cap j} \cdot \ln \left[\frac{p_{i \cap j}}{p_i \cdot p_j} \right]$$

from which, on substituting the numerical data from the table, we obtain $\hat{I} = 0.04$ nats. This value corresponds to the risk score threshold $s_{opt} = 3.23$ on Figures 3 and 4. If we estimate I for further thresholds along the PROC curve shown in Figure 3, the resulting values form a curve with maximum value of $\hat{I} \approx 0.09$ nats at a risk threshold $1 < s < 2$. In this way, expected mutual information can provide an information theoretic perspective on the evaluation of predictive performance for a PROC curve. For further discussion of the qualitative correspondence between PSEP (on a probability scale) and I (on an information scale) as measures of separation, see [14].

4. Discussion

Shiu and Gatsonis [1] provide a comprehensive introduction to the predictive receiver operating characteristic curve for the joint assessment of positive and negative predictive values and its potential application in the evaluation of diagnostic tests. To realize its applicability, the PROC curve needs to become part of what Gatsonis [19] has called 'ROC thinking'—whereby the operational risk score threshold value is considered not just in relation to TPP and FPP (defined conditionally of the actual disease status) but also in relation to predictive values (defined conditionally on the outcome of the forecast). The PROC curve is part of such thinking. The PROC curve provides a format for test evaluation that is focused on predictive values and, in addition, links to the index of separation PSEP [6] and the leaf plot [7,8].

Because the PROC curve often displays complex patterns arising from the way that changes in the risk score threshold affect predictive values, interpreting a PROC curve requires both attention to detail and appropriate contextual data. The following recommendations are based on investigation of the PROC curves as characterized in Table 2, based on binormal ROC curves.

- The PROC curve and the ROC curve should not be regarded as mutually exclusive formats. Both perspectives contribute to test evaluation. In addition, use of both perspectives also serves a pedagogic function. The ROC curve is estimated conditionally on disease status whereas the PROC curve is estimated conditionally on test result. With both frames of reference available for consideration, carefully distinguishing between them becomes a requisite component of the presentation of a test's evaluation statistics.
- Generally (and especially in the case where the background distributional model for case and non-case risk scores may lead to an improper ROC curve) it is useful to include the main diagonal on the graphical plot of the PROC curve. It is also useful to present the full zero-one range of the axes of the plot, even if (e.g., in the case of an empirical analysis) the PROC curve data do not extend over the full range. Standardizing our view of the PROC curve helps in interpretation and comparison of the apparently complex patterns generated by PROC curve analysis.
- Calibrating a PROC curve at intervals with values of the risk score threshold is useful, because the way that the threshold changes along PROC curves is less obviously straightforward than for ROC curves.
- It is always useful to include details of the prevalence value for which a PROC curve has been calculated along with the graphical plot. Availability of a prevalence value allows further calculations (via Bayes' rule) following selection of a test based on a particular risk score threshold on the PROC curve.
- If a test is characterized on the basis of a threshold on the PROC curve, it is then useful to gain a wider perspective by determining TPP (sensitivity) and TNP (specificity) and calculating the leaf plot (for which purpose a spreadsheet is freely available [8]).

PROC curve analysis draws particular attention to the importance of predictive values in the evaluation process, and so in turn brings the index of separation PSEP [6] and the leaf plot [7,8] within the scope of Gatsonis' [15] 'ROC thinking'. Here we have considered PROC curves based on binormal ROC curves; other distributional models are available. When the background distributional models for case and non-case risk scores are continuous, there is a common information theoretic basis for understanding the shape properties of ROC curves [11] that can also contribute to our interpretation of the corresponding PROC curves. There is also an information theoretic approach to the assessment of predictive performance based on PROC curves.

The suggestions presented here are aimed at furthering adoption of the PROC curve and related methods for assessment of predictive values in the context of diagnostic test evaluation. As a set of methods they usefully augment the perspective provided by traditional ROC curve analysis used in isolation. Here, the binormal ROC model has provided the exemplar, but analysis based on risk score frequencies described by other distributional models would be of interest, as would empirical analysis (see, e.g., [20–22]). As yet there are few disease-related PROC analyses in the literature [2]; we expect this to change as understanding of the contribution that the format can make to test evaluation grows.

Funding: This research received no external funding.

Acknowledgments: The author thanks colleagues and reviewers who read the manuscript.

Conflicts of Interest: The author declares no conflict of interest.

References

1. Shiu, S.-Y.; Gatsonis, C. The predictive receiver operating characteristic curve for the joint assessment of the positive and negative predictive values. *Phil. Trans. R. Soc. A* **2008**, *366*, 2313–2333. [CrossRef] [PubMed]
2. Obuchowski, N.A.; Bullen, J. Receiver operating characteristic (ROC) curves: Review of methods with applications in diagnostic medicine. *Phys. Med. Biol.* **2018**, *63*, 07TR01. [CrossRef] [PubMed]
3. Zou, K.H.; Liu, A. Preface. *Phil. Trans. R. Soc. A* **2008**, *366*, 2251–2252. [CrossRef] [PubMed]
4. Sackett, D.L.; Haynes, R.B.; Guyatt, G.H.; Tugwell, P. *Clinical Epidemiology: A Basic Science for Clinical Medicine*, 2nd ed.; Little, Brown Book Group: London, UK, 1991.
5. Pepe, M.S. *The Statistical Evaluation of Medical Tests for Classification and Prediction*; Oxford University Press: Oxford, UK, 2003.
6. Altman, D.G.; Royston, P. What do we mean by validating a prognostic model? *Stat. Med.* **2000**, *19*, 453–473. [CrossRef]
7. Coulthard, M.G. Using urine nitrite sticks to test for urinary tract infection in children aged <2 years: A meta-analysis. *Pediatr. Nephrol.* **2019**, *34*, 1283–1288. [CrossRef] [PubMed]
8. Coulthard, M.G.; Coulthard, T. The leaf plot: A novel way of presenting the value of tests. *Br. J. Gen. Pract.* **2019**, *69*, 205–206. [CrossRef]
9. Bhattacharya, B.; Hughes, G. Symmetry of receiver operating characteristic curves and Kullback-Leibler divergences between the signal and noise populations. *J. Math. Psychol.* **2011**, *5*, 365–367. [CrossRef]
10. Hughes, G.; Bhattacharya, B. Symmetry properties of bi-normal and bi-gamma receiver operating characteristic curves are described by Kullback-Leibler divergences. *Entropy* **2013**, *15*, 1342–1356. [CrossRef]
11. Bhattacharya, B.; Hughes, G. On shape properties of the receiver operating characteristic curve. *Stat. Probabil. Lett.* **2015**, *103*, 73–79. [CrossRef]
12. Hillis, S.L.; Berbaum, K.S. Using the mean-to-sigma ratio as a measure of the improperness of binormal ROC curves. *Acad. Radiol.* **2011**, *18*, 143–154. [CrossRef] [PubMed]
13. Wahl, R.L.; Siegel, B.A.; Coleman, R.E.; Gatsonis, C.G. Prospective multicenter study of axillary nodal staging by positron emission tomography in breast cancer: A report of the staging breast cancer with PET study group. *J. Clin. Oncol.* **2004**, *22*, 277–285. [CrossRef] [PubMed]
14. Hughes, G.; Burnett, F.J. Evaluation of probabilistic disease forecasts. *Phytopathology* **2017**, *107*, 1136–1143. [CrossRef] [PubMed]
15. Lee, W.C. Selecting diagnostic tests for ruling out or ruling in disease: The use of the Kullback-Leibler distance. *Int. J. Epidemiol.* **1999**, *28*, 521–525. [CrossRef] [PubMed]

16. McKeigue, P. Quantifying performance of a diagnostic test as the expected information for discrimination: Relation to the *C*-statistic. *Stat. Methods Med. Res.* **2019**, *28*, 1841–1851. [CrossRef] [PubMed]

17. Hughes, G.; Reed, J.; McRoberts, N. Information graphs incorporating predictive values of disease forecasts. *Entropy* **2020**, *22*, 361. [CrossRef]

18. Hughes, G. Information graphs for epidemiological applications of the Kullback-Leibler divergence. *Meth. Inform. Med.* **2014**, *53*, IV–VI. [CrossRef] [PubMed]

19. Gatsonis, C. Receiver operating characteristic analysis for the evaluation of diagnosis and prediction. *Radiology* **2009**, *253*, 593–596. [CrossRef] [PubMed]

20. Liedenbaum, M.H.; van Rijn, A.F.; de Vries, A.H.; Dekker, H.M.; Thomeer, M.; van Marrewijk, C.J.; Hol, L.; Dijkgraaf, M.G.W.; Fockens, P.; Bossuyt, P.M.M.; et al. Using CT colonography as a triage technique after a positive faecal occult blood test in colorectal cancer screening. *Gut* **2009**, *58*, 1242–1249. [CrossRef] [PubMed]

21. Keihani, S.; Putbrese, B.E.; Rogers, D.M.; Patel, D.P.; Stoddard, G.J.; Hotaling, J.M.; Nirula, R.; Luo-Owen, X.; Mukherjee, K.; Morris, B.J.; et al. Optimal timing of delayed excretory phase computed tomography scan for diagnosis of urinary extravasation after high-grade renal trauma. *J. Trauma Acute Care Surg.* **2019**, *86*, 274–281. [CrossRef] [PubMed]

22. Oehr, P.; Ecke, T. Establishment and characterisation of empirical biomarker predictive value ROC curves (PV-ROC) using results of the UBC®*Rapid* Test in Bladder Cancer. *Entropy* **2020**. submitted.

Article

Establishment and Characterization of an Empirical Biomarker SS/PV-ROC Plot Using Results of the UBC® *Rapid* Test in Bladder Cancer

Peter Oehr [1],* and Thorsten Ecke [2],*

[1]   Faculty of Medicine, Rheinische Friedrich-Wilhelms-Universität Bonn, 53113 Bonn, Germany
[2]   Department of Urology, HELIOS Hospital, 15526 Bad Saarow, Germany
*   Correspondence: peter@oehr.info (P.O.); thorsten.ecke@helios-gesundheit.de (T.E.)

Received: 4 June 2020; Accepted: 29 June 2020; Published: 30 June 2020

Abstract: Background: This investigation included both a study of potential non-invasive diagnostic approaches for the bladder cancer biomarker UBC® *Rapid* Test and a study including comparative methods about sensitivity–specificity characteristic (SS-ROC) and predictive receiver operating characteristic (PV-ROC) curves that used bladder cancer as a useful example. Methods: The study included 289 urine samples from patients with tumors of the urinary bladder, patients with non-evidence of disease (NED) and healthy controls. The UBC® *Rapid* Test is a qualitative point of care assay. Using a photometric reader, quantitative data can also be obtained. Data for pairs of sensitivity/specificity as well as positive/negative predictive values were created by variation of threshold values for the whole patient cohort, as well as for the tumor-free control group. Based on these data, sensitivity–specificity and predictive value threshold distribution curves were constructed and transformed into SS-ROC and PV-ROC curves, which were included in a single SS/PV-ROC plot. Results: The curves revealed TPP-asymmetric improper curves which cross the diagonal from above. Evaluation of the PV-ROC curve showed that two or more distinct positive predictive values (PPV) can correspond to the same value of a negative predictive value (NPV) and vice versa, indicating a complexity in PV-ROC curves which did not exist in SS-ROC curves. In contrast to the SS-ROC curve, the PV-ROC curve had neither an area under the curve (AUC) nor a range from 0% to 100%. Sensitivity of the qualitative assay was 58.5% and specificity 88.2%, PPV was 75.6% and NPV 77.3%, at a threshold value of approximately 12.5 µg/L. Conclusions: The SS/PV-ROC plot is a new diagnostic approach which can be used for direct judgement of gain and loss of predictive values, sensitivity and specificity according to varied threshold value changes, enabling characterization, comparison and evaluation of qualitative and quantitative bioassays.

Keywords: predictive ROC curve; ROC curve; PV-ROC curve; SS-ROC curve; SS/PV-ROC plot; empirical; urinary bladder cancer

---

1. Introduction

In clinical practice for the detection of urinary bladder cancer, the confirmatory gold-standard procedure, cystoscopy, is invasive, costly and time consuming. Thus, there is interest in easy to perform rapid noninvasive bioassays at lower cost for detecting cancer disease in urinary samples from patients with suspected bladder cancer or for follow-up of the disease in bladder cancer patients.

To date, antigens determined by bioassays for urinary bladder cancer are not tumor specific. However, since bladder cancer tissue, compared to normal tissue, often expresses higher concentrations for those antigens, elevated levels can also be found in urine of bladder cancer patients, when compared to the levels of individuals without cancer. This enables the use of such assays for antigen determination in the diagnostics of urinary bladder cancer to a certain extent.

A new type of noninvasive urine-based tumor marker test for detection of urinary bladder cancer is the UBC® *Rapid* Test (Urinary Bladder Cancer Antigen *Rapid* Test). This biomarker test is a visual point of care (POC) test, detecting antigen fragments of cytokeratin 8 and 18 from urine samples only qualitatively. These antigens can also be determined quantitatively using the qualitative POC UBC® *Rapid* Test, combined with a photometric POC reader. With respect to quantitative assay determinations, cytokeratin levels are lower in low-grade tumors and benign urological diseases, compared to high-grade tumors [1,2]. Recent investigations including sensitivity–specificity characteristic (SS-ROC) curves gave evidence of utility for the UBC® *Rapid* Test in detecting CIS (carcinoma in situ) and non-invasive high-risk tumors [3,4].

The aim of this investigation was to evaluate the quality of the non-invasive diagnostic approaches for the qualitative and quantitative bladder cancer biomarker UBC® *Rapid* on the basis of tables and distribution curves for sensitivity, specificity and predictive values. Using the underlying study of bladder cancer as a useful example, the study was furthermore intended to access preliminary information about the utility of a newly developed graph consisting of both a sensitivity–specificity ROC SS-ROC and a predictive value ROC curve (PV-ROC), called a SS/PV-ROC plot. According to the present literature, there seems to be no publications on empirical PV-ROC-curves, and no SS/PV-ROC plot has been published to date.

2. Methods

The study was approved by the local Institutional Review Board of Medical Association Brandenburg (AS 147(bB)/2013).

In total, 289 urine samples were included in this prospective study. Clinical urine samples from 111 patients with tumors of the urinary bladder, 32 patients with non-evidence of disease (NED) and 146 healthy controls were used. Midstream urine was collected in a sterile plastic container and processed subsequently. Urine samples were analyzed by the UBC® *Rapid* Test (concile GmbH, Freiburg/Breisgau, Germany).

All patients with confirmed bladder cancer underwent cystoscopy, bladder ultrasound and transurethral resection of bladder tumor in the case of abnormal findings. Exclusion criteria were any kind of mechanical manipulation (cystoscopy, transrectal ultrasound and catheterization) within 10 days prior to urine sampling. Other exclusion criteria were benign prostate enlargement, urolithiasis other tumor diseases; severe infections; and pregnancy. All these criteria could influence the test to yield false positive results. Table 1 illustrates the characteristics of all bladder cancer patients.

The UBC® *Rapid* Test was performed by qualitative visual estimation of positive/negative results. Presently used POC-assays use qualitative immuno-chromatographic lateral flow assays which develop a concentration-dependent color reaction used as a threshold on a test line. A positive reaction is determined by subjective decision of human operators. The UBC® *Rapid* Test in combination with a POC-reader system enables quantitative determination of a tumor-marker under POC-assay conditions. Photometric readers can transform the concentration-dependent color reaction into quantitative values and represent a new development including objective, quantitative evaluation of POC-assays.

Data for pairs of sensitivity/specificity as well as positive/negative predictive values were created by variation of test threshold values for the whole patient cohort as well as for the tumor-free control group. The cut-offs were only selected to cover the range of the biomarker test used in the study, in order to have several values for plotting the SS-ROC and the PV-ROC curves. There were no clinical selection criteria, or criteria concerning an optimal cut-off. Based on these data, sensitivity–specificity and predictive value distribution curves were constructed and transformed into SS-ROC and PV-ROC curves, and then drawn together on a single SS/PV-ROC plot. In addition, the values for sensitivity, specificity and predictive values of the qualitative POC assay were plotted, each as a single point, for direct assignment to both ROC curves, as well as for estimation of the qualitative test's threshold values.

Table 1. Characteristics of bladder cancer patients.

	Status	*n* (%)
Tumor stage	pTa	61 (55%)
	pT1	14 (12.6%)
	pT2	23 (20.7%)
	pT3	9 (8.1%)
	only CIS	3 (2.7%)
	n.a.	1 (1%)
Grading	G1	26 (23.4%)
	G2	58 (52.3%)
	G3	24 (21.6%)
	G4	1 (0.9%)
	n.a.	2 (1.8%)
EORTC—Risk	Low risk	9
	Intermediate risk	43
	High risk	27
	n.a.	32
Number of tumors in bladder	1	50
	2–7	39
	≥8	8
	Not specified	14
Diameter of tumors in the bladder	$\varnothing < 3$ cm	56
	$\varnothing > 3$ cm	43
	n.a.	12
Primary vs. Recurrent tumors	Primary	58
	Recurrent	52
	n.a.	1
Number of recurrence	1	25
	2	6
	3	6
	≥4	14
	n.a.	1
Gross hematuria	yes	66
	no	45
Alguria	yes	33
	no	78

Explanation of abbreviated medical terminology. Tumor stage: The extent of a cancer in the body. Staging is usually based on the size of the tumor, whether lymph nodes contain cancer and whether the cancer has spread from the original site to other parts of the body; pTa tumors are those neoplasms that are confined to the epithelial layer of the bladder; pT1 tumors are those that invade into the subepithelial connective tissue; CIS (Carcinoma in situ) is a "flat tumor" of the epithelial layer; T2–T4 are muscle-invasive tumors. Tumor grade (G): A description of a tumor based on how abnormal the cancer cells and tissue look under a microscope and how quickly the cancer cells are likely to grow and spread; GI are cancer cells that resemble normal cells and are not growing rapidly; GII are cancer cells that do not look like normal cells and are growing faster than normal cells; GIII are cancer cells that look abnormal and may grow or spread more aggressively. Gross hematuria: Blood in the urine that can be seen with the naked eye. Alguria: Burning sensation when voiding.

Statistical Analysis

According to the tables, all statistical analyses were performed using R version 3.2.3 (R Core Team (2015) [5]. Data are presented descriptively using means and standard deviations for numerical variables and absolute and relative frequencies for categorical variables.

Data evaluation for the curves was conducted using Excel. True and false positive and true and false negative results of the qualitative and quantitative assays were calculated and applied for plotting distribution curves for sensitivity, specificity and positive and negative predictive values, as well as for SS-ROC and PV-ROC curves with respect to their corresponding set threshold values.

3. Results

Table 1 illustrates the data of the patients. Tumor stages pTa and pT1 include non-muscle-invasive (NMI) bladder cancer, while tumor stages pT2 and pT3 include muscle invasive (MI) bladder cancer; only CIS (carcinoma in situ) is a tumor type with a high risk of recurrence and progress. Grading of bladder cancer is stratified from high (G1) to low (G3) differentiation. European Organisation for Research and Treatment of Cancer (EORTC) risk is defined after the definition of the European Association of Urology [6,7]. Table 2 contains the description of study data. Clinical data were evaluated with respect to the UBC® *Rapid* Test. Table 2 shows that pathological concentrations of the UBC® *Rapid* Test are detectable in urine of bladder cancer patients. Pathological levels of the UBC® *Rapid* Test in urine are higher in patients with bladder cancer in comparison to the control group.

Table 2. Description of study data.

		Cancer	NED	NMI-LG	NMI-HG	MI-HG	Control	Total
		($n = 111$)	($n = 32$)	($n = 56$)	($n = 22$)	($n = 33$)	($n = 146$)	($n = 289$)
Age (years)								
	Mean	71.19	68.78	70.80	72.45	73.23	69.61	70.39
	(SD)	(11.46)	(13.14)	(12.35)	(9.41)	(9.31)	(11.94)	(11.71)
	Median	74	70.5	72	75	74	71.5	73
	Range	26 to 92	46 to 88	26 to 92	51 to 92	53 to 88	33 to 93	26 to 93
	n	111	32	56	22	33	146	289
Gender (M, F)	*n* (%)							
F		28 (25.23)	7 (21.88)	14 (25.00)	2 (9.09)	12 (36.36)	46 (31.51)	81 (28.03)
M		83 (74.77)	25 (78.12)	42 (75.00)	20 (90.91)	21 (62.64)	100 (68.49)	208 (71.93)
UBC (µg/L)								
	Mean	53.64	12.37	44.61	109.26	70.85	7.58	30.37
	(SD)	(87.93)	(11.40)	(81.98)	(115.76)	(97.12)	(14.00)	(66.66)
	Median	10.5	6.45	6.15	59.4	20.7	5	5
	Range	5 to 300	5 to 56.5	5 to 300	5 to 300	5 to 300	5 to 166	5 to 300
	n	111	32	56	22	33	146	289

Explanation of abbreviated medical terminology: NED, No evidence of disease according to the Guidelines on Non-Muscle-Invasive Urothelial Carcinoma [5,6]; NMIBC, Non-Muscle-Invasive Bladder Cancer; (TaT1 or carcinoma in situ (CIS); NMI-LG, Non-Muscle-Invasive Low-Grade (Bladder Cancer); NMI-HG, Non-Muscle-Invasive High-Grade (Bladder Cancer); MI-HG, Muscle-Invasive High-Grad (Bladder Cancer).

Figures 1 and 2 show concentration distribution curves calculated from values determined from samples of bladder cancer patients and cancer-free controls. Figure 1 refers to sensitivity and specificity, Figure 2 to PPV and NPV. Both figures confirm that the quantitative UBC® *Rapid* Test can discriminate between bladder cancer patients and cancer-free controls.

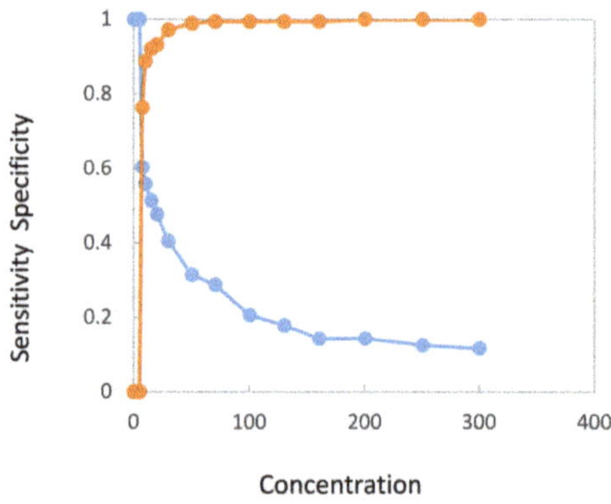

Figure 1. Distribution curves for sensitivity (blue line) and specificity (brown line) from setting various threshold values over the whole range of the UBC® *Rapid* Test concentrations (µg/L).

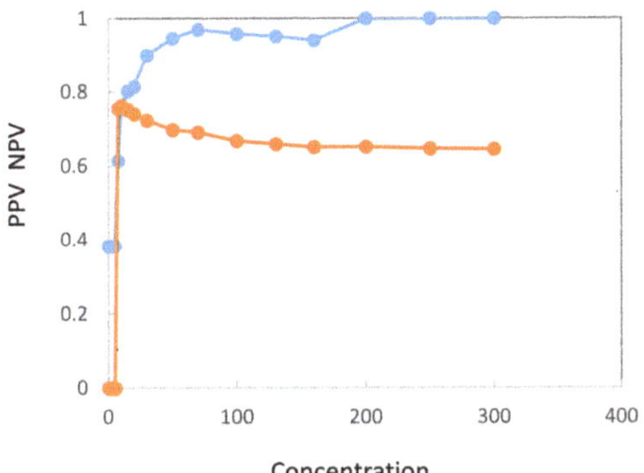

Figure 2. Distribution curves for positive (blue line) and negative (brown line) predictive values from setting various threshold values over the whole range of the UBC® *Rapid* Test concentrations (µg/L).

Figure 3 shows that all results for sensitivity, specificity and positive and negative predictive results of the risk thresholds can be included in a single SS/PV-ROC plot. The curves are calibrated at intervals of the risk score threshold and reveal TPP-asymmetric improper curves, which cross the diagonal from above. The SS-ROC curve (brown line, squares) is related to all created pairs of true positive rates (TPR) against the false positive rates (FPR) at various threshold value settings for specificity and sensitivity, and the PV-ROC curve (blue line, triangles) consists of all created pairs of PPV and NPV as the threshold for test positivity varied. The threshold numbers on the right side of the graph for sensitivity and PPV are not on a calibrated scale. In addition, the corresponding points calculated from the qualitative UBC® *Rapid* Test predictive value POC Test determination for sensitivity/specificity (black circle) and predictive values (black cross) are included in the graph.

Figure 3 demonstrates that, in contrast to the SS-ROC curve, which (generally) contains an area under the curve (AUC) and a full range from 0% to 100%, the PV-ROC curve has neither an AUC nor a range from 0% to 100%.

In addition to Figure 3, a section of this figure was drawn (Figure 4) to provide a detailed description of the course of the PPV curve, as well as to estimate the threshold concentrations of the visually judged quantitative test. This was done by correlating the threshold values of the quantitative assay to points for the sensitivity/specificity, as well as to those of the predictive values visually in this figure, which led to an approximate threshold value of 12.5 µg/L for both sensitivity and PPV. The graph shows a decrease of PPV starting at a threshold value of 70 (empty black triangle) and ending at 160 (empty red triangle), including the threshold values 100 (empty triangle) and 130 (full triangle).

Regarding the course of the predictive values curve, it is evident that two or more distinct values of PPV can correspond to the same value of NPV and vice versa (Figure 4), indicating a complexity in PV-ROC curves which does not exist in SS-ROC curves. Evaluation of the PV-ROC curve showed an (unexpected) decrease of PPV values at threshold values of 70, 100, 130 and 160.

The sensitivity of the qualitative assay at the threshold of 12.5 µg/L was 58.5%, the specificity was 88.2%, the PPV was 75.6% and the NPV was 77.3%. Figures 3 and 4 show that the values were located close to the respective curves, which confirms a good agreement of results. Visual estimation of the threshold concentration for the quantitative assay seemed to be equivalent to the threshold value range from 10.0 to 12.5 µg/L. At a threshold value of 10 µg/L, the values for the quantitative assay were 55.8, 88.8, 75.6 and 76.3 µg/L. The highest 1-NPV was 0.234 µg/L, close to the NPV threshold of the quantitative assay at a value of 0.227 µg/L.

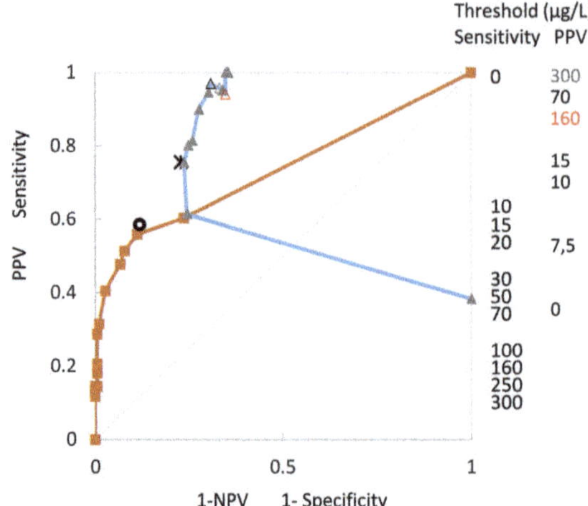

Figure 3. SS-ROC and PV-ROC curves, which were included in a single SS/PV-ROC plot. The curves are calibrated at intervals of the risk score threshold and reveal TPP-asymmetric improper curves, which cross the diagonal from above. The SS-ROC curve is shown as a brown line with squares, the PV-ROC curve as a blue line with triangles. Thresholds on each of the curves can be noted by reading horizontally across from the appropriate column of values on the right-hand side of the plot to the corresponding curve. In addition, the corresponding points calculated from the qualitative UBC® *Rapid Test* predictive value POC test determination for sensitivity/specificity (black circle) and predictive values (black cross) are included in the graph.

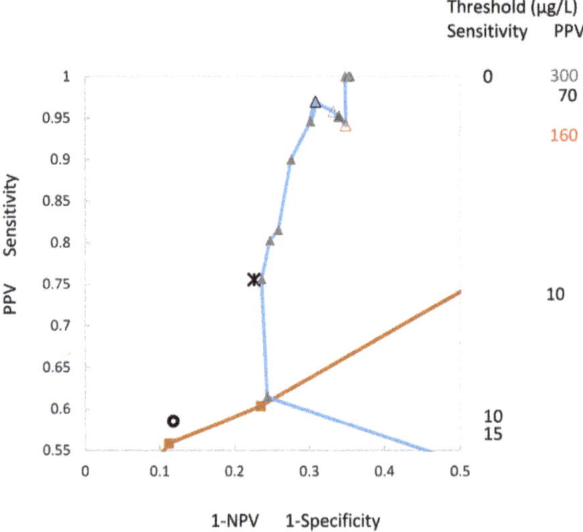

Figure 4. This figure is a section of Figure 3, which can be used to estimate the threshold value concentration for the qualitative test, derived from data for sensitivity (black circle) and PPV (black star), found to be approximately 12.5 µg/L. The graph illustrates decreasing PPV-values starting at a threshold value of 70 (empty black triangle) and ending at 160 (empty red triangle), including the threshold values 100 (empty triangle) and 130 (full triangle). The curves are calibrated at intervals of the risk score threshold. Thresholds on each of the curves can be noted by reading horizontally across from the appropriate column of values on the right-hand side of the plot to the corresponding curve.

4. Discussion

The SS/PV-ROC plot may become a good tool to judge patient values received from a biomarker determination from a more clinical perspective, within the context of the variable parameters seen in the SS-ROC and PV-ROC curves when both are considered together. Furthermore, it is a valuable approach to characterize quantitative biomarker assays and to compare them with others, including qualitative assays. However, evaluation of such diagnostics seems to be a complex procedure and clinicians will need assistance in learning how to deal with this approach with respect to the complexity of PV-ROC curves. Recently, articles published by Hughes [8] and Benish [9] provided valuable information and aid in dealing with this task (Benish does not discuss PV-ROC). A detailed evaluation of the findings in the presented article would include a discussion via information theory. In this article, which is primarily focused on establishment of the empirical SS/PV-ROC plot, a detailed analysis via information theory is not included. Instead, the reader is referred to the articles of Hughes [8] and Benish [9] who reported on mutual information as a metric for predictive performance for PV-ROC and SS-ROC, respectively.

The investigation presented in this article includes both a study of the potential non-invasive diagnostic approaches for the qualitative and quantitative cancer biomarker UBC® *Rapid* and a study including empirical SS-ROC and PV-ROC curves, using bladder cancer as an example. The results are presented in a single SS/PV-ROC plot for direct characterization, comparison and evaluation of two clinically applied bioassays.

One purpose of this study was to evaluate the clinical usefulness of the UBC® *Rapid* Test for diagnosis of bladder cancer with a focus on patients with non-muscle-invasive high-grade tumors (NMI-HG) of the urinary bladder compared with healthy individuals. The results of the present study show that cytokeratin concentrations determined by the UBC® *Rapid* Test measured by POC reader are statistically significant for patients with bladder cancer compared with healthy controls. Similar results were shown by Pichler et al. [10] and Styrke et al. [11].

The other purpose of this study was to use the results of a quantitative biomarker assay for bladder cancer in order to establish empirical PV-ROC curves and combine them with SS-ROC curves for sensitivity and specificity. Thus, conventional characterization, evaluation and comparability of bioassays could be applied at a broader scale to use this new tool to improve clinical diagnostics.

Concerning sensitivity and specificity for biomarkers, the first ROC curves were published in 1981 by Oehr et al. [12] for different patient groups affected with cancer of the breast, lung, urinary bladder and testis, in comparison to groups of healthy persons or patient groups with benign diseases. Within this first approach, ROC curves were established in order to directly compare different markers or different test systems independent of the correspondent marker concentrations.

Theoretical predictive value ROC curves including a study of the effect of the positivity threshold on the pair of PPV and NPV of tests were first published by Shiu and Gatsonis in 2008 [13], defining the curves mathematically, discussing the geometric patterns of these curves and describing methods for evaluating a test's predictive performance. According to the authors, it is essential "to study and attempt to characterize the geometric properties of PROC curves before undertaking an investigation of how the curves can be used to evaluate the performance of a diagnostic test" (PROC is a synonym of PV-ROC). To the best of our knowledge, empirical predictive ROC curves have not yet been published for bioassays by other authors, and accordingly this would also be valid for the SS/PV-ROC plot presented in this study.

The underlying investigation of bioassays included the qualitative UBC® *Rapid* urine-based point-of-care (POC) test, which can also be evaluated quantitatively by combining it with a reader system. The first evaluation of both the qualitative and the quantitative UBC® *Rapid* Tests was published by Ritter et al. [2]. In this study, the quantitative UBC® *Rapid* Test showed similar results when compared to the quantitative determination. The sensitivity of the quantitative assay was 55.7%, the specificity was 81.0%, the PPV was 56.7% and the NPV was 80.4%. The results for the quantitative assay were 60.7%, 70.1%, 46.8% and 79.3%, respectively. According to the quantitative assay, the optimal threshold value was calculated to be 12.3 µg/L, using the optimal threshold value obtained by receiver operating characteristic analysis for the quantitative assay according to the highest Youden index. However, a threshold for the qualitative assay was not known and could not be included for comparison. For comparing two assays, technically using the same POC test as a base for antigen detection, it would be optimal to use the same threshold value because at different thresholds the deviation of the results will normally increase, as can be seen according to threshold changes in the SS/PV-ROC plot in Figure 3. Without knowledge of both threshold values, however, the resulting diagnostic values can only reflect the degree of deviation, but such data cannot be interpreted in the same way as results from a direct comparison under (broadly similar) defined conditions.

With respect to the study presented in this article, the resulting diagnostic values found by comparison of the qualitative and quantitative assays are much closer to each other. Sensitivity of the qualitative assay was 58.5%, specificity 88.2%, PPV 75.6% and NPV 77.3%. In comparison, at the threshold concentration value of 10 µg/L, sensitivity of the quantitative assay was 55.8%, the specificity was 88.8%, the PPV was 75.6%, and the NPV was 77.3%. These results give evidence that both assays show a high agreement. The comparison could be made by prior estimation of the threshold by use of the SS/PV-ROC plot, and the outcome of this approach might be taken as an example for its utility.

The reason for including the thresholds for the resulting values of sensitivity and PPV in Figures 3 and 4 is that developing ROC curves from cut-off distribution curves for sensitivity/specificity or predictive values involves loss of information about the threshold concentrations. Adding the threshold values into the SS/PV-ROC plot again is regarded as important supplementary information which supports the reader in understanding and interpretation of SS/PV-ROC plots. It is important to know how the thresholds for the curves change and that the changes are not gradual but dynamic. As for publications using the SS/PV-ROC plot, it is recommended to use this approach.

The optimal threshold is presently calculated by most authors of tumor marker studies according to the highest Youden index. Ritter et al. [2] published an optimal cutoff at ≥12.3 µg/L. Styrke et al. [11] calculated an optimal threshold value at ≥8.1 µg/L, resulting in a sensitivity of 70.8%, specificity of 61.4%, PPV of 71.3% and NPV of 60.8%. Pichler et al. [10] reported the best cutoff at a threshold value of 6.7 ng/mL. The sensitivity, specificity, PPV and NPV of the visually evaluated qualitative UBC® *Rapid* Test were 61.3%, 77.3%, 65.5% and 73.9%, respectively. For the quantitative UBC® *Rapid* Test, sensitivity, specificity, PPV and NPV were 64.5%, 81.8%, 71.4% and 76.6%, respectively. This accumulation of different optimal threshold values in diagnostics and follow-up of bladder cancer patients seems to be confusing. The reason for differences in optimal threshold values may be related to different clinical states of the included patients and/or controls.

To find the appropriate threshold, regarding the clinical situation for requesting an examination of a patient, deriving threshold decisions from an SS/PV-ROC plot might be an alternative solution. At present, the calculation of the highest Youden index disregards the predictive values, and the low specificity of the published optimal values might lead to an increased number of false positive values, which could involve unnecessary invasive diagnostics in clinical follow-up of patients with suspected cancer.

Concerning the establishment of the SS/PV-ROC plot for the present study, an artifact appears to have evolved when plotting the empirical PPV-ROC at high values in a study with low case numbers. As illustrated in Figure 4, there was an (unexpected) decrease of PPV values at threshold values of 70, 100, 130 and 160. This was due to the fact that in case of all calculations for this threshold range there was always only a single case result for false positive values (FPR) in the calculations for PPV. There was no effect on NPV, since FPR values are not included in the calculations. In case a study includes a higher number of cases, this artifact effect might decrease or disappear. Specificity results regarding the mentioned FPR values were not markedly affected because the results for specificity were 99.4% in all cases.

The present study of the SS/PV-ROC plot is regarded as a first step, and application of this approach in daily clinical work is still regarded a goal. As we understand more about the characteristics of SS/PV-ROC plots, there will be opportunities for the re-analysis of existing datasets in order to gain a more detailed understanding of the operation of risk thresholds. The study by Styrke et al. [11] is one such example.

At present, evaluation of empirical PV-ROC curves remains a difficult task. Unlike the SS-ROC, the form of PV-ROC is dependent on prevalence of cases in the dataset, and this has some impact on the extent to which any particular SS/PV-ROC can be generalized. PPV and $1-$NPV increase as prevalence increases [8]. With respect to urinary bladder cancer, prevalence is known to differ in men and woman. Here, for the purpose of illustration, we treated the data as homogeneous. However, we note that, where sources of heterogeneity can be identified statistically within a dataset, this might call for separate PV-ROC curves for the different sub-sets. This is beyond the scope of the present article but worth noting as a topic for further research.

Support is necessary to understand the information theoretic perspective on evaluation, as well as to provide recommendations with a view to aiding understanding and interpretation of the sometimes-complex patterns generated by PV-ROC curves, their correlations with SS-ROC and their correlations to other related statistical methods, including estimation of prevalence and the leaf plot. Furthermore, to obtain agreement on a standardized PV-ROC curve evaluation, it is important to make future empirical studies by different authors or institutions comparable. Recently published articles [8,9] will help to develop this new path relating to the diagnostics of bioassays for cancer and provide a base in other fields of science for general application of the SS/PV-ROC plot.

Author Contributions: Conceptualization, methodology, analysis and writing—original draft preparation of PV-ROC and SS-ROC curves, SS/PV-ROC plot and clinical discussion, P.O. Conceptualization, methodology, analysis and writing—original draft preparation of clinical data T.E. P.O. and T.E. share first authorship. All authors have read and agreed to the published version of the manuscript.

Funding: The test materials were sponsored by concile GmbH, Freiburg/Breisgau, Germany and IDL Biotech, Bromma, Sweden.

Conflicts of Interest: The authors declare no conflict of interest. The funders had no role in the design of the study, in the collection, analyses, or interpretation of data; in the writing of the manuscript, or in the decision to publish the results.

References

1. Schroeder, G.L.; Lorenzo-Gomez, M.F.; Hautmann, S.H.; Toma, M.; Lorenzo Gomez, M.F.; Friedrich, M.G.; Jaekel, T.; Michl, U.; Schroeder, G.L.; Huland, H.; et al. A side by side comparison of cytology and biomarkers for bladder cancer detection. *J. Urol.* **2004**, *172*, 1123–1126. [CrossRef] [PubMed]
2. Ritter, R.; Hennenlotter, J.; Kuhs, U.; Ritter, R.; Hennenlotter, J.; Kühs, U.; Hofmann, U.; Aufderklamm, S.; Blutbacher, P.; Deja, A.; et al. Evaluation of a new quantitative point-of-care test platform for urine-based detection of bladder cancer. *Urol. Oncol.* **2014**, *32*, 337–344. [CrossRef] [PubMed]
3. Ecke, T.H.; Weiß, S.; Stephan, C.; Hallmann, S.; Barski, D.; Otto, T.; Gerullis, H. UBC(R) Rapid Test for detection of carcinoma in situ for bladder cancer. *Tumour. Biol.* **2017**, *39*, 1010428317701624. [CrossRef] [PubMed]
4. Ecke, T.H.; Weiß, S.; Stephan, C.; Hallmann, S.; Arndt, C.; Barski, D.; Otto, T.; Gerullis, H. UBC((R)) Rapid Test-A Urinary Point-of-Care (POC) Assay for Diagnosis of Bladder Cancer with a focus on Non-Muscle Invasive High-Grade Tumors: Results of a Multicenter-Study. *Int. J. Mol. Sci.* **2018**, *19*, 3841. [CrossRef] [PubMed]
5. R Core Team. *R: A Language and Environment for Statistical Computing;* R Foundation for Statistical Computing: Vienna, Austria, 2015. Available online: http://www.R-project.org/ (accessed on 21 April 2019).
6. Sylvester, R.J.; van der Meijden, A.P.M.; Oosterlinck, W.; Witjes, J.A.; Bouffioux, C.; Denis, L.; Newling, D.W.W.; Kurth, K. Predicting recurrence and progression in individual patients with stage Ta T1 bladder cancer using EORTC risk tables: A combined analysis of 2596 patients from seven EORTC trials. *Eur. Urol.* **2006**, *49*, 466–477. [CrossRef] [PubMed]

7. Babjuk, M.; Burger, M.; Compérat, E.M.; Gontero, P.; Mostafid, A.H.; Palou, J.; van Rhijn, B.W.G.; Rouprêt, M.; Shariat, S.F.; Sylvester, R. European Association of Urology Guidelines on Non-muscle-invasive Bladder Cancer (TaT1 and Carcinoma In Situ)—2019 Update. *Eur. Urol.* **2019**, *76*, 639–657. [CrossRef] [PubMed]
8. Hughes, G. On the binormal predictive receiver operating characteristic curve for the joint assessment of positive and negative predictive values. *Entropy* **2020**, *22*, 593. [CrossRef]
9. Benish, W.A. A Review of the Application of Information Theory to Clinical Diagnostic Testing. *Entropy* **2020**, *22*, 97. [CrossRef]
10. Pichler, R.; Tulchiner, G.; Fritz, J.; Schaefer, G.; Horninger, W.; Heidegger, I. Urinary UBC Rapid and NMP22 Test for Bladder Cancer Surveillance in Comparison to Urinary Cytology: Results from a Prospective Single-Center Study. *Int. J. Med. Sci.* **2017**, *14*, 811–819. [CrossRef] [PubMed]
11. Styrke, J.; Henriksson, H.; Ljungberg, B.; Hasan, M.; Silfverberg, I.; Einarsson, R.; Malmström, P.U.; Sheriff, A. Evaluation of the diagnostic accuracy of UBC((R)) Rapid in bladder cancer: A Swedish multicentre study. *Scand. J. Urol.* **2017**, *51*, 293–300. [CrossRef] [PubMed]
12. Oehr, P.D.H.; Altmann, R. Evaluation and characterisation of tumor associated antigens by conversion of inverse-distribution function values into ROC curves. *Tumordiagnostik* **1981**, *2*, 283.
13. Shiu, S.Y.; Gatsonis, C. The predictive receiver operating characteristic curve for the joint assessment of the positive and negative predictive values. *Philos. Trans. A Math. Phys. Eng. Sci.* **2008**, *366*, 2313–2333. [CrossRef] [PubMed]

Article

Mutual Information as a Performance Measure for Binary Predictors Characterized by Both ROC Curve and PROC Curve Analysis

Gareth Hughes [1,*], Jennifer Kopetzky [2] and Neil McRoberts [2]

[1] SRUC (Scotland's Rural College), The King's Buildings, Edinburgh EH9 3JG, UK
[2] Department of Plant Pathology, University of California, Davis, CA 95616, USA;
jlkopetzky@ucdavis.edu (J.K.); nmcroberts@ucdavis.edu (N.M.)
* Correspondence: gareth.hughes@sruc.ac.uk

Received: 10 August 2020; Accepted: 24 August 2020; Published: 26 August 2020

Abstract: The predictive receiver operating characteristic (PROC) curve differs from the more well-known receiver operating characteristic (ROC) curve in that it provides a basis for the evaluation of binary diagnostic tests using metrics defined conditionally on the outcome of the test rather than metrics defined conditionally on the actual disease status. Application of PROC curve analysis may be hindered by the complex graphical patterns that are sometimes generated. Here we present an information theoretic analysis that allows concurrent evaluation of PROC curves and ROC curves together in a simple graphical format. The analysis is based on the observation that mutual information may be viewed both as a function of ROC curve summary statistics (sensitivity and specificity) and prevalence, and as a function of predictive values and prevalence. Mutual information calculated from a 2 × 2 prediction-realization table for a specified risk score threshold on an ROC curve is the same as the mutual information calculated at the same risk score threshold on a corresponding PROC curve. Thus, for a given value of prevalence, the risk score threshold that maximizes mutual information is the same on both the ROC curve and the corresponding PROC curve. Phytopathologists and clinicians who have previously relied solely on ROC curve summary statistics when formulating risk thresholds for application in practical agricultural or clinical decision-making contexts are thus presented with a methodology that brings predictive values within the scope of that formulation.

Keywords: diagnostic test; mutual information; prevalence; PROC curve; positive predictive value; negative predictive value; ROC curve; sensitivity; specificity

---

1. Introduction

Receiver operating characteristic (ROC) curves and predictive receiver operating characteristic (PROC) curves are graphical formats with application in the determination of threshold values for proxy variables used in disease risk assessment when it is, for whatever reason, deemed inappropriate to use the gold standard. The work described in the present article concerns graphical threshold determination for binary predictors based on 2 × 2 prediction-realization tables. In crop protection decision making, binary tests are disease predictors that provide a probabilistic risk assessment of, for example, epidemic vs. no epidemic, or treatment required vs. no treatment required. Context for the work described here is provided by four previous articles; in chronological order of publication, Vermont et al. [1], Shiu and Gatsonis [2], Reibnegger and Schrabmair [3] and Hughes [4]. Vermont et al. [1] described general strategies of threshold determination for both ROC curves and PROC curves. Shiu and Gatsonis [2] described PROC curves and discussed a probabilistic measure of performance. Reibnegger and Schrabmair [3] described ROC curves and discussed both probabilistic and information theoretic measures of performance. Hughes [4] described both ROC curves and PROC curves and briefly discussed both probabilistic and information theoretic measures of performance for the latter.

Both ROC curves and PROC curves are based on graphical plots of conditional probabilities. In the case of the more well-known ROC curve, the probabilities are conditioned on the actual (gold standard) disease status. For the PROC curve, the probabilities are conditioned on the outcome of the test. The shape of an ROC curve is independent of disease prevalence, whereas the shape of a PROC curve varies with prevalence. Performance measures for both ROC and PROC curves are metrics that are deployed to search for a suitable balance of the conditional probabilities on which the plots are based. Much more work has been done on describing performance measures for ROC curves than for PROC curves, reflecting the historical levels of application of the curves in the evaluation of disease predictors. The work discussed here is presented as a unifying approach to the description of performance measures for both types of curve.

To illustrate this approach, we first extend the scope of [3], a study of performance measures for ROC curves, by calculating the corresponding PROC curves. This then provides a context for a discussion of performance measures as characterized in [2–4] in a range of ROC curves and the corresponding PROC curves. In particular we investigate the properties of the information theoretic performance measure mutual information, applied to both ROC curves and PROC curves. The work of Vermont et al. [1] is of interest in that although it appears to be one of the earliest studies of the application of both ROC and PROC curves to the problem of probabilistic risk assessment, it has not always been cited in the subsequent literature. Thus, we will integrate a discussion of [1] with our analysis of the findings of the present study.

The methodology described here is applicable to the development of binary prediction tools in phytopathology and also in clinical medicine. In particular, we show that the adoption of an information theoretic approach to performance measurement allows the choice of an appropriate risk score threshold to take both ROC curve and PROC curve characteristics into account in a single analysis.

2. Methods

2.1. Background to ROC Curves and PROC Curves

The present analysis of ROC curves and PROC curves uses the same starting point as a previous study of some performance measures for ROC curves [3]. However, it is helpful at the outset to place the analysis in the context of the kind of phytopathological studies in which these graphical formats find application for the evaluation of disease predictors in practice.

In crop protection decision making, an ROC curve is based on the analysis of a data set that typically comprises two observations derived from agronomic data collected during the growing season from each of a set of experimental crops, untreated for the disease in question. One observation is the gold standard disease assessment, often a measure of disease intensity, yield, or quality, made at the end of the growing season. The other observation is a risk score, based on data collected earlier in the season. The risk score provides a basis for crop protection decision making because in practice, a gold standard observation would come too late for application in decision making. Risk scores are typically calibrated so that higher scores are indicative of greater probability of a disease outbreak, or of the need for a disease management intervention. The methods we describe here assume that this data set of gold standard observations and their corresponding risk scores is already available for analysis. For further information on the assembly of such a data set, see Hughes [5] for background on methods for the calculation of risk scores from agronomic data, and Yuen et al. [6] and Twengström et al. [7] for an example of the experimentation that underlies the necessary agronomic data collection.

Crops are classified as cases ('c') or non-cases ('nc'), based, respectively, on whether or not the gold standard end-of-season assessment is indicative of economically significant damage. We may then calculate histograms of risk scores separately for the c and nc crop categories. Now, consider the introduction of a threshold on the risk score scale. Scores above the threshold are designated '+', indicative of (predicted) need for a crop protection intervention. Scores at or below the threshold are designated '−', indicative of (predicted) no need for a crop protection intervention.

The proportion of + predictions made for c crops is referred to as the true positive proportion (TPP or sensitivity) written $p_{+|c}$ in conditional probability notation. The complementary false negative proportion (FNP) is written $p_{-|c}$. Similarly, the proportion of + predictions made for nc crops is referred to as the false positive proportion (FPP), written $p_{+|nc}$. The complementary true negative proportion (TNP or specificity) is written $p_{-|nc}$. Thus, sensitivity and specificity are metrics defined conditionally on actual disease status. The ROC curve, which has become a familiar device in crop protection decision support following the pioneering work of Jonathan Yuen and colleagues [6,7], is a graphical plot of probabilities $p_{+|c}$ (sensitivity) against $p_{+|nc}$ (1 − specificity) derived by systematically varying the position of the threshold on the risk score scale and plotting the resulting probabilities over a range of risk scores.

In practice, the application of this analysis depends on the adoption of a particular threshold risk score for use in a given crop protection context. The variable that characterizes the risk score together with the adopted threshold on the risk score scale characterize a classification rule that may be referred to as a (binary) test ('predictor' is synonymous). Since the values of sensitivity and specificity are linked, a disease predictor based on a particular threshold must represent values chosen in order to achieve an appropriate balance; see Madden [8] for discussion. The considerations underlying adoption of a particular threshold risk score for use in a given crop protection context are beyond the scope of this article.

While sensitivity and specificity are of interest in characterizing a test, they are of limited significance in terms of the way we consider test results in the context of crop protection decision making. This is because they are metrics conditioned on the actual disease status which, in a practical decision-making context, we do not know. If we begin with a disease prevalence denoted p_c, often what we would really like to know is the predicted probability after a + test result, denoted $p_{c|+}$. To obtain this and similar probabilities, we apply Bayes' Rule.

Generally, we can write $i = +, -$ (for the predictions) and $j = c, nc$ (for the realizations). The p_i for a prediction either of intervention required ($i = +$) or of intervention not required ($i = -$) can be written as $p_i = p_{i|c} \cdot p_c + p_{i|nc} \cdot p_{nc}$ from the Law of Total Probability. The p_j for case ($j = c$, prevalence) or non-case ($j = nc$) status, such that $p_{nc} = 1 - p_c$, are taken as Bayesian prior probabilities (i.e., before the test is used to make a prediction). From Bayes' Rule, $p_{i|j} \cdot p_j = p_{j|i} \cdot p_i$, so we have $p_{c|+} = (p_{+|c} \cdot p_c)/p_+$ (positive predictive value, PPV) and the complement $p_{nc|+} = 1 - p_{c|+}$. Here, PPV refers to correct predictions of the need for a crop protection intervention; the complement 1 − PPV refers to incorrect predictions of the need for an intervention. We also have $p_{nc|-} = (p_{-|nc} \cdot p_{nc})/p_-$ (negative predictive value, NPV) and the complement $p_{c|-} = 1 - p_{nc|-}$. Here, NPV refers to correct predictions of no need for an intervention; the complement 1 − NPV refers to incorrect predictions of no need for an intervention. The predictive values are Bayesian posterior probabilities, calculated after obtaining the prediction. We note that the positive and negative predictive values are metrics conditioned on the test outcomes. Also, unlike sensitivity and specificity, which are independent of disease prevalence, the positive and negative predictive values vary with prevalence. The PROC curve is a graphical plot of probabilities $p_{c|+}$ (PPV) against $p_{c|-}$ (1 − NPV).

2.2. Analytical Scenarios and the Calculation of ROC Curves and Corresponding PROC Curves

Reibnegger and Schrabmair [3] described four scenarios "with quite different distributional characteristics". Each scenario comprised a pair of statistical probability distributions, modelling the separate (normalized) histograms of risk scores for c and nc subject categories. Here, we begin with the same four scenarios (Table 1).

In Table 1, each scenario's pair of distributions implicitly describes a parametric ROC curve. However, Reibnegger and Schrabmair [3] did not make these ROC curves explicit. Instead they used each pair of distributions as the basis for sampling c and nc data sets of various sizes. Their simulation study of ROC curve performance measures was based on the resulting sample data. Understandably, then, Reibnegger and Schrabmair [3] had no need to discuss the underlying parametric ROC curves and their properties. Here, however, these curves provide a basis for further analysis, so we explicitly calculate the ROC curve for each scenario (Figure 1) and discern its properties. An important reason for using the parametric ROC curves, rather than adopting the simulation approach of [3], is that we wish to be able to discuss the shape properties of the ROC and corresponding PROC curves for each scenario. The parametric ROC curves provide us with a non-varying baseline for this purpose. Visually, the curve for Scenario 4 passes noticeably closer to the top left-hand corner of the plot than the others, the curve for Scenario 2 stays noticeable further from the top left-hand corner, while the curves for Scenarios 1 and 3 are intermediate (Figure 1). By visual inspection, none of these ROC curves appears markedly asymmetrical.

Table 1. The four analytical scenarios [i,ii].

Scenario	Distribution of c	Distribution of nc
1 [iii]	Lognormal; mean = 2.5, s.d. = 0.3	Lognormal; mean = 2.0, s.d. = 0.4
2 [iv]	Chi-squared; d.f. = 10	Chi-squared; d.f. = 7
3	Inverse gamma; shape = 3	Inverse gamma; shape = 6
4	Weibull; shape = 10, scale = 20	Chi-squared; d.f. = 6

[i] Notation: c, cases; nc, non-cases; s.d., standard deviation; d.f., degrees of freedom. [ii] See Figure 1 in [3] for a graphical illustration of these scenarios. Each distribution was plotted over the range from 1 to 30 on the horizontal axis. [iii] See [9] for further discussion of the bi-lognormal receiver operating characteristic (ROC) curve. [iv] See [10] for further discussion of the bi-chi-squared ROC curve.

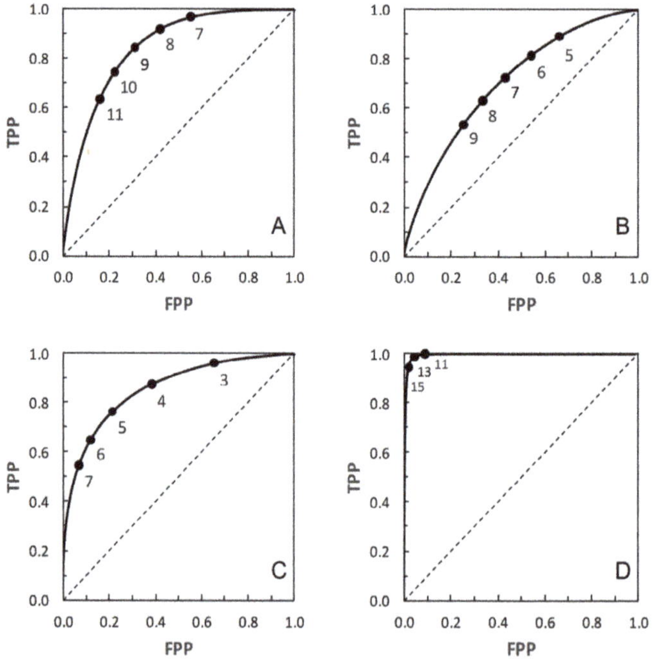

Figure 1. ROC curves for: (**A**) Scenario 1. (**B**) Scenario 2. (**C**) Scenario 3. (**D**) Scenario 4. See Table 1 for details. Risk score thresholds are calibrated in units of 1 unit on a 1 to 30 scale, following [3]. The risk score threshold increases along the curve from the top right-hand corner to the bottom left-hand corner. On each curve a subset of thresholds is indicated.

ROC curves are often described in terms of being "proper" or "improper". A proper ROC curve has a negative second derivative (i.e., decreasing slope) over the whole range; such a proper ROC curve never crosses the main diagonal of the plot [11]. However, an ROC curve that does not cross the diagonal may still be improper [11]. From the literature, Scenario 2 provides a proper ROC curve [9], and it appears from [10] that Scenario 1 provides an improper curve. We found no information relating to the curves for Scenarios 3 and 4. For the purpose of the present study, it is of more interest whether or not an ROC curve crosses the diagonal than whether it is strictly defined as proper or improper, so all we can really draw for certain from the literature is that the ROC curve in Figure 1B does not cross the main diagonal.

Having described the ROC curves, the first element of further analysis is to calculate the corresponding PROC curves for each of the four scenarios. The required probabilities can be obtained by adopting a value of p_c (prevalence), systematically varying the position of the threshold on the risk score scale to obtain values of $p_{+|c}$ (TPP) and $p_{+|nc}$ (FPP = 1 − TNP), then calculating PPV and 1 − NPV via Bayes' Rule. For each scenario, a PROC curve is calculated for each of nine prevalence values, from $p_c = 0.1$ to 0.9 at intervals of 0.1 (Figures 2–5).

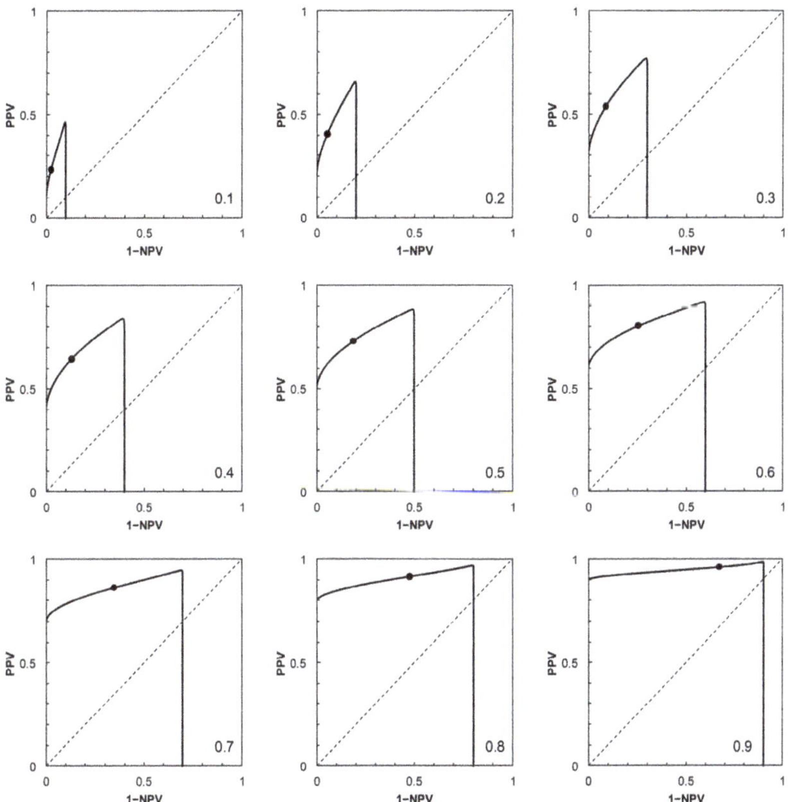

Figure 2. Scenario 1: Predictive receiver operating characteristic (PROC) curves corresponding to the ROC curve in Figure 1A. Each panel is labelled with the prevalence value at which the graph was calculated. For reference to Figure 1A, the threshold risk score at 9 is marked on each graph. Threshold risk scores increase along the curves, starting from the vertical axis (where 1 − NPV = 0), crossing the main diagonal (at which point PPV = 1 − NPV = prevalence) from above, and continuing the horizontal axis (where PPV = 0). NPV: negative predictive value, PPV: positive predictive value.

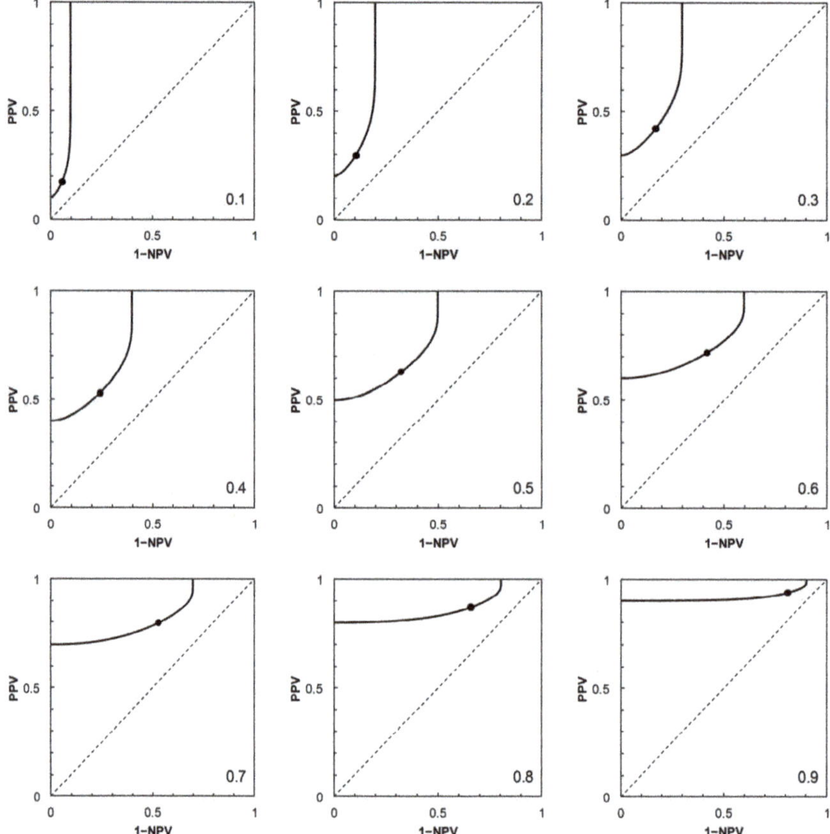

Figure 3. Scenario 2: PROC curves corresponding to the ROC curve in Figure 1B. Each panel is labelled with the prevalence value at which the graph was calculated. For reference to Figure 1B, the threshold risk score at 7 is marked on each graph. Threshold risk scores increase along the curves, starting from the vertical axis (where $1 - \text{NPV} = 0$) and continuing to the upper horizontal of the plot (where $\text{PPV} = 1$) without crossing the main diagonal.

As noted in [2], the shapes of PROC curves can appear rather complicated. There is not, as yet, an accepted vocabulary for discussion of the shapes of PROC curves. Here, we offer a descriptive account, based on [2,4]. The PROC curves in Figures 3 and 4, corresponding to ROC curves in Figure 1B (Scenario 2) and Figure 1C (Scenario 3) respectively, do not cross the main diagonal of the PROC plot. Since we know from [4] that where a PROC crosses the diagonal, it does so at the same risk score threshold as the corresponding ROC curve, this suggests that neither ROC curve crosses the diagonal. We know this definitively to be the case for Scenario 2, based on a proper ROC curve.

The PROC curves in Figures 2 and 5, corresponding to ROC curves in Figure 1A (Scenario 1) and Figure 1D (Scenario 4) respectively, cross the main diagonal of the PROC plot. Qualitatively, the shape of these PROC curves resembles that of Figure 2B in [4]. Starting at the left-hand vertical (PPV) axis of the plot, the risk score threshold increases along the curve. The curve cuts the main diagonal of the plot from above, then continues until meeting the horizontal ($1 - \text{NPV}$) axis. Now consider the ROC curves in Figure 1A (for corresponding PROC curves in Figure 2) and Figure 1D (for corresponding PROC curves in Figure 5). From [4], we know that these ROC curves must also cross the diagonal (in fact, they must cross at the same risk score threshold as the corresponding PROC curve). Starting in the top right-hand corner of the ROC plot (FPP = 1, TPP = 1), the risk score threshold increases along the curve. The curve cuts the main diagonal of the plot from above, then continues to the bottom left-hand corner of the plot (FPP = 0, TPP = 0). The point where the ROC curve cuts the diagonal is close to the bottom left-hand corner of the plot in Figure 1A,D, so is not obvious from visual inspection.

At the point where an ROC curve cuts the main diagonal of the plot, TPP = 1 − FPP, and we know that the positive and negative likelihood ratios (LR+ and LR−, respectively) are both equal to 1. Now, via the odds form of Bayes' Rule (i.e., posterior odds = prior odds × LR(+ or − as appropriate)), the posterior odds of c (given either a + or − test result) is equal to the prior odds of c; and similarly the posterior odds of nc (given either a + or − test result) is equal to the prior odds of nc. Converting these odds back to probabilities, we have $p_{c|+} = p_{c|-} = p_c$ and $p_{nc|+} = p_{nc|-} = p_{nc}$. In words, the result means that application of a test based on a threshold positioned on the main diagonal of an ROC plot is uninformative because it results in no revision of prior probabilities to new posteriors. This is a well-known observation; we include it here in order to compare the corresponding observation for a PROC curve. The points where the corresponding PROC curves cut their respective diagonals are (Figures 2 and 5) visually much clearer. We note that when the PROC curve crosses the diagonal of the plot, it does so at the point (1 − NPV, PPV), where both these conditional (posterior) probabilities are equal to the prior, p_c. So we can see directly that a test based on a threshold positioned on the main diagonal of an PROC plot is, by definition, uninformative.

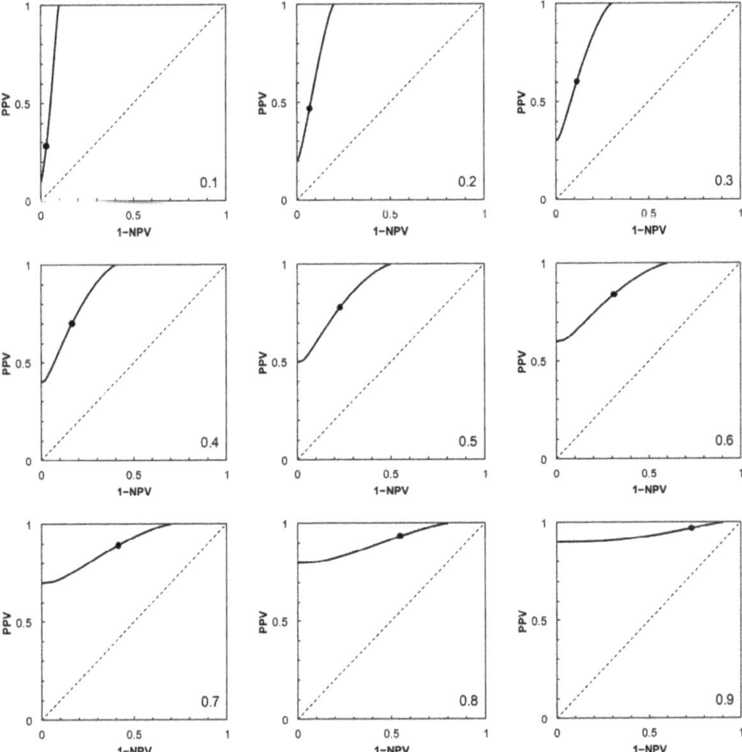

Figure 4. Scenario 3: PROC curves corresponding to the ROC curve in Figure 1C. Each panel is labelled with the prevalence value at which the graph was calculated. For reference to Figure 1C, the threshold risk score at 5 is marked on each graph. Threshold risk scores increase along the curves, starting from the vertical axis (where 1 − NPV = 0) and continuing to the upper horizontal of the plot (where PPV = 1) without crossing the main diagonal.

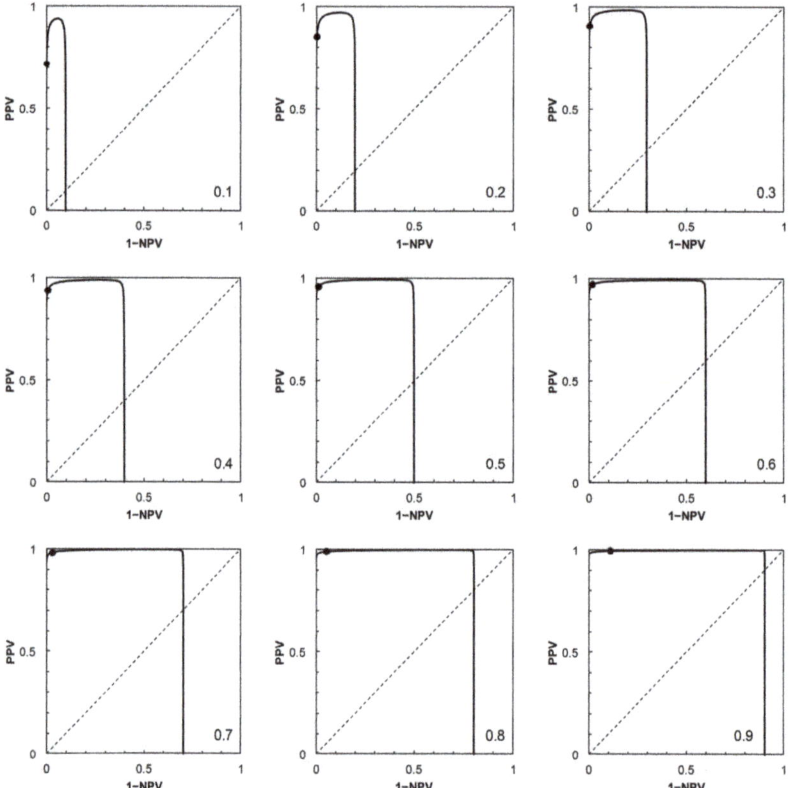

Figure 5. Scenario 4: PROC curves corresponding to the ROC curve in Figure 1D. Each panel is labelled with the prevalence value at which the graph was calculated. For reference to Figure 1D, the threshold risk score at 13 is marked on each graph. Threshold risk scores increase along the curves, starting from the vertical axis (where $1 - \mathrm{NPV} = 0$), crossing the main diagonal (at which point $\mathrm{PPV} = 1 - \mathrm{NPV} =$ prevalence) from above, and continuing to the horizontal axis (where $\mathrm{PPV} = 0$).

2.3. Performance Measures for ROC Curves and Corresponding PROC Curves

Performance measures for ROC and PROC curves are metrics that summarize the consequences of different choices about the position of the threshold on the risk score scale. Thus they provide methods for identification of what Reibnegger and Schrabmair [3] called the "optimum binary cut-off threshold". In [3] three such methods for ROC curves are considered in a simulation study: a probability-scale metric, an information-scale metric, and a metric based on logistic regression. Here we consider further the first two of these, but do not pursue their logistic regression analysis.

For ROC curves, Reibnegger and Schrabmair [3] calculated the probability-scale metric Youden's index [12], where the index $J = \mathrm{TPP} + \mathrm{TNP} - 1 = \mathrm{TPP} - \mathrm{FPP}$. J was originally proposed as a generic index for rating diagnostic tests, without reference to ROC curves. For a geometrical interpretation of J in the context of a test with TPP and FPP described by an ROC curve, consider two points on the ROC plot. The first is a point on the ROC curve positioned at a value TPP on the vertical axis; the second a point vertically below the first, positioned on the main diagonal of the plot (where $\mathrm{TPP} = \mathrm{FPP}$). The vertical distance between the two points is thus $\mathrm{TPP} - \mathrm{FPP}$. J can thus be thought of as the vertical distance between the curve and the main diagonal on an ROC plot at a given value of TPP. Reibnegger and Schrabmair sought the optimum risk score threshold on an ROC curve by systematically varying the threshold and observing the value at which J was maximized. In practice, a search for the maximum value of J would only need to consider thresholds where the ROC curve was above the main diagonal of the plot.

Now consider the equivalent geometrical examination of two points on a PROC plot. The first point is on the PROC curve positioned at a given value of PPV on the vertical axis (and, in practice, above the main diagonal of the plot); the second is a point vertically below the first, positioned on the main diagonal of the plot (where PPV = 1 − NPV). The vertical distance between the two points is thus calculated as PPV − (1 − NPV) = PPV + NPV − 1. This probability-scale metric was discussed in the context of the evaluation of diagnostic tests by Altman and Royston [13], who referred to it as PSEP. Note that Altman and Royston's discussion was generic. It concerned neither ROC curves nor PROC curves. In the present context, one could seek the optimum risk score threshold on an PROC curve by systematically varying the threshold and observing the value at which PSEP was maximized. These geometrical interpretations of the performance measures J (as applied to ROC curves) and PSEP (as applied to PROC curves) are both illustrated in Figure 6. The maximum values of J and of PSEP occur at different risk score thresholds.

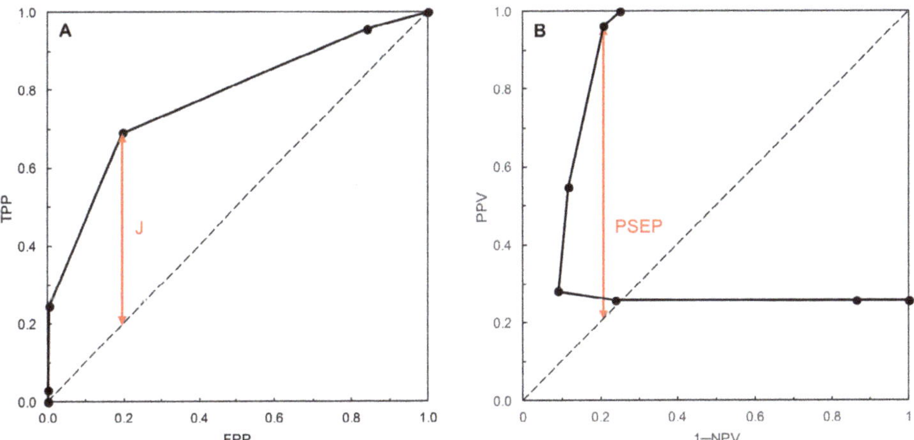

Figure 6. (**A**) The ROC curve is based on the normal distribution, with *c* subjects being N~(1.72, 0.42) and *nc* subjects N~(1.27, 0.27) (see [1] for details). Qualitatively, the shape of this improper ROC curve resembles that of Figure 1C in [4]. The risk score threshold increases along the ROC curve from the top right-hand corner of the plot to the bottom left-hand corner, crossing the main diagonal from below close to the top right-hand corner. The approximate maximum value of J = 0.494 (correct to 3 d.p.) occurs at a risk score threshold of 1.5. (**B**) The corresponding PROC curve was calculated as outlined in [4], with prevalence set to 180/702 = 0.256 (see [1]). Qualitatively, the shape of this PROC curve resembles that of Figure 2C in [4]. The risk score threshold increases along the PROC curve from the right-hand upright of the plot (where 1 − NPV = 1) to the upper horizontal (where PPV = 1), crossing the main diagonal from below at 1 − NPV = PPV = 0.256 (prevalence). The approximate maximum value of PSEP = 0.754 (correct to 3 d.p.) occurs at a risk score threshold of 2.0. Risk score thresholds on both curves are calibrated in units of 0.5 on a −10 to +10 scale (resulting data points may overlap).

We note that the metric *r* = (1 − PPV) + (1 − NPV) = 1 − PSEP [4] was discussed as a performance measure for PROC curves by Shiu and Gatsonis [2] (without reference to PSEP). It is a measure of distance (but not the shortest distance) between a given point on a PROC curve and the point (0, 1) in the top left-hand corner of the plot, with minimum value denoted *r**. In passing, we note that the ROC curve analogue of *r* is 1 − J = (1 − TPP) + (1 − TNP). We did not find any discussion of the use of this metric as a performance measure in the literature. The distance metrics J (and its complement) (for ROC curves) and PSEP and *r* (for PROC curves), and other metrics derived from them, have application in graphical determination of thresholds, as discussed in, for example, [1] (see Strategies 5 and 6) and [14].

We turn now to the information-scale metric mutual information (denoted here I). In the present context, mutual information is the expected value of the amount of information gained by application of a diagnostic test. Metz et al. [15] and McNeil et al. [16] appear to have described the first applications of I in the particular context of ROC curve analysis. As with J and PSEP, I is not defined specifically for such application [17]. Reibnegger and Schrabmair [3] sought the optimum risk score threshold on an ROC curve by systematically varying the threshold and observing the value at which I was maximized. Here we extend this approach to include the study of both ROC and PROC curves. Hughes [4] briefly discussed I as a potential performance measure for PROC curves.

Starting from a generic 2×2 prediction-realization table (Table 2), and working in natural logarithms, we obtain mutual information I via:

$$I = \sum_{i=+,-} \sum_{j=c,nc} p_{i \cap j} \cdot \ln\left\{\frac{p_{i \cap j}}{p_i \cdot p_j}\right\} \tag{1}$$

from which, on substituting the appropriate numerical data, we may calculate the required estimates of I in nats. In the present study, the calculation of I via Equation (1) was carried out on systematically varying the risk score threshold over the range 1 to 30 (in increments of 1 unit, along the calculated ROC curves for each scenario shown in Figure 1). In order to apply the results to the corresponding PROC curves (Figures 2–5), these calculations were carried out using nine different prior probabilities (prevalence values) over the range 0.1–0.9 in increments of 0.1.

We note at this stage that Equation (1) can be viewed either from an ROC curve perspective (i.e., in terms of sensitivity and specificity and their complements) or from a PROC curve perspective (i.e., in terms of predictive values). For the ROC perspective, we rewrite Equation (1) as:

$$\begin{aligned}
I = {} & p_{+|c} \cdot p_c \cdot \ln\left\{\frac{p_{+|c}}{p_{+|c} \cdot p_c + p_{+|nc} \cdot p_{nc}}\right\} \\
& + p_{+|nc} \cdot p_{nc} \cdot \ln\left\{\frac{p_{+|nc}}{p_{+|c} \cdot p_c + p_{+|nc} \cdot p_{nc}}\right\} \\
& + p_{-|c} \cdot p_c \cdot \ln\left\{\frac{p_{-|c}}{p_{-|c} \cdot p_c + p_{-|nc} \cdot p_{nc}}\right\} \\
& + p_{-|nc} \cdot p_{nc} \cdot \ln\left\{\frac{p_{-|nc}}{p_{-|c} \cdot p_c + p_{-|nc} \cdot p_{nc}}\right\}
\end{aligned} \tag{2}$$

in nats, which is Equation (2) from [15] written in the notation of the current article. Here mutual information is written as a function of sensitivity and specificity (and their complements) and the prevalence values for cases and non-cases. For the PROC perspective, we rewrite Equation (1) as:

$$I = \sum_{i=+,-} p_i \sum_{j=c,nc} p_{j|i} \cdot \ln\left\{\frac{p_{j|i}}{p_j}\right\} \tag{3}$$

in nats, which is Equation (4) from [18] written in the current notation. Here, mutual information is written as the information obtained from a specific test outcome (either $+$ or $-$) averaged over both c and nc subjects (this is relative entropy), then averaged over both $+$ and $-$ outcomes. Both [15] and [18] worked in base 2 logarithms rather than natural logarithms. To convert from natural logarithms to base 2 logarithms, divide by $\ln(2) \approx 0.693$ (in which case the units are bits).

Table 2. The prediction-realization table for a test with two categories of realized (actual) status (c, nc) and two categories of prediction ($+$, $-$). In the body of the table are the joint probabilities.

Prediction (i)	Realization (j)		
	c	nc	Row Sums
$+$	$p_{+\cap c}$	$p_{+\cap nc}$	p_+
$-$	$p_{-\cap c}$	$p_{-\cap nc}$	p_-
Column Sums	p_c	p_{nc}	1

3. Results

An immediate consequence of the fact that Equation (1) can be viewed either from the perspective of an ROC curve (Equation (2)) or a PROC curve (Equation (3)) is that the mutual information calculated for a given 2×2 prediction-realization table applies to the same risk score threshold on both curves. Thus, mutual information as a performance measure for binary predictors characterized by both ROC and PROC analysis has the same value at the same risk score threshold on both curves. Having obtained this result, we do not pursue the separate probability metrics J (for ROC curves) and PSEP (for PROC curves) further. We focus instead on the information metric I, applicable to both curves.

It is tests based on the part of the ROC curve above the main diagonal of the plot that are of interest in the context of diagnostic decision making. Here, $p_{+|c} > p_{+|nc}$, which implies $p_{c|+} > p_c$ and $p_{nc|-} > p_{nc}$ [4]. And as noted above, we know from [4] that for an ROC curve that crosses the main diagonal of the ROC plot, the corresponding PROC curve crosses the main diagonal of the PROC plot at the same threshold risk score. Looking first at Equation (2), recall that $p_c + p_{nc} = 1$, and that at the point where the ROC curve crosses the diagonal, $p_{+|c} = p_{+|nc}$ and $p_{-|c} = p_{-|nc}$. Thus at that point, each of the four terms in curly brackets in Equation (2) is equal to 1, and as $\ln\{1\} = 0$, $I = 0$ nats. Looking now at Equation (3), recall that where the PROC curve crosses the diagonal of the plot, we have $p_{c|+} = p_{c|-} - p_c$ and $p_{nc|+} - p_{nc|-} = p_{nc}$. So in Equation (3), we again have four terms in curly brackets, each term equal to 1 at the point where the PROC curve crosses the diagonal, so again we have $I = 0$ nats. This result confirms that at the risk score threshold where an ROC curve and the corresponding PROC curve cross the main diagonal of their respective plots, characterizing an uninformative predictor, the mutual information I is zero nats.

We now return to the scenarios outlined in Table 1. These are arbitrary in the sense that they represent plausible statistical simulacra of data used in the context of diagnostic test evaluation, rather than any specific disease diagnostic scenario. So the results presented here (Figures 7–10) are of interest mainly in terms of their qualitative characteristics. Note, in particular, that in the examples presented there is always a single maximum value of I (referred to here as I_{max}) over the range of threshold risk scores, whatever the shapes of the ROC and PROC curves. Somoza and Mossman [19] also observed this in a study based on bi-normal ROC curves. The threshold risk score for I_{max} decreases slowly with increasing prior probability, as noted in Reibnegger and Schrabmair's simulation study [3].

For each of Figures 7–10, each of the nine panels shows how I varies with risk score threshold at a specified prior probability. I_{max} refers to the maximum value of I for a particular panel. Clearly there is variation in I_{max} over the set of panels in each of Figures 7–10. Recall that in Figures 7–10, each panel applies both to an ROC curve from Figure 1A–D respectively and to a PROC curve from the corresponding panel from Figures 2–5 respectively. The values of I_{max} obtained in this way characterize an information-scale specification of the optimum risk score threshold at a specified prevalence for an ROC curve as discussed by [3], which is shown here to apply also to the corresponding PROC curves.

Metz et al. [15] were not directly concerned with characterizing the optimum risk score threshold on an ROC curve. Instead, their application of I_{max} was as measure of the "system quality" attributable to a device used in diagnostic decision making and described by an ROC curve, for the purpose of comparison with other such devices. Nevertheless, the calculations of mutual information in [15] are the same as those required for application in characterizing ROC curve thresholds [3], and those presented here with application further extended to characterizing PROC curve thresholds.

Metz et al. [15] pointed out a distinction between I_{max} and the global "information capacity" of a system. Information capacity, which we refer to here as channel capacity (denoted C) is the maximum value of I at a given risk score threshold taken over all values of prevalence. A (binary) "channel", in this case, is represented quantitatively by data from a numerical version of a 2 × 2 table such as Table 2. Now, for example, suppose we obtain from Figures 7–10 the risk score thresholds at which the largest value of I_{max} is observed for each scenario. These thresholds occur at 9 (Scenario 1, Figure 7), 7 (Scenario 2, Figure 8), 5 (Scenario 3, Figure 9), and 13 (Scenario 4, Figure 10). The corresponding largest observed values of I_{max} for each respective specified risk score threshold are then $I_{max} = 0.154$ nats (Figure 7), $I_{max} = 0.046$ nats (Figure 8), $I_{max} = 0.158$ nats (Figure 9) and $I_{max} = 0.568$ nats (Figure 10). We note in passing that these values of I_{max} reflect our earlier visual description of the ROC curves for the four scenarios in terms of the relative proximity of their paths to the top left-hand corner of the plot (Figure 1).

What we cannot say without further analysis is that these estimates of I_{max} are in the vicinity of C. While the calculation of C from a general prediction-realization table requires application of an iterative algorithm, there is a relatively simple analytical solution available in the case of a channel represented by a 2 × 2 table [20,21]. From this, using the same thresholds as above, we obtain for Scenario 1, $C = 0.155$ nats; for Scenario 2, $C = 0.046$ nats; for Scenario 3, $C = 0.158$ nats; and for Scenario 4, $C = 0.569$ nats (all to 3 d.p.). We find that the maximum value of I_{max}, obtained graphically at specified thresholds from Figures 7–10 for each of the four scenarios, is an approximation of the corresponding value of C. Thus calculation of the maximum value of I_{max} at a specified threshold can provide an estimate of what Metz et al. [15] called information capacity, furnishing an upper limit to their information theoretic measure of system quality. This result was unforeseen by Metz et al. [15].

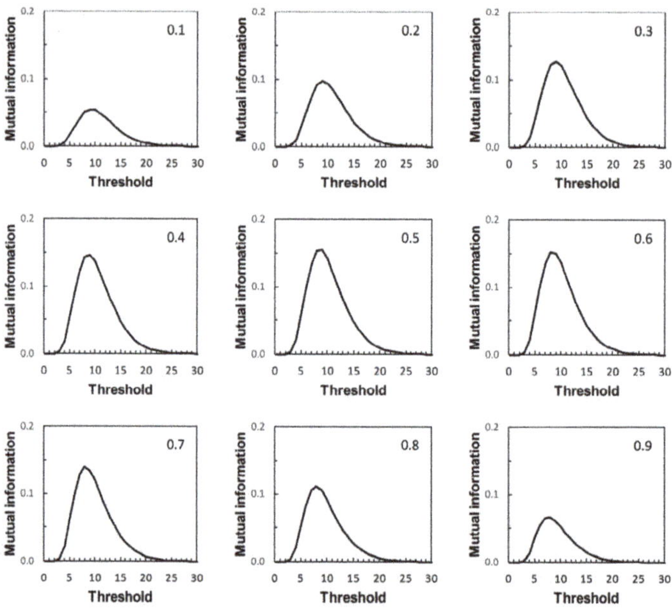

Figure 7. Scenario 1: variation of mutual information with risk score threshold. The calculated values of mutual information apply at risk score thresholds on the ROC curve in Figure 1A and at the same risk score thresholds on the corresponding PROC curves in Figure 2. Each panel is labelled with the prevalence value at which the graph was calculated. The vertical axis scales on Figures 7–10 differ.

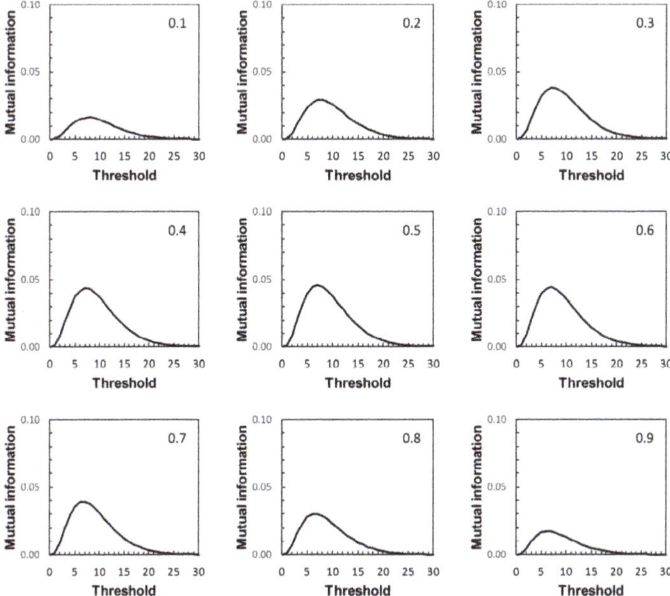

Figure 8. Scenario 2: variation of mutual information with risk score threshold. The calculated values of mutual information apply at risk score thresholds on the ROC curve in Figure 1B and at the same risk score thresholds on the corresponding PROC curves in Figure 3. Each panel is labelled with the prevalence value at which the graph was calculated. The vertical axis scales on Figures 7–10 differ.

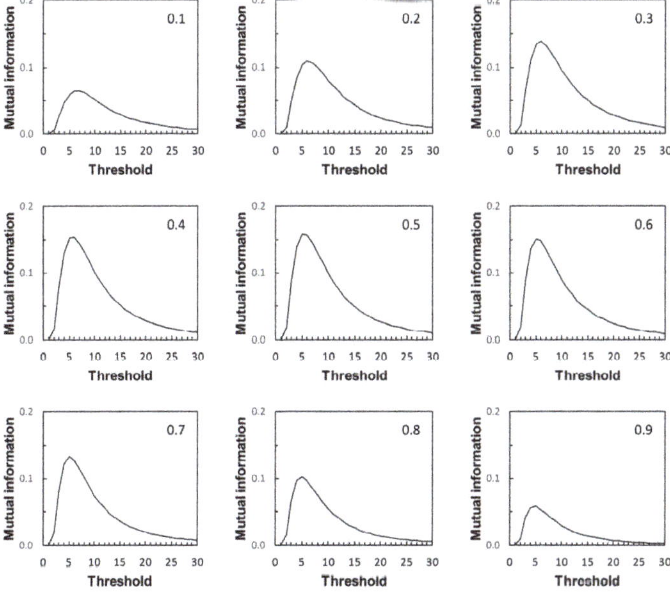

Figure 9. Scenario 3: variation of mutual information with risk score threshold. The calculated values of mutual information apply at risk score thresholds on the ROC curve in Figure 1C and at the same risk score thresholds on the corresponding PROC curves in Figure 4. Each panel is labelled with the prevalence value at which the graph was calculated. The vertical axis scales on Figures 7–10 differ.

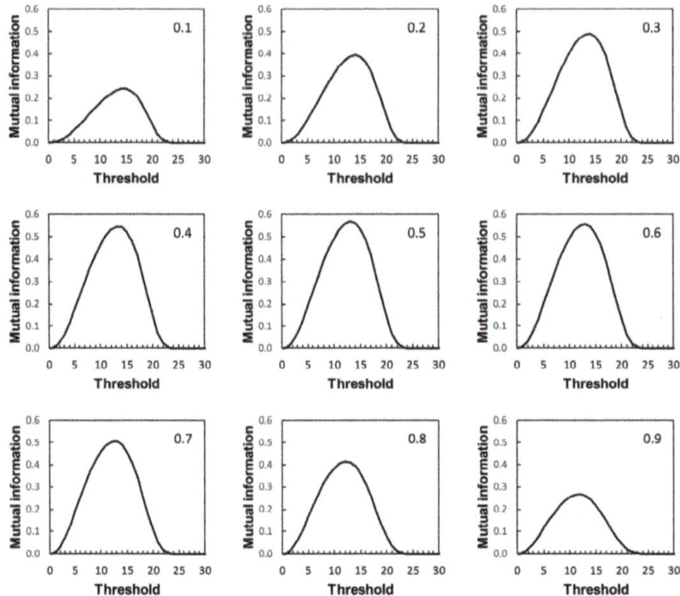

Figure 10. Scenario 4: variation of mutual information with risk score threshold. The calculated values of mutual information apply at risk score thresholds on the ROC curve in Figure 1D and at the same risk score thresholds on the corresponding PROC curves in Figure 5. Each panel is labelled with the prevalence value at which the graph was calculated. The vertical axis scales on Figures 7–10 differ.

4. Discussion

Vermont et al. [1], concluding their study of the roles of ROC curves and PROC curves in the context of graphical methods for diagnostic threshold determination, wrote as follows: "we do not feel that it is possible to choose a segmentation threshold by only using the ROC curve of a variable when this threshold must be used for diagnostic purposes; the PROC curves are less attractive, more chaotic and imprecise than the ROC curves but can help to select or reject certain threshold choice strategies". Much the same point—that the complex patterns of PROC curves made their implementation difficult—was later made by Shiu and Gatsonis [2]. The question thus arises as to how we may realize the advantages of PROC curves in application (that is to say, how to make them more attractive) in the face of apparent presentational difficulties. Answering this question would facilitate use of PROC curve analysis to augment what we can learn from the application of ROC curve analysis, not to substitute for it.

Because of the dependence of PROC curves on prevalence, we displayed an array of PROC curves corresponding to the ROC curve on which each scenario was based (Figures 2–5). When calibrating predictive values for a predictor initially based on an ROC curve, there is potential application for an array of PROC curves such as shown in each of Figures 2–5 if consideration of more than one prevalence value is deemed necessary. For example, it was noted in [22] that the prevalence of bladder cancer is known to differ between subgroups of males and females. In such a situation, an array of PROC curves for different prevalence values may allow a preview of the likely extent of differences between the curves for each of the subgroups. A similar situation may arise in crop protection decision making with a predictor based on an ROC curve. For example, a predictor may be used in separate locations where geographical and/or climatic differences result in subgroups with differing disease prevalence [23].

Vermont et al. [1] discussed strategies for threshold determination based on probability measures; sensitivity and specificity for ROC curves, predictive values for PROC curves. We have discussed examples of such measures; J [12] for ROC curves and its analogue PSEP [13] for PROC curves. Probability measures require separate calculation and interpretation of performance measures for ROC curve analysis and for PROC curve analysis. Mutual information is an information theoretic performance measure that has had application in the analysis of ROC curves, for example [3,15,16]. We have studied the concurrent application of mutual information to the analysis of ROC curves and their corresponding PROC curves. The important new result of our study is that mutual information is a performance measure that is applicable to the analysis of both ROC curves and PROC curves. In particular, for a given prevalence, mutual information calculated at a specified risk score threshold on an ROC curve (using Equation (2)) is the same as mutual information calculated at the same risk score threshold on a PROC curve (using Equation (3)). In our study this result applied to scenarios based on proper, improper, and unspecified-type ROC curves. It is also applicable to empirical ROC and PROC analysis, as for example in [22].

The presentation of this result is noteworthy. We begin with an ROC curve, the graphical plot of sensitivity (TPP) against $1 - $ specificity ($1 - $ TNP $=$ FPP) (e.g., Figure 1). This curve is independent of prevalence. However, a PROC curve, the graphical plot of positive predictive value (PPV) against $1 - $ negative predictive value ($1 - $ NPV), is not independent of prevalence. Thus, in our study, we calculate PROC curves corresponding to an ROC curve for a range of prevalence values, from 0.1 to 0.9 at intervals of 0.1. Then, in each of Figures 2–5, we present an array of nine PROC curves for each ROC curve shown in Figure 1. Now we can calculate mutual information for risk score thresholds from 1 to 30 at intervals of 1 unit (thus following the methodology of [3]). These mutual information values apply to risk score thresholds along the ROC curve and to the same thresholds along the corresponding array of PROC curves. Thus, if we describe a scenario for description of a diagnostic device in terms of an ROC curve and a set of likely prevalence values in which the device may be operational, we can present an array of graphical plots of mutual information against risk score threshold as a performance measure that applies both to the ROC curve and the corresponding PROC curves (e.g., Figures 7–10).

If we set out to integrate ROC curve analysis and PROC curve analysis into a strategy for graphical threshold determination [1], an array such as shown in each of Figures 7–10 provides an information theoretic basis on which to meet this objective. We note that the threshold at which I_{max} is indicated in the appropriate panel of an array (for the specified prevalence) is not prescriptive. It provides guidance towards the choice of an appropriate threshold, taking into consideration data on both sensitivity and specificity (via the ROC curve) and predictive values (via the PROC curve). Values of sensitivity, specificity (and so J) and predictive values (and so PSEP) in the vicinity of the threshold identified by I_{max} can always be investigated if required.

Drawing mutual information contours calculated at a specified prevalence onto ROC space [15] is another way in which to present the information theoretic analysis of an ROC curve. However, this approach does not allow for integration of an analysis of the corresponding PROC curves into the same graphic. Nor, we believe, does this contour plot depict I_{max} as clearly as a graph of mutual information against risk score threshold. Metz et al. [15] were concerned with measuring and comparing system quality via mutual information, specifically by calculating I_{max} from an ROC curve by means of Equation (2) applied at a given prevalence. Any one panel from an array of graphical plots of mutual information against risk score threshold (e.g., Figures 7–10) fulfils this objective for a particular prevalence value. In addition, the maximum value of I_{max} at a specified risk score threshold across an array, independent of prevalence, is an estimate of channel capacity C.

There is little doubt that the complexity of PROC curves [1,2] is an obstacle to their application in assessment of the performance of binary predictors. Equally, few would disagree that predictive values, alongside sensitivity and specificity, should have a role to play in characterizing predictor performance. We have shown that adoption of an information theoretic performance measure, mutual information, in a graphical format that plots the variation of mutual information over an appropriate range of risk score thresholds, allows integration of ROC curve analysis and PROC curve analysis. So the undoubted difficulties of interpretation that the PROC graph's complexity presents may be avoided, while retaining the benefits of considering predictive values alongside ROC curve characteristics in the evaluation of predictor performance.

Author Contributions: The authors contributed equally to this research. G.H. wrote the first draft manuscript, which J.K. and N.M. reviewed and edited. All authors have read and agreed to the published version of the manuscript.

Funding: This research received no external funding. Work by NM on this paper falls under the objectives of USDA-NIFA Hatch project CA-D-PPA-2131-H.

Conflicts of Interest: The authors declare no conflict of interest.

References

1. Vermont, J.; Bosson, J.L.; Francois, P.; Robert, C.; Rueff, A.; Demongeot, J. Strategies for graphical threshold determination. *Comput. Methods Progr. Biomed.* **1991**, *35*, 141–150. [CrossRef]
2. Shiu, S.-Y.; Gatsonis, C. The predictive receiver operating characteristic curve for the joint assessment of the positive and negative predictive values. *Philos. Trans. R. Soc. A* **2008**, *366*, 23132333. [CrossRef] [PubMed]
3. Reibnegger, G.; Schrabmair, W. Optimum binary cut-off threshold of a diagnostic test: Comparison of different methods using Monte Carlo technique. *BMC Med. Inform. Decis. Mak.* **2014**, *14*, 99. [CrossRef] [PubMed]
4. Hughes, G. On the binormal predictive receiver operating characteristic curve for the joint assessment of positive and negative predictive values. *Entropy* **2020**, *22*, 593. [CrossRef]
5. Hughes, G. The evidential basis of decision making in plant disease management. *Annu. Rev. Phytopathol.* **2017**, *55*, 41–59. [CrossRef] [PubMed]
6. Yuen, J.; Twengström, E.; Sigvald, R. Calibration and verification of risk algorithms using logistic regression. *Eur. J. Plant Pathol.* **1996**, *102*, 847–854. [CrossRef]
7. Twengström, E.; Sigvald, R.; Svensson, C.; Yuen, J. Forecasting Sclerotinia stem rot in spring sown oilseed rape. *Crop Prot.* **1998**, *17*, 405–411. [CrossRef]
8. Madden, L.V. Botanical epidemiology: Some key advances and its continuing role in disease management. *Eur. J. Plant Pathol.* **2006**, *115*, 3–23. [CrossRef]
9. Pundir, S.; Amala, R. Detecting diagnostic accuracy of two biomarkers through a bivariate log-normal ROC curve. *J. Appl. Stat.* **2015**, *12*, 2671–2685. [CrossRef]
10. Hillis, S.L.; Berbaum, K.S. Using the mean-to-sigma ratio as a measure of the improperness of binormal ROC curves. *Acad. Radiol.* **2011**, *18*, 143–154. [CrossRef]
11. Dorfman, D.D.; Berbaum, K.S.; Metz, C.E.; Length, R.V.; Hanley, J.A.; Abu Dagga, H. Proper receiver operating characteristic analysis: The bigamma model. *Acad. Radiol.* **1997**, *4*, 138–149. [CrossRef]
12. Youden, W.J. Index for rating diagnostic tests. *Cancer* **1950**, *3*, 32–35. [CrossRef]
13. Altman, D.G.; Royston, P. What do we mean by validating a prognostic model? *Stat. Med.* **2000**, *19*, 453–473. [CrossRef]
14. Hua, J.; Tian, L. A comprehensive and comparative review of optimal cut-points selection methods for diseases with multiple ordinal stages. *J. Biopharm. Stat.* **2020**, *30*, 46–68. [CrossRef] [PubMed]
15. Metz, C.E.; Goodenough, D.J.; Rossmann, K. Evaluation of receiver operating characteristic curve data in terms of information theory, with applications in radiography. *Radiology* **1973**, *109*, 297–303. [CrossRef]
16. McNeil, B.J.; Keeler, E.; Adelstein, S.J. Primer on certain elements of medical decision making. *N. Engl. J. Med.* **1975**, *293*, 211–215. [CrossRef]
17. Benish, W.A. Mutual information as an index of diagnostic test performance. *Methods Inf. Med.* **2003**, *42*, 260–264. [CrossRef]
18. Benish, W.A. A review of the application of information theory to clinical diagnostic testing. *Entropy* **2020**, *22*, 97. [CrossRef]
19. Somoza, E.; Mossman, D. Comparing and optimizing diagnostic tests: An information-theoretical approach. *Med. Decis. Mak.* **1992**, *12*, 179–188. [CrossRef]
20. Silverman, R.A. On binary channels and their cascades. *IRE Trans. Inf. Theory* **1955**, *1*, 19–27. [CrossRef]

21. Benish, W.A. The channel capacity of a diagnostic test as a function of test sensitivity and test specificity. *Stat. Methods Med. Res.* **2015**, *24*, 1044–1052. [CrossRef] [PubMed]

22. Oehr, P.; Ecke, T. Establishment and characterization of an empirical biomarker SS/PV-ROC plot using results of the UBC® *Rapid* Test in bladder cancer. *Entropy* **2020**, *22*, 729. [CrossRef]

23. Duttweiler, K.B.; Gleason, M.L.; Dixon, P.M.; Sutton, T.B.; McManus, P.S.; Monteiro, J.E.B.A. Adaptation of an apple sooty blotch and flyspeck warning system for the Upper Midwest United States. *Plant Dis.* **2008**, *92*, 1215–1222. [CrossRef] [PubMed]

 *entropy*

Article

Information Graphs Incorporating Predictive Values of Disease Forecasts

Gareth Hughes [1],*, Jennifer Reed [2] and Neil McRoberts [2]

[1]   SRUC, The King's Buildings, Edinburgh EH9 3JG, UK

[2]   Department of Plant Pathology, University of California, Davis, CA 95616, USA; reedje8@gmail.com (J.R.);
      nmcroberts@ucdavis.edu (N.M.)

*   Correspondence: gareth.hughes@sruc.ac.uk

Received: 10 January 2020; Accepted: 13 March 2020; Published: 20 March 2020

Abstract: Diagrammatic formats are useful for summarizing the processes of evaluation and comparison of forecasts in plant pathology and other disciplines where decisions about interventions for the purpose of disease management are often based on a proxy risk variable. We describe a new diagrammatic format for disease forecasts with two categories of actual status and two categories of forecast. The format displays relative entropies, functions of the predictive values that characterize expected information provided by disease forecasts. The new format arises from a consideration of earlier formats with underlying information properties that were previously unexploited. The new diagrammatic format requires no additional data for calculation beyond those used for the calculation of a receiver operating characteristic (ROC) curve. While an ROC curve characterizes a forecast in terms of sensitivity and specificity, the new format described here characterizes a forecast in terms of relative entropies based on predictive values. Thus it is complementary to ROC methodology in its application to the evaluation and comparison of forecasts.

Keywords: probability; forecast; likelihood ratio; positive predictive value; negative predictive value; diagnostic information; relative entropy

---

1. Introduction

Forecasting using two categories of actual status and two categories of forecast is common in many scientific and technical applications where evidence-based risk assessment is required as a basis for decision-making, including plant pathology and clinical medicine. The statistical evaluation of probabilistic disease forecasts often involves the calculation of metrics defined conditionally on actual disease status. For the purpose of disease management decision making, metrics defined conditionally on forecast outcomes (i.e., predictive values) are also of interest, although these are less frequently reported. Here we introduce a new diagrammatic format for disease forecasts with two categories of actual status and two categories of forecast. The format displays relative entropies, functions of predictive values that characterize expected information provided by disease forecasts. Our aims in introducing a new diagrammatic format are two-fold. First, we wish to highlight that performance metrics conditioned on forecast outcomes have a useful role in the overall evaluation of diagnostic tests and disease forecasters; second, bearing in mind the first aim, we wish to demonstrate that performance metrics based on information theoretic quantities can help distinguish characteristics of such tests and forecasters that may not be apparent from probability-scale metrics. The new diagrammatic format we introduce is intended to provide a generic approach that can applied in any suitable context.

Diagrammatic formats are useful for summarizing the processes of evaluation and comparison of disease forecasts in plant pathology and other disciplines where decisions about a subject must often be taken based on a proxy risk variable rather than knowledge of a subject's actual status. The receiver operating characteristic (ROC) curve [1] is one such well-known format. In plant pathology,

ROC curves are widely applied to characterize disease forecasters in terms of probabilities defined conditionally on actual disease status. Calculating the new diagrammatic format that we describe here has the same data requirements as the calculation of the ROC curve, but relates to relative entropy, an information theoretic metric that quantifies the expected amount of diagnostic information consequent on probability revision from prior to posterior arising from application of a disease forecaster. That is to say, it depicts (functions of) probabilities defined conditionally on the forecast. Even when the full underlying ROC curve data are not available, the new format can be constructed simply from ROC curve summary statistics.

The new diagrammatic format is linked analytically to other formats in ways that may not always be obvious simply from the resulting diagrams. We describe other formats and the links between them and the new format, using example data from a previously published study. In a general discussion, we consider the complementarity of metrics defined conditionally on the actual disease status and metrics defined conditionally on the outcome of the forecast.

2. Methods

We discuss information graphs for disease forecasters with two categories of actual status for subjects and two categories of forecast. In the present article, the terms 'forecast' and 'prediction' are synonymous. We place our discussion in the context of plant pathology, but the information graphs we describe likely have wider application. We are not concerned here with the detailed experimental and analytical methodology that underlies the development of disease forecasters. Readers seeking a description of such work are referred to Yuen et al. [2], Twengström et al. [3], and Yuen and Hughes [4], for example. Rather, we will describe some graphical methods for the comparison and evaluation of forecasters, and will outline some terminology and notation accordingly.

We need forecasters for support in crop protection decision making because the stage of the growing season at which disease management decisions are taken is usually much earlier than an assessment of actual (or 'gold standard') disease status could be made. For the purpose of development of a forecaster, two disease assessments are made on each of a series of experimental crops during the growing season. The actual status of each crop is characterized by an assessment of yield, or of disease intensity, at the end of the growing season. Crops are classified as cases ('c') or non-cases ('nc'), based on whether or not the gold standard end-of-season assessment indicates economically significant damage, respectively. Because the end-of-season assessment takes place too late to provide a basis for crop protection decision-making, an earlier assessment of disease risk is made, at a stage of the growing season when appropriate action can still be taken, if necessary. This earlier risk assessment may take the form of observation of a single variable that provides a risk score for the crop in question, or observation of a set of variables that are then combined to provide a risk score [5]. The risk score is a proxy variable, related to the actual status of the crop, that can be obtained at an appropriately early stage of the growing season for use in crop protection decision-making. Risk scores are usually calibrated so that higher scores are indicative of greater risk.

Now, consider the introduction of a threshold on the risk score scale. Scores above the threshold are designated '+', indicative of (predicted) need for a crop protection intervention. Scores at or below the threshold are designated '−', indicative of (predicted) no need for a crop protection intervention. The considerations underlying the adoption of a specific threshold risk score for use in a particular crop protection setting are beyond the scope of this article. Madden [6] discusses this in connection with an example data set that we consider in more detail below. In all settings, an adopted threshold characterizes the operational classification rule that is used as a basis for predictions of the need or otherwise for a crop protection intervention. The variable that characterizes the risk score together with the adopted threshold risk score that characterizes the operational classification rule together characterize what we may refer to as a (binary) 'test' ('forecaster' and 'predictor' are synonymous). A prediction-realization table [7] encapsulates the cross-classified experimental data underlying such a test. The data provide estimates of probabilities as shown in Table 1. Then, from Table 1 via Bayes'

Rule, we can write $\hat{p}_{i \cap j} = [\hat{p}_{j \cap i}] = \hat{p}_{i|j} \cdot \hat{p}_j = \hat{p}_{j|i} \cdot \hat{p}_i$, with $i = +, -$ (for the predictions) and $j = c, nc$ (for the realizations). The $\hat{p}_j$ are taken as the Bayesian prior probabilities of case ($j = c$) or non-case ($j = nc$) status, such that $\hat{p}_{nc} = 1 - \hat{p}_c$. Note also that the $\hat{p}_i$ for intervention required ($i = +$) and intervention not required ($i = -$) can be written as $\hat{p}_i = \hat{p}_{i|c} \cdot \hat{p}_c + \hat{p}_{i|nc} \cdot \hat{p}_{nc}$ via the Law of Total Probability.

Table 1. The prediction-realization table for a test with two categories of realized (actual) status (c, nc) and two categories of prediction ($+$, $-$). In the body of the table are the joint probabilities.

	Realization		
Prediction	c	nc	**Row Sums**
$+$	$\hat{p}_{+\cap c}$	$\hat{p}_{+\cap nc}$	$\hat{p}_+$
$-$	$\hat{p}_{-\cap c}$	$\hat{p}_{-\cap nc}$	$\hat{p}_-$
Column sums	$\hat{p}_c$	$\hat{p}_{nc}$	1

The posterior probability of (gold standard) case status (c) given a $+$ prediction on using a test is $p_{c|+}$, referred to as the *positive predictive value*. Here, this refers to correct predictions of the need for a crop protection intervention; the complement $p_{nc|+} = 1 - p_{c|+}$ refers to incorrect predictions of the need for an intervention. The posterior probability of (gold standard) non-case (nc) status given a $-$ prediction on using a test is $p_{nc|-}$, referred to as the *negative predictive value*. Here, this refers to correct predictions of no need for an intervention; the complement $p_{c|-} = 1 - p_{nc|-}$ refers to incorrect predictions of no need for an intervention. If we think of p_j ($j = c, nc$) as representing the Bayesian prior probabilities (i.e., before the test is used to make a prediction), the $p_{j|i}$ ($i = +, -$) then represent the corresponding posteriors (i.e., after obtaining the prediction). Predictive values are metrics defined conditionally on forecast outcomes.

The proportion of $+$ predictions made for cases is referred to as the true positive proportion, or *sensitivity*, and provides an estimate of the conditional probability $p_{+|c}$. The complementary false negative proportion is an estimate of $p_{-|c}$. The proportion of $+$ predictions made for non-cases is referred to as the false positive proportion, and provides an estimate of $p_{+|nc}$. The complementary true negative proportion, or *specificity*, is an estimate of $p_{-|nc}$. *Sensitivity* and *specificity* are metrics defined conditionally on actual disease status. The ROC curve, which has become a familiar device in crop protection decision support following the pioneering work of Jonathan Yuen and colleagues [2,3], is a graphical plot of *sensitivity* against $1-specificity$ for a set of possible binary tests, based on the disease assessments made during the growing season and derived by varying the threshold on the risk score scale. Since *sensitivity* and *specificity* values are linked, a disease forecaster based on a particular threshold represents values chosen to achieve an appropriate balance [8].

3. Results

3.1. Biggerstaff's Analysis

We denote the likelihood ratio of a $+$ prediction as L_+, estimated by:

$$\hat{L}_+ = \frac{\hat{p}_{+\mid c}}{\hat{p}_{+\mid nc}} \tag{1}$$

(in words, the expression on the RHS is the true positive proportion divided by the false positive proportion or *sensitivity/(1–specificity)*). We denote the likelihood ratio of a $-$ prediction as L_-, estimated by:

$$\hat{L}_- = \frac{\hat{p}_{-\mid c}}{\hat{p}_{-\mid nc}} \tag{2}$$

(in words, the expression on the RHS is the false negative proportion divided by the true negative proportion or *(1–sensitivity)/specificity*). Likelihood ratios are properties of a predictor (i.e., they are independent of prior probabilities) [9]. Values $L_+ > 1$ and $0 < L_- < 1$ are the minimum requirements for a useful binary test; within these ranges, larger positive values of L_+ and smaller positive values of L_- are desirable. L_+ characterizes the extent to which a + prediction is more likely from c crops than from nc crops; L_- characterizes the extent to which a − prediction is less likely from c crops than from nc crops.

Now, working in terms of odds (o) rather than probability (p) (with $o = p/(1-p)$), we can write versions of Bayes' Rule, for example:

$$\hat{o}_{c|+} = \hat{o}_c \cdot \hat{L}_+ \tag{3}$$

and:

$$\hat{o}_{c|-} = \hat{o}_c \cdot \hat{L}_-. \tag{4}$$

Thus, a + prediction increases the posterior odds of c status relative to the prior odds by a factor of $\hat{L}_+$ and a − prediction decreases the posterior odds of c status relative to the prior odds by a factor of $\hat{L}_-$. Biggerstaff [10] used Equations (3) and (4) to make pairwise comparisons of binary tests (with both tests applied at the same prior odds), premised on the availability only of the sensitivities and specificities corresponding to the two tests' operational classification rules (for example, when considering tests for application based on their published ROC curve summary statistics, *sensitivity* and *specificity*).

At this point, we refer to a previously published phytopathological data set [11] in order to illustrate our analysis. Note, however, that the analysis we present is generic, and is not restricted to application in one particular pathosystem. Table 2 summarizes data for five different scenarios, based in essence on five different normalized prediction-realization tables, derived from the original data set and discussed previously in [6] in the context of decision making in epidemiology.

Table 2. Example data set. See [6,11] for full details.

Scenario	$\hat{p}_c$	$\hat{p}_{+|c}$	$\hat{p}_{-|nc}$	$\hat{p}_{c|+}$	$\hat{p}_{nc|-}$
A	0.36	0.833	0.844	0.75	0.90
B	0.05	0.833	0.844	0.22	0.99
C	0.05	0.390	0.990	0.67	0.97
D	0.85	0.833	0.844	0.97	0.47
E	0.85	0.944	0.656	0.94	0.67

$\hat{p}_c$: prior probability of an epidemic or for the need for a control intervention, estimated by disease prevalence. $\hat{p}_{+|c}$: estimated probability of an actual epidemic being correctly predicted on using a test (as defined by a prediction-realization table). Referred to as *sensitivity*. $\hat{p}_{-|nc}$: estimated probability of an actual non-epidemic being correctly predicted on using a test (as defined by a prediction-realization table). Referred to as *specificity*. $\hat{p}_{c|+}$: estimated posterior probability of an epidemic given that one is predicted on using a test (as defined by a prediction-realization table). Referred to as *positive predictive value*. $\hat{p}_{nc|-}$: estimated posterior probability of no epidemic given that one is not predicted on using a test (as defined by a prediction-realization table). Referred to as *negative predictive value*.

Recall that we are interested in probability (or odds) revision calculated on the basis of a forecast. For illustration, we first consider the pairwise comparison of the tests derived from Scenario B (reference) and Scenario C (comparison) made at $\hat{p}_c = 0.05$ (Table 2). Madden [6] gives a detailed comparison based on knowledge of the full ROC curve derived from field experimentation. Biggerstaff's analysis essentially represents an attempt to reverse engineer a similar comparison based only on knowledge of the tests' published sensitivities and specificities. Scenario B yields *sensitivity* = 0.833 and *specificity* = 0.844, so we have $\hat{L}_+ = 5.333$ and $\hat{L}_- = 0.198$. Scenario C yields *sensitivity* = 0.390 and *specificity* = 0.990, so we have $\hat{L}_+ = 39.000$ and $\hat{L}_- = 0.616$. Thus, Scenario C's test is superior in terms of $\hat{L}_+$ values but inferior in terms of $\hat{L}_-$ values (even though its *sensitivity* is lower and *specificity* higher than that of the reference test). As long as we restrict ourselves to pairwise comparisons of binary tests at the same prior probability we have a simple analysis that leads, via calculation of likelihood ratios, to an evaluation of tests made on the basis of Bayesian posteriors (directly in terms of posterior odds, but these are easily converted to posterior probabilities if so desired). The diagrammatic version of this comparison is shown in Figure 1. The likelihood ratios graph comprises two single-point ROC curves. A similar analysis for Scenario D (reference) and Scenario E (comparison) (Figure 2) shows that Scenario E's test is inferior in terms of $\hat{L}_+$ values but superior in terms of $\hat{L}_-$ values (even though its *sensitivity* is higher and *specificity* lower than that of the reference test).

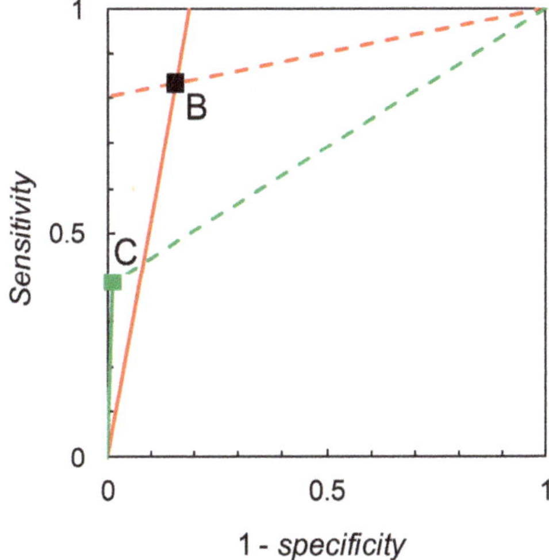

Figure 1. Biggerstaff's likelihood ratios graph for Scenario B (reference) and Scenario C (comparison). The graph for Scenario B consists of a single point at $1-specificity = 0.156$, $sensitivity = 0.833$ (see Table 2). The solid red line through $(0, 0)$ and $(0.156, 0.833)$ has slope $= sensitivity/(1-specificity) = 5.333 = \hat{L}_+$. The dashed red line through $(0.156, 0.833)$ and $(1, 1)$ has slope $= (1-sensitivity)/specificity = 0.198 = \hat{L}_-$. The graph for Scenario C consists of a single point at $1-specificity = 0.01$, $sensitivity = 0.39$ (see Table 2). The solid green line through $(0.01, 0.39)$ and $(1, 1)$ has slope $= sensitivity/(1-specificity) = 39.0 = \hat{L}_+$. The dashed green line through $(0.156, 0.833)$ and $(1, 1)$ has slope $= (1-sensitivity)/specificity = 0.616 = \hat{L}_-$.

Referring back to Table 2, the likelihood ratios, and corresponding graphs, for Scenarios A, B and D would be numerically identical. It is in this context that the information theoretic properties of likelihood ratios graphs (not pursued by Biggerstaff) are of interest. To elaborate further, we will require an estimate of the prior probability $\hat{p}_c$. This is beyond what Biggerstaff's analysis allowed, but it is not so unlikely that such an estimate might be available. For example, a $\hat{p}_c$ value is provided for any test for which a numerical version of the prediction-realization table (see Table 1) is accessible.

For information quantities, the specified unit depends on the choice of logarithmic base; bits for log base 2, nats for log base e, and hartleys (abbreviation: Hart) for log base 10 [12]. Our preference is to use base e logarithms, symbolized ln, where we need derivatives, following Thiel [7]. In this article, we will also make use of base 10 logarithms, symbolized $\log_{10}$, where this serves to make our presentation straightforwardly compatible with previously published work, specifically that of Johnson [13]. To convert from hartleys to nats, divide by $\log_{10}(e)$; or to convert from nats to hartleys, divide by ln(10). When logarithms are symbolized just by log, as immediately following, this indicates use of a generic format such that specification of a particular logarithmic base is not required until the formula in question is used in calculation.

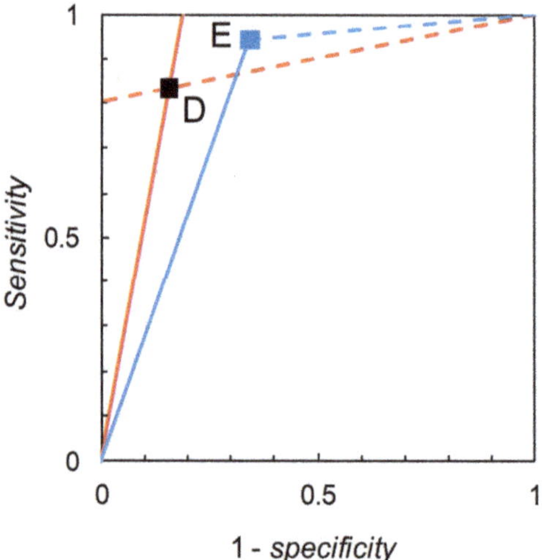

Figure 2. Biggerstaff's likelihood ratios graph for Scenario D (reference) and Scenario E (comparison). The graph for Scenario D consists of a single point at 1–*specificity* = 0.156, *sensitivity* = 0.833 (see Table 2). The solid red line through (0, 0) and (0.156, 0.833) has slope = *sensitivity*/(1–*specificity*) = 5.333 = $\hat{L}_+$. The dashed red line through (0.156, 0.833) and (1, 1) has slope = (1–*sensitivity*)/*specificity* = 0.198 = $\hat{L}_-$. The graph for Scenario E consists of a single point at 1–*specificity* = 0.344, *sensitivity* = 0.944 (see Table 2). The solid blue line through (0, 0) and (0.344, 0.944) has slope = *sensitivity*/(1–*specificity*) = 2.744 = $\hat{L}_+$. The dashed blue line through (0.344, 0.944) and (1, 1) has slope = (1–*sensitivity*)/*specificity* = 0.085 = $\hat{L}_-$.

We start with disease prevalence as an estimate of the prior probability $\hat{p}_c$ of need for a crop protection intervention, and seek to update this by application of a predictor. The information required for certainty (i.e., when the posterior probability of need for an intervention is equal to one) is then $\log(1/\hat{p}_c)$ denominated in the appropriate information units. However, a predictor typically does not provide certainty, but instead updates $\hat{p}_c$ to $\hat{p}_{c|i} < 1$. The information still required for certainty is then $\log(1/\hat{p}_{c|i})$ in the appropriate information units. We see from $\log(1/\hat{p}_c) - \log(1/\hat{p}_{c|i}) = \log(\hat{p}_{c|i}/\hat{p}_c)$ that the term $\log(\hat{p}_{c|i}/\hat{p}_c)$ represents the information content of prediction i in relation to actual status c in the appropriate information units. Provided the prediction is correct (i.e., in this case, $i = +$), the posterior probability is larger than the prior, and thus information content of the *positive predictive value* is > 0. In general, the information content of correct predictions is > 0. Predictions that result in a posterior unchanged from the prior have zero information content and incorrect predictions have information content < 0.

Here, we consider the information content of a particular forecast, averaged over the possible actual states. These quantities are *expected* information contents, often referred to as relative entropies. For a binary test:

$$\hat{I}_+ = \sum_{c,nc} \hat{p}_{j|+} \cdot \log\left[\frac{\hat{p}_{j|+}}{\hat{p}_j}\right] \tag{5}$$

for the forecast $i = +$ and:

$$\hat{I}_- = \sum_{c,nc} \hat{p}_{j|-} \cdot \log\left[\frac{\hat{p}_{j|-}}{\hat{p}_j}\right] \tag{6}$$

for the forecast $i = -$. Relative entropies measure expected information consequent on probability revision from prior $\hat{p}_j$ to posterior $\hat{p}_{j|i}$ after obtaining a forecast. Relative entropies are ≥ 0, with equality only if the posterior probabilities are the same as the priors. Larger values of both $\hat{I}_+$ and $\hat{I}_-$ are preferable, as being indicative of forecasts that, on average, provide more diagnostic information.

We can write the relative entropies $\hat{I}_+$ and $\hat{I}_-$ in terms of *sensitivity*, *specificity* and (constant) prior probability. Working here in natural logarithms, and recalling that $\hat{p}_{-\,|\,c} = 1 - \hat{p}_{+\,|\,c}$, $\hat{p}_{-\,|\,nc} = 1 - \hat{p}_{+\,|\,nc}$, and $\hat{p}_{nc} = 1 - \hat{p}_c$ we have:

$$
\begin{aligned}
\hat{I}_+ \;=\; & \frac{\hat{p}_{+\,|\,c}\cdot\hat{p}_c}{\hat{p}_{+\,|\,c}\cdot\hat{p}_c+\hat{p}_{+\,|\,nc}\cdot\hat{p}_{nc}}\cdot\ln\!\left[\frac{\hat{p}_{+\,|\,c}}{\hat{p}_{+\,|\,c}\cdot\hat{p}_c+\hat{p}_{+\,|\,nc}\cdot\hat{p}_{nc}}\right]\\
+\,& \frac{\hat{p}_{+\,|\,nc}\cdot\hat{p}_{nc}}{\hat{p}_{+\,|\,c}\cdot\hat{p}_c+\hat{p}_{+\,|\,nc}\cdot\hat{p}_{nc}}\cdot\ln\!\left[\frac{\hat{p}_{+\,|\,nc}}{\hat{p}_{+\,|\,c}\cdot\hat{p}_c+\hat{p}_{+\,|\,nc}\cdot\hat{p}_{nc}}\right]
\end{aligned}
\tag{7}
$$

in nats and:

$$
\begin{aligned}
\hat{I}_- \;=\; & \frac{\hat{p}_{-\,|\,c}\cdot\hat{p}_c}{\hat{p}_{-\,|\,c}\cdot\hat{p}_c+\hat{p}_{-\,|\,nc}\cdot\hat{p}_{nc}}\cdot\ln\!\left[\frac{\hat{p}_{-\,|\,c}}{\hat{p}_{-\,|\,c}\cdot\hat{p}_c+\hat{p}_{-\,|\,nc}\cdot\hat{p}_{nc}}\right]\\
+\,& \frac{\hat{p}_{-\,|\,nc}\cdot\hat{p}_{nc}}{\hat{p}_{-\,|\,c}\cdot\hat{p}_c+\hat{p}_{-\,|\,nc}\cdot\hat{p}_{nc}}\cdot\ln\!\left[\frac{\hat{p}_{-\,|\,nc}}{\hat{p}_{-\,|\,c}\cdot\hat{p}_c+\hat{p}_{-\,|\,nc}\cdot\hat{p}_{nc}}\right]
\end{aligned}
\tag{8}
$$

again in nats. Now we can use these formulas to plot sets of iso-information contours for constant relative entropies $\hat{I}_+$ and $\hat{I}_-$ on the graph with axes *sensitivity* and $1 -$ *specificity*, for given prior probabilities. From Equation (7) we obtain:

$$
\frac{\mathrm{d}(\hat{p}_{+\,|\,c})}{\mathrm{d}(\hat{p}_{+\,|\,nc})} = \frac{\hat{p}_{+\,|\,c}}{\hat{p}_{+\,|\,nc}}
\tag{9}
$$

the solution of which is the straight line $\hat{p}_{+\,|\,c} = a\cdot\hat{p}_{+\,|\,nc}$, which yields $a = \hat{L}_+$. From Equation (8) we obtain:

$$
\frac{\mathrm{d}(\hat{p}_{+\,|\,c})}{\mathrm{d}(\hat{p}_{+\,|\,nc})} = \frac{1-\hat{p}_{+\,|\,c}}{1-\hat{p}_{+\,|\,nc}}
\tag{10}
$$

the solution of which is the straight line $\hat{p}_{+\,|\,c} = (1-b) + b\cdot\hat{p}_{+\,|\,nc}$, which yields $b = \hat{L}_-$. Thus, we find that iso-information contours for $\hat{I}_+$ and $\hat{I}_-$ are straight lines on the graph with axes *sensitivity* and $1 -$ *specificity*, i.e., Biggerstaff's likelihood ratios graph (see Figure 3).

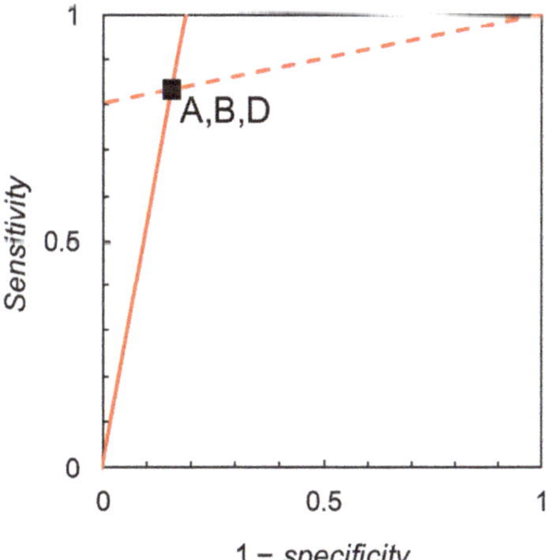

Figure 3. Biggerstaff's likelihood ratios graphs for Scenarios A, B and D (Table 2). The slopes of the lines are the likelihood ratios $\hat{L}_+ = 5.333$ and $\hat{L}_- = 0.198$, calculated from Table 2. Analysis shows that the lines themselves are also iso-information contours for the expected information contents of $+$ and $-$ forecasts. However, the calculated values of these expected information contents depend on the prior probability as well as on *sensitivity* and *specificity*. Making use of the available data on the prior probabilities allows us to calculate relative entropies in order to distinguish analytically between scenarios, but the likelihood ratios graph does not distinguish visually between scenarios with the same *sensitivity* and *specificity*.

Now consider Scenarios A, B and D; from the data in Table 2, we calculate likelihood ratios $\hat{L}_+ = 5.333$ and $\hat{L}_- = 0.198$ for all three scenarios (these are the slopes of the lines shown in Figure 3). However, the three scenarios differ in their prior probabilities: $\hat{p}_c = 0.36, 0.05, 0.85$ for A, B, and D respectively. This situation may arise in practice when a test is developed and used in one geographical location, and then subsequently evaluated with a view to application in other locations where the disease prevalence is different. The difference in test performance is reflected by the relative entropy calculations. For Scenario A, we calculate relative entropies $\hat{I}_+ = 0.315$ and $\hat{I}_- = 0.179$ (both in nats, these characterize the lines shown in Figure 3 interpreted as iso-information contours for the expected information contents of + and − forecasts respectively). For Scenario B, we calculate $\hat{I}_+ = 0.171$ and $\hat{I}_- = 0.024$ nats. For Scenario D, $\hat{I}_+ = 0.076$ and $\hat{I}_- = 0.289$ nats. Thus we may view Biggerstaff's likelihood ratios graph from an information theoretic perspective. While likelihood ratios are independent of prior probability, relative entropies are functions of prior probability. There is further discussion of relative entropies, including calculations for Scenarios C and E, in Section 3.3.

3.2. Johnson's Analysis

Johnson [13] suggested transformation of the likelihood ratios graph (e.g., Figures 1–3), such that the axes of the graph are denominated in log likelihood ratios. At the outset, note that Johnson works in base 10 logarithms and that this choice is duplicated here, for the sake of compatibility. Thus, although Johnson's analysis is not explicitly information theoretic, where we use it as a basis for characterizing information theoretic quantities, these quantities will have units of hartleys. Note also that Johnson calculates $\left|\log_{10}\hat{L}_+\right|$ and $\left|\log_{10}\hat{L}_-\right|$ but here we take account of the signs of the log likelihood ratios. Fosgate's [14] correction of Johnson's terminology is noted, although this does not affect our analysis at all.

From Equation (3), we write:

$$\log_{10}\hat{o}_c|_+ = \log_{10}\hat{o}_c + \log_{10}\hat{L}_+ \tag{11}$$

and from Equation (4):

$$\log_{10}\hat{o}_c|_- = \log_{10}\hat{o}_c + \log_{10}\hat{L}_- \tag{12}$$

with $\log_{10}\hat{L}_+ > 0$ (larger positive values are better) and $\log_{10}\hat{L}_- < 0$ (larger negative values are better) for any useful test. As previously, the objective is to make pairwise comparisons of binary tests (with both tests applied at the same prior odds), premised on the availability only of the sensitivities and specificities corresponding to the two tests' operational classification rules.

With Scenario B as the reference test and Scenario C as the comparison test, we find Scenario C's test is superior in terms of $\log 10\hat{L}_+$ values but inferior in terms of $\log 10\hat{L}_-$ values (Figure 4). With Scenario D as the reference test and Scenario E as the comparison test, we find Scenario E's test is inferior in terms of $\log 10\hat{L}_+$ values, but superior in terms of $\log 10\hat{L}_-$ (Figure 4). Moreover, we find that the transformed likelihood ratios graph still does not distinguish visually between Scenarios A, B and D (Figure 4). Thus, the initial findings from the analysis of the scenarios in Table 2 are the same as previously.

Now, as with Biggerstaff's [10] original analysis, we seek to view Johnson's analysis from an information theoretic perspective. As before, we will require an estimate of the prior probability $\hat{p}_c$. After some rearrangement, we obtain from Equation (11):

$$\log_{10}\left[\frac{\hat{p}_c|_+}{\hat{p}_c}\right] - \log_{10}\left[\frac{\hat{p}_{nc}|_+}{\hat{p}_{nc}}\right] = \log_{10}\hat{L}_+ \text{Hart} \tag{13}$$

where $\log_{10}[\hat{p}_c|_+ /\hat{p}_c]$ (> 0) and $\log_{10}[\hat{p}_{nc}|_+ /\hat{p}_{nc}]$ (< 0) on the LHS are information contents (as outlined in Section 3.1) with units of hartleys. From Equation (12):

$$\log_{10}\left[\frac{\hat{p}_c|_-}{\hat{p}_c}\right] - \log_{10}\left[\frac{\hat{p}_{nc}|_-}{\hat{p}_{nc}}\right] = \log_{10}\hat{L}_-\text{Hart} \tag{14}$$

where $\log_{10}[\hat{p}_c|_- /\hat{p}_c]$ (< 0) and $\log_{10}[\hat{p}_{nc}|_- /\hat{p}_{nc}]$ (> 0) on the LHS are information contents, again with units of hartleys. Thus, we recognize that $\log_{10}$ likelihood ratios also have units of hartleys. Figure 5 shows the information theoretic characteristics of Johnson's analysis when data on priors are incorporated, by drawing $\log_{10}$-likelihood contours on a graphical plot that has information contents on the axes.

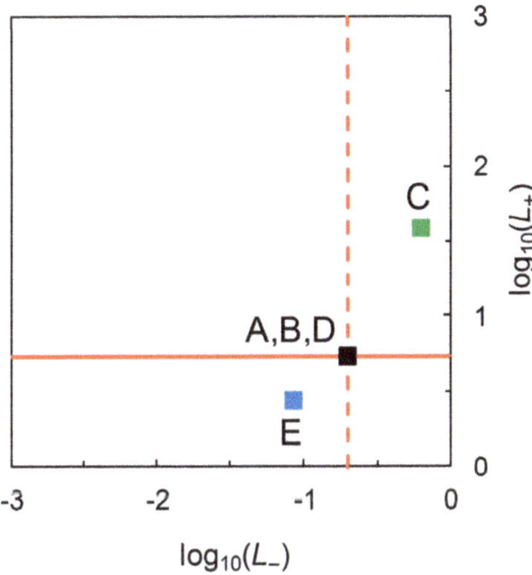

Figure 4. A version of Johnson's $\log_{10}$ likelihood ratios diagram for data from Table 2. Here $\log_{10}\hat{L}_+ = 0.727$ and $\log_{10}\hat{L}_- = -0.704$ for Scenarios A, B and D (■). For Scenario C (■), $\log_{10}\hat{L}_+ = 1.591$ and $\log_{10}\hat{L}_- = -0.208$. For Scenario E (■), $\log_{10}\hat{L}_+ = 0.438$ and $\log_{10}\hat{L}_- = -1.071$. Valid comparisons (i.e., for scenarios with equal prior probabilities) are Scenario B (reference) with Scenario C (comparison) and Scenario D (reference) with Scenario E (comparison).

In Figure 5, both the $\log_{10}\hat{L}_+$ and $\log_{10}\hat{L}_-$ contours always have slope = 1. As the decompositions characterized in Equations (13) and (14) show, any (constant) $\log_{10}$ likelihood ratio is the sum of two information contents. Looking at the "north-west" corner of Figure 5 and taking Scenarios A, B, and D from Table 2 as examples, we have $\log_{10}[\hat{p}_c|_+ /\hat{p}_c] = 0.642, 0.319, 0.056$ Hart and $\log_{10}[\hat{p}_{nc}|_+ /\hat{p}_{nc}] = -0.085, -0.408, -0.671$ Hart for $\hat{p}_c = 0.05$ (B), 0.36 (A), 0.85 (D), respectively. In each case, Equation (13) yields $\log_{10}\hat{L}_+ = 0.727$ Hart. Looking at the "south-east" corner of Figure 5, again taking Scenarios A, B, and D from Table 2 as examples, we have $\log_{10}[\hat{p}_{nc}|_- /\hat{p}_{nc}] = 0.498, 0.148, 0.018$ Hart and $\log_{10}[\hat{p}_c|_- /\hat{p}_c] = -0.207, -0.556, -0.687$ Hart for $\hat{p}_{nc} = 0.15$ (D), 0.64 (A), 0.95 (B), respectively. In each case, Equation (14) yields $\log_{10}\hat{L}_- = -0.704$ Hart. Thus we have an information theoretic perspective on Johnson's analysis when data on priors are available, and this time one that separates Scenarios A, B and D visually (Figure 5).

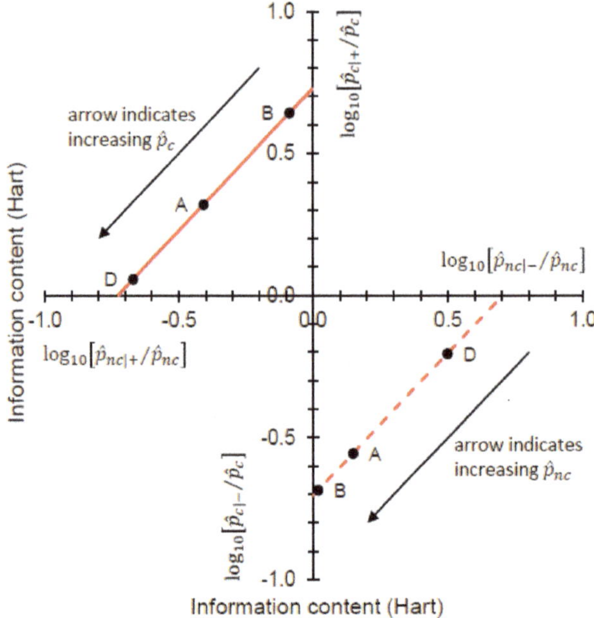

Figure 5. The "north-west" region of the figure is characterized by Equation (13), so relates to + predictions (which are correct for *c* subjects and incorrect for *nc* subjects). $\mathrm{Log}_{10}L_+$ contours are always straight lines with slope = 1. The solid red line indicates the contour for $\log_{10}\hat{L}_+ = 0.727$ Hart, corresponding to Scenarios A, B, and D (Table 2). A correct + prediction has a large information content when $\hat{p}_c$ is small (B), and a small information content is when $\hat{p}_c$ is large (D) (the arrow indicates the direction of increasing $\hat{p}_c$ along the contour). As the information content $\log_{10}[\hat{p}_c|_+ /\hat{p}_c]$ (on the vertical axis) becomes decreasingly positive, the information content $\log_{10}[\hat{p}_{nc}|_+ /\hat{p}_{nc}]$ (on the horizontal axis) becomes increasingly negative. The "south-east" region of the figure is characterized by Equation (14), so relates to − predictions (which are correct for *nc* subjects and incorrect for *c* subjects). $\mathrm{Log}_{10}L_-$ contours are always straight lines with slope = 1. The dashed red line indicates the contour for $\log_{10}\hat{L}_- = -0.704$ Hart, corresponding to Scenarios A, B, and D (Table 2). A correct − prediction has a large information content when $\hat{p}_{nc}$ is small (D), and a small information content is when $\hat{p}_{nc}$ is large (B) (the arrow indicates the direction of increasing $\hat{p}_{nc}$ along the contour, $\hat{p}_{nc} = 1 - \hat{p}_c$). As the information content $\log_{10}[\hat{p}_{nc}|_- /\hat{p}_{nc}]$ (on the horizontal axis) becomes decreasingly positive, the information content $\log_{10}[\hat{p}_c|_- /\hat{p}_c]$ (on the vertical axis) becomes increasingly negative.

3.3. A New Diagrammatic Format

Biggerstaff's [10] diagrammatic format for binary predictors allows an information theoretic interpretation once the data on prior probabilities have been incorporated. This distinguishes predictors with the same likelihood ratios analytically, but not visually. Johnson's [13] transformed version of Biggerstaff's diagrammatic format also allows an information theoretic interpretation once data on prior probabilities are incorporated. This approach distinguishes predictors with the same likelihood ratios both analytically and visually, but does not contribute to the comparison and evaluation of predictive values of disease forecasters.

We now return to the information theoretic interpretation of Biggerstaff's likelihood ratios graph (and revert to working in natural logarithms for continuity with previous analysis based on Figure 3). In Figure 3, the likelihood ratios are the slopes of the lines on the graphical plot. The lines themselves are relative entropy contours, the value of which depends on prior probability. We can now visually separate scenarios that have the same likelihood ratios but different relative entropies (e.g., A, B, D in

Table 2) by calculating the graph with relative entropies $\hat{I}_+$ and $\hat{I}_-$ on the axes of the plot (Figure 6). If we consider the predictor based on Scenario A as the reference, then the predictor based on Scenario B falls in the region of Figure 6 indicating comparatively less information is provided by both + and − predictions, while the predictor based on Scenario D falls in the region indicating comparatively less diagnostic information is provided by + predictions but comparatively more by − predictions.

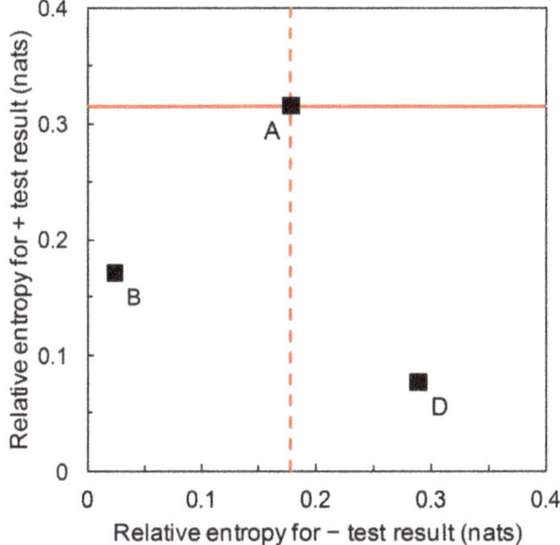

Figure 6. Scenario A: from the data in Table 2, we calculate relative entropies $\hat{I}_+ = 0.315, \hat{I}_- = 0.179$ (both in nats) ($\hat{p}_c = 0.36$) (Equations (3) and (4)). Similarly, for Scenario B we calculate $\hat{I}_+ = 0.171, \hat{I}_- = 0.024$ nats ($\hat{p}_c = 0.05$) and for Scenario D, $\hat{I}_+ = 0.076, \hat{I}_- = 0.289$ nats ($\hat{p}_c = 0.85$).

There is an alternative view of the diagrammatic format presented in Figure 6. Scenarios A, B and D all have the same likelihood ratios, $\mathring{L}_+ = 5.333$ and $\mathring{L}_- = 0.198$ (see Figure 3). What differs between scenarios is the prior probability $\hat{p}_c$. We can remove the gridlines indicating the relative entropies for Scenario A and plot the underlying prior probability contour (Figure 7). In Figure 7, starting at the origin and moving clockwise, prior probability increases as we move along the contour. The contour has maximum points with respect to both the horizontal axis and the vertical axis. The maximum value of the contour with respect to the horizontal axis is:

$$\hat{p}_c = \frac{\hat{p}_{+|nc}\cdot\left[\hat{p}_{+|c}\cdot\left(\ln\left[\frac{\hat{p}_{+|c}}{\hat{p}_{+|nc}}\right]-1\right)+\hat{p}_{+|nc}\right]}{\left[\hat{p}_{+|c}-\hat{p}_{+|nc}\right]^2} \tag{15}$$

and the maximum value of the contour with respect to the vertical axis is:

$$\hat{p}_c = \frac{\hat{p}_{-|nc}\cdot\left[\hat{p}_{-|c}\cdot\left(\ln\left[\frac{\hat{p}_{-|c}}{\hat{p}_{-|nc}}\right]-1\right)+\hat{p}_{-|nc}\right]}{\left[\hat{p}_{+|c}-\hat{p}_{+|nc}\right]^2}. \tag{16}$$

The corresponding values of $\hat{I}_+$ and $\hat{I}_-$, respectively, can then be calculated by substitution into Equations (7) and (8). The two maxima (together with the origin) divide the prior probability contour into three monotone segments (see Figure 7). As $\hat{p}_c$ increases, we observe a segment where $\hat{I}_+$ and $\hat{I}_-$ are both increasing (this includes Scenario B), then one where $\hat{I}_+$ is decreasing and $\hat{I}_-$ is increasing, this includes Scenario A), and then one where $\hat{I}_+$ and $\hat{I}_-$ are both decreasing (this includes Scenario D).

From Figure 7, we see that for the predictor based on Scenarios A, B and D, a + prediction provides most diagnostic information around prior probability $0.2 < \hat{p}_c < 0.3$. A − prediction provides most diagnostic information around prior probability $0.7 < \hat{p}_c < 0.8$. Recall that this contour describes performance (in terms of diagnostic information provided) for predictors with *sensitivity* = 0.833 and *specificity* = 0.844 (Table 2) (alternatively expressed as likelihood ratios $\hat{L}_+$ = 5.333 and $\hat{L}_-$ = 0.198). No additional data beyond *sensitivity* and *specificity* are required in order to produce this graphical plot; that is to say, by considering the whole range of prior probability we remove the requirement for any particular values. The point where the contour intersects the main diagonal of the plot is where $\hat{I}_+ = \hat{I}_-$. In this case, we find that $\hat{I}_+ = \hat{I}_-$ at prior probability $\approx$ 0.5 (Figure 7). At lower prior probabilities, + predictions provide more diagnostic information than − predictions, while at higher prior probabilities, the converse is the case. This contour's balance of relative entropies at prior probability $\approx$ 0.5 is noteworthy because it is not necessarily the case that there is always scope for such balance.

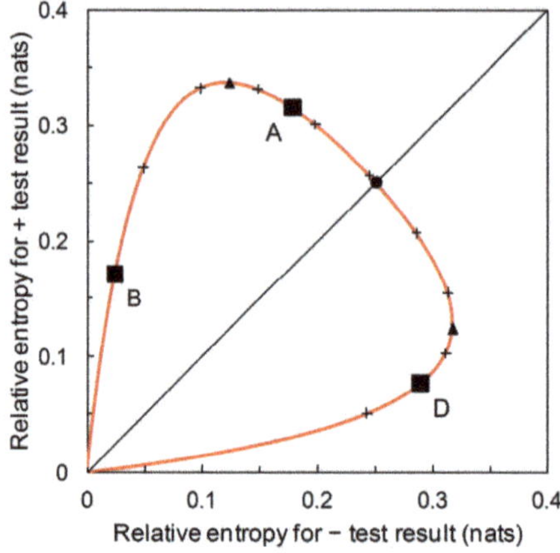

Figure 7. The prior probability $\hat{p}_c$ contour for Scenarios A, B, and D (solid red line). The contour is calibrated at 0.1 intervals of $\hat{p}_c$, clockwise from the origin, 0.1 to 0.9 (+ symbol on curve). Scenarios B ($\hat{p}_c$ = 0.05), A ($\hat{p}_c$ = 0.36), and D ($\hat{p}_c$ = 0.85) as characterized in Table 2 are indicated (■). Also indicated on the prior probability contour: maximum $\hat{I}_+$ = 0.337 nats (▲) ($\hat{p}_c$ = 0.245), maximum $\hat{I}_-$ = 0.317 nats (▲) ($\hat{p}_c$ = 0.749), $\hat{I}_+ = \hat{I}_-$= 0.251 nats (●) ($\hat{p}_c$ = 0.513).

Recall from Section 3.1 that we start with disease prevalence as an estimate of the prior probability $\hat{p}_c$ of need for a crop protection intervention. The information required (from a predictor) for certainty is then $\log(1/\hat{p}_c)$ denominated in the appropriate information units. This is the amount of information that would result in a posterior probability of need for an intervention equal to one. Similarly, $\log(1/\hat{p}_{nc})$, denominated in the appropriate information units, is the amount of information that would result in a posterior probability of no need for an intervention equal to one. We can plot the contour for these information contents on the diagrammatic format of Figure 7. This contour, illustrated in Figure 8, indicates the upper limit for the performance of any binary predictor. No phytopathological data are required to calculate this contour.

The diagrammatic format of Figure 7 (for Scenarios A, B and D) can accommodate prior probability contours for other Scenarios (i.e., for predictors based on different *sensitivity* and *specificity* values). For example, Figure 9 shows, in addition, the prior probability contours for the predictors based on Scenario C (with *sensitivity* = 0.39 and *specificity* = 0.99) and on Scenario E (with *sensitivity* = 0.944 and *specificity* = 0.656). We observe that a predictor based on Scenario C's *sensitivity* and *specificity* values potentially provides a large amount of diagnostic information from a + prediction, but over a very narrow range of prior probabilities. Scenario C itself represents one such predictor. The amount of diagnostic information from − predictions is very low over the whole range of prior probabilities. A predictor based on Scenario E's *sensitivity* and *specificity* values potentially provides a large amount of diagnostic information from − predictions over a narrow range of prior probabilities. Scenario E itself represents one such predictor. The amount of diagnostic information from + predictions remains low over the whole range of prior probabilities.

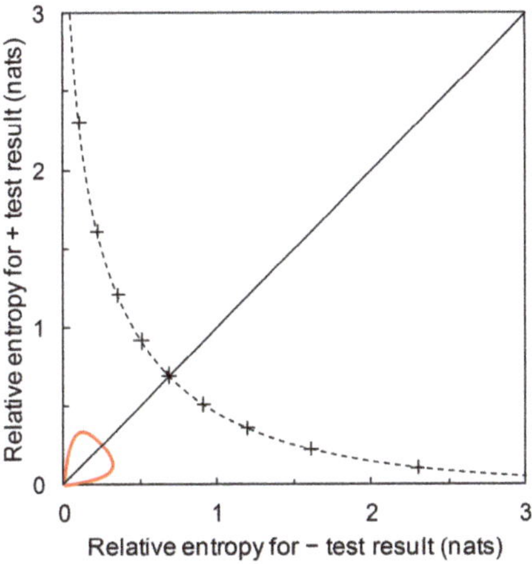

Figure 8. The dashed curve is the prior probability $\hat{p}_c$ contour showing the upper limit for performance of any binary predictor. The contour is calibrated at 0.1 intervals of $\hat{p}_c$ from upper left to lower right, 0.1 to 0.9 (+ symbol on curve). The maximum relative entropy for a + test result increases indefinitely as $\hat{p}_c$ approaches 0 while the maximum relative entropy for a − test result increases indefinitely as $\hat{p}_c$ approaches 1. The prior probability contour for Scenarios A, B, and D from Figure 7 (solid red line) is also shown, for reference (note the rescaled axes).

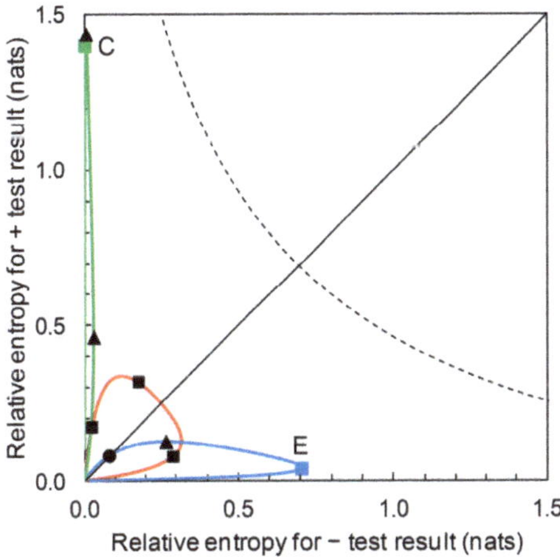

Figure 9. The prior probability contours for Scenarios C (solid green line) and E (solid blue line). Starting at the origin, the green prior probability contour passes through points (clockwise from origin): Scenario C, $\hat{I}_+ = 1.399$, $\hat{I}_- = 0.004$ (prior = 0.05) (■); maximum $\hat{I}_+ = 1.436$ (prior = 0.073) (▲); maximum $\hat{I}_- = 0.029$ (prior = 0.580) (▲). This contour does not coincide with the main diagonal of the plot other than at the origin. Starting at the origin, the blue prior probability contour passes through points (clockwise from origin): $\hat{I}_+ = \hat{I}_- = 0.080$ (●) (prior = 0.109); maximum $\hat{I}_+ = 0.126$ (prior = 0.337) (▲); Scenario E, $\hat{I}_+ = 0.039$, $\hat{I}_- = 0.700$ (prior = 0.850) (■); maximum $\hat{I}_- = 0.701$ (prior = 0.842) (point obscured from view). The prior probability contour for Scenarios A, B, and D (solid red line) is included here for reference; clockwise from origin, points marked ■ indicate Scenarios B, A and D (see Figure 7 for details). The dashed curve shows the contour indicating the upper limit for performance of a binary predictor (see Figure 8 for details). Note the changes in the scales on the axes compared with Figures 7 and 8.

4. Discussion

Diagrammatic formats have the potential to aid interpretation in the evaluation and comparison of disease forecasts. Biggerstaff's [10] likelihood ratios graph is a particularly interesting example. This graph uses the format of the ROC curve, as widely applied in exhibiting and explaining *sensitivity* and *specificity* for binary tests. However, while *sensitivity* and *specificity* are defined conditionally on actual disease status, the likelihood ratios graph is used to compare tests on the basis of predictive values, defined conditionally on the forecast (when tests are applied at the same prior probability). As Biggerstaff notes, one is less interested in *sensitivity* and *specificity* when it comes to the application of a test, because the conditionality is in the wrong order. The predictive values, or some functions of them, are also important, and ideally one would be able use these when assessing test performance in application (Figures 1 and 2).

Altman and Royston [15] discussed this idea in some detail and proposed PSEP as a metric for use in the assessment of predictor performance (in the binary case, PSEP = *positive predictive value + negative predictive value* − 1). Hughes and Burnett [16] later used an information theoretic analysis (including a diagrammatic representation) to show how PSEP was related to both the *Brier score* [17] and the information theoretic *divergence score* [18] methods of assessing predictor performance. In the current article, further analysis shows that Biggerstaff's likelihood ratios graph has underlying information theoretic properties that specifically relate to predictive values. The lines on the likelihood ratios graph are relative entropy contours, quantifying the expected information consequent on revising the prior probability of disease to the posterior probability after obtaining a forecast. However, the likelihood ratios graph does not visually distinguish relative entropy contours when predictors that have the same ROC curve summary statistics (sensitivities and specificities, or equivalently, likelihood ratios for both + and − predictions) are compared at different prior probabilities (Figure 3). A modified diagrammatic format that does so would therefore be of interest.

Johnson [13] provides a modified format, with log likelihood ratios on the axes of the graph (Figure 4), and suggests various possible advantages of this format. Our further analysis again shows that this modified format has underlying information theoretic properties. These properties relate to the statistical decomposition of log likelihood ratios (Figure 5; see also [5] for further discussion) but do not appear to be straightforwardly helpful as an aid to interpretation in the evaluation and comparison of disease forecasters based on predictive values.

Benish [19] applied information graphs for relative entropy to evaluate and compare clinical diagnostic tests. Here we derive relative entropies from Biggerstaff's likelihood ratios graph and present the results in a new diagrammatic format, with relative entropies for + and − predictions on the axes of the graph. Compared with the likelihood ratios graph, this visually distinguishes between predictors that have the same ROC curve summary statistics when compared at different (known) prior probabilities (Figure 6). So, referring to the scenarios listed in Table 2 with likelihood ratios $\hat{L}_+ = 5.333$ and $\hat{L}_- = 0.198$ (i.e., A, B, and D) we see that Scenario A has the highest relative entropy for a + prediction, then B, then D. Scenario D has the highest relative entropy for a − prediction, then A, then B. Recall that relative entropies are functions of the predictive values.

Suppose now that our aim is not to compare predictor performance in particular scenarios, but to evaluate performance over the range of possible scenarios. We can use our new format not just to plot relative entropies for a comparison of predictor performance for various known prior probability (disease prevalence) scenarios (Figure 6), but to also draw the contour showing how relative entropies change as prior probability of disease varies over the range from zero to one (Figure 7). This diagrammatic format requires no particular prior probabilities for calculation, only the ROC curve summary statistics. In the same way that the ROC curve relates to all predictors (by *sensitivity* and *specificity*) until a particular operational threshold is set, Figure 7 relates to all predictors (by relative entropies based on predictive values) until a particular prior probability value is specified. Maximum relative entropy points on the contour are calculable analytically in this format. Moreover, we can include the contours for predictors with different summary statistics. Figure 9 shows the contour that includes the predictor based on Scenario C and the contour that includes the predictor based on Scenario E, in addition to the contour that includes predictors based on Scenarios A, B and D from Figure 7. In this diagrammatic format, we can easily see the difference between contours that include predictors with high performance (in terms of relative entropies) in a narrow range of applicability (in terms of prior probabilities) when compared with a contour that balances predictor performance with a wider range of applicability. Unless we wish to evaluate and/or compare particular scenarios—in which case, not unreasonably, estimates of the corresponding prior probability (disease prevalence) values are required—producing the contour plot (Figures 7 and 9) has no data requirements beyond those for producing the ROC curve.

Figures 8 and 9 include the contour showing the upper limit for performance of a binary predictor. This upper limit serves as a qualitative visual calibration of predictor performance, rather in the way that we look at an ROC curve in relation to the upper left-hand corner of the ROC plot (where *sensitivity* and *specificity* are both equal to one). The contour cuts the main diagonal of the plot at prior probability $\hat{p}_c = 0.5$, when $\ln(1/\hat{p}_c) = \ln(2) = 0.693$ nats (Figure 8). This is the amount of information required to be certain of a binary outcome when the prior probability is equal to 0.5. However, the amount of information required to be certain of an outcome is not of any great practical significance in crop protection decision making. Rather than seeking certainty, a realistic objective is to develop predictors that provide enough information to enable better decisions, on average, than would be made with reliance only on prior probabilities. Thus we need to be able to consider predictor performance in terms of predictive values.

Perhaps the most important instrument available to the developer of a binary predictor is the placement of the threshold on the risk score scale [2,3,6,8]. This determines a predictor's *sensitivity* and *specificity*, and consequently the likelihood ratios for + and − predictions. However, this does not guarantee predictor performance in terms of predictive values. ROC curve analysis and diagrammatic formats that characterize predictive values (or functions of them) are therefore complementary aspects of predictor evaluation and comparison. For example, the appropriate placement of the threshold on the risk score scale may be informed by knowledge of disease prevalence for the scenario in which the predictor is intended for application. This in turn affords an evaluation of likely performance—in terms of predictive values—for the predictor in operation. Sometimes, however, we may wish to compare predictors' likely performances—perhaps in a novel scenario—when we are simply a potential user of the predictors in question, having had no development input but with access to the predictors' ROC curve summary statistics. In both settings, the diagrammatic formats we have discussed have potential application. They lead to information graphs that are visually distinct but analytically linked. All give extra insight via the predictive values of disease forecasts.

Author Contributions: Conceptualization, G.H., J.R. and N.M.; Formal analysis, G.H., J.R. and N.M.; Methodology, G.H., J.R. and N.M.; Writing–original draft, G.H.; Writing–review & editing, J.R. and N.M. All authors have read and agreed to the published version of the manuscript.

Funding: This research received no external funding.

Conflicts of Interest: The authors declare no conflict of interest.

References

1. Swets, J.A.; Dawes, R.M.; Monahan, J. Better decisions through science. *Sci. Am.* **2000**, *283*, 70–75. [CrossRef] [PubMed]
2. Yuen, J.; Twengström, E.; Sigvald, R. Calibration and verification of risk algorithms using logistic regression. *Eur. J. Plant Pathol.* **1996**, *102*, 847–854. [CrossRef]
3. Twengström, E.; Sigvald, R.; Svensson, C.; Yuen, J. Forecasting Sclerotinia stem rot in spring sown oilseed rape. *Crop Prot.* **1998**, *17*, 405–411. [CrossRef]
4. Yuen, J.E.; Hughes, G. Bayesian analysis of plant disease prediction. *Plant Pathol.* **2002**, *51*, 407–412. [CrossRef]
5. Hughes, G. The evidential basis of decision making in plant disease management. *Annu. Rev. Phytopathol.* **2017**, *55*, 41–59. [CrossRef] [PubMed]

6. Madden, L.V. Botanical epidemiology: Some key advances and its continuing role in disease management. *Eur. J. Plant Pathol.* **2006**, *115*, 3–23. [CrossRef]

7. Theil, H. *Economics and Information Theory*; North-Holland: Amsterdam, The Netherlands, 1967.

8. Makowski, D.; Denis, J.-B.; Ruck, L.; Penaud, A. A Bayesian approach to assess the accuracy of a diagnostic test based on plant disease measurement. *Crop Prot.* **2008**, *27*, 1187–1193. [CrossRef]

9. Go, A.S. Refining probability: An introduction to the use of diagnostic tests. In *Evidence-Based Medicine: A Framework for Clinical Practice*; Friedland, D.J., Go, A.S., Ben Davoren, J., Shilpak, M.G., Bent, S.W., Subak, L.L., Mendelson, T., Eds.; McGraw-Hill/Appleton & Lange: New York, NY, USA, 1998; pp. 11–33.

10. Biggerstaff, B.J. Comparing diagnostic tests: A simple graphic using likelihood ratios. *Stat. Med.* **2000**, *19*, 649–663. [CrossRef]

11. De Wolf, E.D.; Madden, L.V.; Lipps, P.E. Risk assessment models for wheat Fusarium head blight epidemics based on within-season weather data. *Phytopathology* **2003**, *93*, 428–435. [CrossRef] [PubMed]

12. Harremoës, P. *Entropy*—New editor-in-chief and outlook. *Entropy* **2009**, *11*, 1–3. [CrossRef]

13. Johnson, N.P. Advantages to transforming the receiver operating characteristic (ROC) curve into likelihood ratio co-ordinates. *Stat. Med.* **2004**, *23*, 2257–2266. [CrossRef] [PubMed]

14. Fosgate, G.T. Letter to the editor. *Stat. Med.* **2005**, *24*, 1287–1288. [CrossRef] [PubMed]

15. Altman, D.G.; Royston, P. What do we mean by validating a prognostic model? *Stat. Med.* **2000**, *19*, 453–473. [CrossRef]

16. Hughes, G.; Burnett, F.J. Evaluation of probabilistic disease forecasts. *Phytopathology* **2017**, *107*, 1136–1143. [CrossRef] [PubMed]

17. Brier, G.W. Verification of forecasts expressed in terms of probability. *Mon. Weather Rev.* **1950**, *78*, 1–3. [CrossRef]

18. Weijs, S.V.; van Nooijen, R.; van de Giesen, N. Kullback-Leibler divergence as a forecast skill score with classic reliability-resolution-uncertainty decomposition. *Mon. Weather Rev.* **2010**, *138*, 3387–3399. [CrossRef]

19. Benish, W.A. The use of information graphs to evaluate and compare diagnostic tests. *Methods Inform. Med.* **2002**, *41*, 114–118. [CrossRef]

Article

Canine Olfactory Detection of a Non-Systemic Phytobacterial Citrus Pathogen of International Quarantine Significance

Timothy Gottwald [1,*], Gavin Poole [1,2], Earl Taylor [1], Weiqi Luo [1,2], Drew Posny [1,2], Scott Adkins [1], William Schneider [3] and Neil McRoberts [4]

[1] U.S. Department of Agriculture, Agricultural Research Service, Fort Pierce, FL 34945, USA; gavin.poole@usda.gov (G.P.); earl.taylor@usda.gov (E.T.); weiqi.luo@usda.gov (W.L.); Drew.posny@usda.gov (D.P.); scott.adkins@usda.gov (S.A.)
[2] Center for Integrated Pest Management, North Carolina State University, Raleigh, NC 27695, USA
[3] F1K9 LLC, Palm Coast, FL 34797, USA; wschneider@f1-k9.com
[4] Plant Pathology Department, University of California Davis, Davis, CA 95616, USA; nmcroberts@ucdavis.edu
*   Correspondence: citrusdoc1@gmail.com

Received: 19 August 2020; Accepted: 27 October 2020; Published: 9 November 2020

Abstract: For millennia humans have benefitted from application of the acute canine sense of smell to hunt, track and find targets of importance. In this report, canines were evaluated for their ability to detect the severe exotic phytobacterial arboreal pathogen *Xanthomonas citri* pv. *citri* (Xcc), which is the causal agent of Asiatic citrus canker (Acc). Since Xcc causes only local lesions, infections are non-systemic, limiting the use of serological and molecular diagnostic tools for field-level detection. This necessitates reliance on human visual surveys for Acc symptoms, which is highly inefficient at low disease incidence, and thus for early detection. In simulated orchards the overall combined performance metrics for a pair of canines were 0.9856, 0.9974, 0.9257 and 0.9970, for sensitivity, specificity, precision, and accuracy, respectively, with 1–2 s/tree detection time. Detection of trace Xcc infections on commercial packinghouse fruit resulted in 0.7313, 0.9947, 0.8750, and 0.9821 for the same performance metrics across a range of cartons with 0–10% Xcc-infected fruit despite the noisy, hot and potentially distracting environment. In orchards, the sensitivity of canines increased with lesion incidence, whereas the specificity and overall accuracy was >0.99 across all incidence levels; i.e., false positive rates were uniformly low. Canines also alerted to a range of 1–12-week-old infections with equal accuracy. When trained to either Xcc-infected trees or Xcc axenic cultures, canines inherently detected the homologous and heterologous targets, suggesting they can detect Xcc directly rather than only volatiles produced by the host following infection. Canines were able to detect the Xcc scent signature at very low concentrations (10,000× less than 1 bacterial cell per sample), which implies that the scent signature is composed of bacterial cell volatile organic compound constituents or exudates that occur at concentrations many fold that of the bacterial cells. The results imply that canines can be trained as viable early detectors of Xcc and deployed across citrus orchards, packinghouses, and nurseries.

Keywords: early detection; Asiatic citrus canker; latent class; information theory; field diagnostic; scent signature; direct assay; deployment

1. Introduction

Asiatic citrus canker (ACC) is a fruit, foliar, and twig lesion disease that has significant international, national, and local quarantine implications and has been the focus of multiple extensive eradication programs. The disease is caused by the bacterium *Xanthomonas citri* pv. *citri* (Xcc). When pathogen infection is severe, leaf drop and up to 69% crop loss can occur due to fruit drop [1]. The bacteria disperse via meteorological events ranging from gentle rain to tropical storms and hurricanes. The more severe the meteorological event, the more effective the inoculum dispersal [2,3]. Hurricanes and tropical storms have been associated with long-range dissemination as well as local increase of the bacterium [2,4–6]. Xcc inoculum can also be transmitted mechanically by machinery and humans if the foliage is wet. Inoculum begins to exude from Xcc lesions within 1–5 min of becoming wet [4,6–8]. The maximum concentration of bacteria is exuded within the first 1–2 h period following the beginning of the rainfall event or wetting of the lesion, although inoculum is produced continuously at a lower concentration for the duration of the wetting period or storm [2–4]. Infection takes place when inoculum-laden water passes through stomata of foliage, fruit, or green twigs. Infection can occur through wounds as well, and it is highly exacerbated by the Asian citrus leafminer (*Phyllocnistis citrella* Stainton), whose feeding galleries create a labyrinth of wounds that expose susceptible leaf mesophyll tissues to splashed inoculum, greatly increasing the probability of infection by Xcc [9,10].

After infection occurs, bacteria propagate within the plant tissues, eventually forming small blister-like protrusions that become visible 5–7 days post-infection with close examination augmented by 10× magnification. Within 12–14 days, these protrusions erupt through the epidermis, forming 1–2 mm light brown erumpent lesions. As the lesions age, they darken to brown and develop water-soaked margins with a surrounding chlorotic halo [7]. The lesion center develops a raised spongy and corky appearance on adaxial and abaxial foliar surfaces. Older lesions can reach 1 cm in diameter and can coalesce to form mass infections.

Pathogen detection in the field is almost exclusively by human visual inspection. Confirmation of Xcc infection can be accomplished by serological or Polymerase Chain Reaction (PCR) assays, both of which require infected tissue. Therefore, such assays are of minimal use for a field survey of a pathogen that causes only local lesions, i.e., non-systemic infections. Visual inspection is tedious, labor intensive, and highly variable due to the elusiveness of Xcc lesions, especially when high in the canopy. Since mature citrus trees can have in excess of 100,000 leaves, finding initial infections visually in such a large canopy is challenging and uncertain. As a consequence, multiple infection cycles must occur before symptoms are of sufficient prevalence to permit visual detection. During a major epidemic in Florida (1996–2006), the detection time for Xcc by trained regulatory inspectors averaged 106 days post-infection [11]. In one regulatory exercise, 14 teams of two inspectors per team examined the same infected orchard in succession. No two teams found the same infected trees, each successive team found new infected trees previously undetected by prior teams, and no team found all known infections. [Riley, unpublished results]. Since it is highly improbable that all infected trees in an orchard will be found by visual inspection, the true incidence of infection cannot be determined, and thus, calculating the detection accuracy of visual inspection is not possible.

The unpromising results for human visual detection suggest the need for the deployment of a detection technology with much better performance. Genetic analyses indicate that modern-day canines (*Canis lupus familiaris*) were the first domesticated animal, arising from two separate wolf (*Canis lupus*) populations ≈ 15,000 years ago [12–17]. The mammalian (including canids) olfactory system is antediluvian, having evolved from early chemotactic receptors during the Precambrian over 600 million years ago. Acute olfaction evolved to enhance finding food, mates, detecting danger, avoiding predators, etc. [18]. An overarching advantage to the human use of canine olfaction is that all canids non-destructively interrogate their environment for a scent signature of interest. This scent signature is composed of a specific volatile organic compound (VOC) or complex of VOCs. In contrast, commonly used molecular or biochemical assays often require destructively sampling a small proportion of the host or environment and are specific and do not detect complex VOC composites.

Wolves employ acute vision and hearing when prey is in close proximity. However, when tracking prey, often over great distances, wolves resort to olfactory cues to locate widely dispersed and often low-density prey [19], as is true for domestic canines [20]. To find rare targets of human interest, we exploit the tracking expertise and acute sense of smell handed down to modern canines from wolves and their evolutionary predecessors. Nonetheless, scent detection is pervasive through the animal kingdom. Other vertebrates and invertebrate animals have been explored recently in the burgeoning science of scent detection research with diverse practical applications [21]. However, canines dominate as olfactory detector tools because of their unsurpassed domestication and easily deployed tracking skills.

Canines are used broadly to detect and locate a wide array of organic and inorganic odors, (e.g., explosives, drugs, accelerants, pollutants, toxins, pesticides), tracking humans and game animals, finding cadavers, and matching the scent of criminal perpetrators to crime scenes [22]. Canines are also proficient at medical human disease detection, especially cancer (malignant melanoma, non small-cell, small-cell lung, breast, prostrate, bladder, ovarian, and colorectal cancers) often with equivalent or superior sensitivity to current medical assays and can detect epileptic seizures prior to onset [22–25].

Relative to agricultural diseases, canines were recently trained extensively to detect *Candidatus* Liberibacter asiaticus (CLas), which is the causal pathogen of citrus Huanglongbing with greater than 99%, 96%, and 92% accuracy in field trials, commercial citrus orchards and citrus in residential properties, respectively [26]. In the same study, canines detected CLas in the Asian citrus psyllid, *Diaphorina citri*, that vectors the bacteria and in bacterial co-cultures devoid of plant or animal host cells. Additionally, canines have been trained to detect plum pox virus in commercial peach (*Prunus persica*) orchards [Gottwald unpublished] and the fungal pathogen *Raffaelea lauricola*, which is the cause of Laurel wilt disease of avocado (*Persea americana*) [27]. Thus, canines have been demonstrated to detect bacterial, viral, and fungal pathogens in plant hosts.

Canine detection of the plant pathogens as indicated above represents a novel extension of previous detection targets. For drugs, explosives, pollutants, etc., canines are trained to detect specific VOC scent signatures of the target compounds. For biologicals such as plants, plant parts, humans, and other animals, the canines are trained to recognize the volatilome, e.g., unique VOC complex, emitted by the target. Conversely, when trained to detect pathogenic organisms, the canines must detect one organism (bacterium, virus, or fungus) within another organism, i.e., the host plant or animal. The three plant pathogens discussed above, CLas, plum pox virus and *R. lauricola*, are all systemic within the vascular system of the host, and therefore, they can be completely or incompletely distributed within the plant as the infection progresses. In contrast, Xcc is non-systemic and causes only local lesions, which can range in incidence from a single lesion to thousands per tree and can be rare, highly aggregated, or diffused within the host.

In this study, we demonstrate proof of concept that canines can discriminate and detect Xcc and can be trained as viable detectors of the pathogen in agricultural environments. Here we document the detection of Xcc, an exotic bacterial pathogen by use of canine olfactory surveillance in orchards and packinghouses. Finally, we determine that canines can directly detect the target bacterium *in planta* and *in vitro*.

2. Materials and Methods

2.1. Initial Sensitization Training of Canines for Xcc Scent Signature Recognition

Eleven canines of various breeds were utilized for the various trials throughout the course of the study (Table 1).

Table 1. Canines trained to detect *Xanthomonas citri* pv.*citri* (Xcc) scent signature and utilized in this study.

Canine	Breed	Trial(s)
Kimba	Belgian Malinois	Proof of concept studies/scent transfer unit
Tank	German Shepherd	Proof of concept studies/scent transfer unit
NDD-1	Beagle	Proof of concept studies/grapefruit seedlings
NDD-2	Beagle	Proof of concept studies/grapefruit seedlings
NDD-3	German Pointer	Proof of concept studies/grapefruit seedlings
NDD-4	Labrador retriever	Initial orchard detection/grapefruit leaves
Juice	Labrador retriever	Simulated orchards, Packinghouse, Commercial orchards, Spatial heterogeneity
Bady	German Shepherd	Lesion age, Lesion incidence, Abscised/senescing leaves, spatial heterogeneity
Maxi	German Shepherd	Lesion age, Lesion incidence, Abscised/senescing leaves, spatial heterogeneity
Mi	German Shepherd/Belgian Malinois	Direct bacterial detection, Bacterial dilution
Ti	German Shepherd	Direct bacterial detection, Bacterial dilution

At the onset of this study, Xcc was considered a quarantine pathogen in Florida [3,6,7,11], and neither the bacteria nor infected plant material could be transported to new locations for experimental purposes. Therefore, we used a Scent Transfer Unit (STU) (Model STU-100, Tolhurst Big "T" Enterprises, Lockport, New York, NY, USA, 14094) vacuum device commonly used in canine detection work to draw air at a constant rate ($\approx$300 L/min) to collect volatile samples containing the "scent signature" of the target, in this case Xcc-infected plants, and deposit the scent onto a cotton "scent pad" [28]. Since the scent pad does not contain the pathogen (Xcc is only splash dispersed, generally by rain and not dispersed in dry air) nor any potentially infected plant material, it can be safely transported to non-endemic areas for testing. Scent pad samples were collected from non-infected and Xcc-infected Ruby Red grapefruit (*Citrus paradisii*) trees using the STU. A sterile 12.7 $\times$ 22.9 cm (5×9 inch) cotton pad was placed into the STU for 10 min by setting the STU within the canopy of a non-infected or Xcc-infected tree 5–10 cm from leaves, fruit and branches while air was pulled through the pad. Following each 10 min sampling period, scent pads were aseptically removed, placed in a volatile-proof plastic bag (K-pak), and heat sealed. The STU was disinfected with ethyl alcohol between each use. Multiple heat-sealed samples from infected and non-infected trees were segregated, placed in zip lock plastic bags, and stored at -20 °C until transported (Figure 1A–C).

Canine detection companies are for-profit ventures, and therefore, they are often reticent to undertake non-profitable and time-consuming basic research. Fortuitously, a canine detection company in California was willing to dedicate two canines to do an initial testing of canine detection of Xcc. In preliminary studies, it was not feasible to bring canines to the Xcc-infected sites due to quarantine constraints. Therefore, scent pad samples were collected in South Florida, sealed in a styrofoam shipping box with ice packs, and transported overnight to the California training facility.

One Belgian Malinois (Kimba) and one German Shepard (Tank), previously trained for criminology scent detection, were trained to the Xcc scent signature by imprinting, during which the canine was introduced to the target and other neutral scents. Relative to scent signature training, when the canine becomes interested in/reacts to the correct target, the experience is immediately encouraged with verbal and play rewards [18]. In our case, canines were imprinted to the scent signature presented on Xcc-infected scent pads to discriminate the scent from scent pads without the scent signature. To take advantage of the candid passion to search and track, a mixed population of mostly non-infected with a few Xcc-infected scent pad samples were arrayed in a series of rows of metal cans placed outdoors on metal stands (Figure 1D). Correctly alerting on the Xcc-infected pad placed in the bottom of a can resulted in the canine receiving verbal praise and a few moments of play with the handler and a ball or "kong" (hard rubber ball on a short rope). Initial training used non-frozen scent pads. Canines subsequently detected both non-frozen and frozen scent pads from Xcc-infected trees. Thereafter, scent pads were stored frozen until use in training and assessment.

Figure 1. Initial training of canines for detection of the phytobacterial arboreal pathogen *Xanthomonas citri* pv. *citri* (Xcc), the causal agent of Asiatic citrus canker (Acc). (**A**) Xcc-infected red grapefruit fruit and (**B**) foliage. (**C**) Scent Transfer Unit (STU) used to draw in Xcc volatiles and deposit on cotton scent collection pad. (**D**) Canine "Kimba" training by interrogating a row of metal cans containing Xcc-positive and negative scent pads. (**E,F**) Detector canine NDD-1 and NDD-3 alerting on boxes containing Xcc-infected foliage at the USDA, APHIS, National Detector Dog Training Center.

Following initial scent training, a series of preliminary studies were conducted in which canines interrogated scent pads exposed to Xcc-infected red grapefruit for periods of 1, 5, 10, or 30 min versus pads exposed to non-infected trees for similar times, to determine if there was a lower threshold of exposure necessary to create scent pad training materials. Additionally, once the scent pads were removed from the heat-sealed plastic bags, the canines interrogated them every few days to determine the temporal viability of the scent pads as a training tool.

2.2. Training Canines for Detection of Xcc-Infected Plant Material

During 2004–5, in collaboration with the U.S. Department of Agriculture, Animal and Plant Health Inspection Service (USDA, APHIS), National Detector Dog Training Center (NDDTC) in Orlando, Florida, we trained canines to detect Xcc-infected Duncan grapefruit seedlings. At the USDA, ARS laboratory in Fort Pierce, Florida, quarantine greenhouse, seedlings were infected by placing drops of Xcc inoculum, $\approx 10^6$ to 10^8 cfu/mL prepared from pure cultures on young leaves and using a sterile needle to wound the leaf lamina, causing 5–10 wounds/leaf of five leaves/seedling. Seedlings were incubated in the greenhouse for ≈ 2–3 weeks until symptomatic and then transported under regulatory permit in sealed containers to the NDDTC in Orlando along with non-infected seedlings as training materials where they were maintained until use. At the NDDTC, two beagles (NDD-1 and NDD-2), one German pointer (NDD-3), and one Labrador retriever (NDD-4) were trained to recognize the Xcc scent signature by a similar imprinting/reward method as described above. Scent detection proficiency was tested by placing excised branches or whole potted Xcc-infected or non-infected trees in closed cardboard boxes and arraying the boxes on the floor of the training facility (Figure 1E,F). In a second NDDTC study, the trained Labrador was transported to an Xcc-infected orchard in Indian River County, Florida. To avoid disease quarantine issues, leaves were collected from Xcc-infected and non-infected trees and placed into separate sealed plastic pipe containers with holes drilled in the sides to allow the volatiles but not infected plant material to escape for canine interrogation. The canine interrogated the containers, which were placed on the ground at the edge of the orchard. When the trial ended, the infected and non-infected plant materials were removed from the containers, discarded in the infected field, and the containers were disinfected. Due to quarantine issues, intermittent and uncoordinated testing by NDDTC personnel in absence of the authors, and uncoordinated replacement of canines assigned to the study, data were not consistently taken with the exception of an overall assessment of performance by NDDTC personnel.

In all subsequent studies, we collaborated with three commercially certified canine detector companies who specialize in trained detector canines for military, police and domestic clientele, for the detection of an array of targets including explosives, drugs, bed bugs, etc. This was important because professional canine handlers are themselves trained not to give any voluntary or involuntary cues to the canines while testing their performance. A single handler was used for all replications of most studies to avoid bias or variation due to handler. These companies trained nine canines (7 were used in this study) for the detection of Xcc via imprinting on scent pads from Xcc-infected trees or Xcc-infected grapefruit seedlings until canines gained proficiency at differentiating Xcc-infected from non-infected samples (Table 1).

2.3. Assessment of Canine Performance

To measure the performance of canines individually and as a group, a binary classification test was performed on the data from each of the studies below and standard diagnostic accuracy statistics (latent-class metrics) were calculated:

True Positive (TP) correct canine alert on Xcc-positive target
True Negative (TN) correct rejection, no canine alert on Xcc-negative target
False Positive (FP) incorrect canine alert on Xcc-negative target, Type I error
False Negative (FN) incorrect canine rejection of Xcc-positive target, Type II error
Sensitivity (SEN) or true positive rate,

$$SEN = TP/(TP + FN)$$

Specificity (SPE) or True Negative Rate

$$SPE = TN/(FP + TN)$$

Precision or Positive Predictive Value (PPV)

$$PPV = TP/(TP + FP)$$

Negative Predictive Value (NPV)

$$NPV = TN/(TN + FN)$$

False Positive Rate (FPR)

$$FPR = FP/(FP + TN)$$

False Negative Rate (FNR)

$$FNR = FN/(FN + TP)$$

False Discovery Rate (FDR)

$$FDR = FP/(TP + FP)$$

Accuracy (ACC)

$$ACC = (TP + TN)/n$$

where n = total number of samples assessed in each trial.

The above metrics are commonly used to evaluate the performance of diagnostic tests in medicine and other endeavors, including the evaluation of canine detectors [26,29–32]. No single performance metric can capture all aspects of canine detection accuracy. Throughout this study, we report the metrics listed above, with a particular focus on sensitivity, specificity, and accuracy. In addition to those standard metrics for test evaluation, we also examined the diagnostic capability of the canines using three graphical approaches that either depend directly on information quantities or which have direct connections to information theoretic concepts.

To illustrate the use of these methods, we focus on a single set of performance statistics derived from assessment of the impact of disease prevalence in samples on the detection performance of the canines (see Section 3.5 below). Using the summary statistics for the overall performance for canines Bady and Maxi in this set of experiments, we constructed likelihood ratio graphs [33,34], a leaf plot based on the PPV and NPV for each canine, and a "loop" plot illustrating the expected mutual information for the positive and negative diagnoses for each animal [34].

2.4. Detection of Xcc in Simulated New Plantings

Ruby Red Grapefruit seedlings were inoculated using a needless syringe tightly appressed to the abaxial laminar surface of 1/2 to 2/3 expanded leaves and forcing inoculum ($\approx 10^5$ cfu/mL) into the lamina via pressure injection infiltration; then, they were allowed to develop citrus canker symptoms for 4 weeks prior to canine interrogation [35]. Non-inoculated trees were arrayed in a 100 tree, 10×10 grid, with ≈ 3 m between trees within row and between rows, and with 1 to 10 Xcc-infected trees randomly placed within the grid. Canines interrogated each tree in the grid in sequence via a serpentine pattern up and down the rows. Three detector canines associated with two collaborating commercial canine training companies assessed the randomized arrays of grapefruit trees.

For the first collaborating company, a single canine, a Labrador retriever "Juice", interrogated the simulated orchard with 2, 4, 5, 7, or 10 Xcc-infected trees randomly placed within the 100-tree grid. Each of the incidence levels was replicated ten times. Tree placement was re-randomized between each replication. The experiment was repeated during four separate months, July, September, December, and May to examine the effect of seasonality. Due to commercial handler availability, not all incidence levels were examined each month (Figure 2A–C). For the second collaborating company, two canines, German shepherds, "Bady" and "Maxi", each interrogated the simulated orchard with 1 to 6 Xcc-infected trees randomly placed within the 100-tree grid (Figure 2D,E). Tree placement was re-randomized between each replication, and care was taken to ensure that only technicians setting up each grid replicate knew the positions of Xcc-infected and non-infected trees, i.e., both handlers and canines were unaware of true positive and true negative target positions. This same "blind" test methodology was implemented through all subsequent trials. However, when a canine alerted correctly on a true positive, the technician confirmed the correct detection to the handler so the handler could appropriately reward the canine. The experiment was conducted over a two-month duration, depending upon commercial handler availability. Each canine interrogated a grid of each Xcc-incidence level at least twice. Canine alerts were recorded, and latent class metrics were calculated to assess canine performance.

Figure 2. Canine detection of *Xanthomonas citri* pv. *citri* (Xcc) in simulated and commercial orchards. (**A**) Xcc-infected, potted red grapefruit inserted into ground—Xcc lesions indicated by red arrows. Detector canine "Juice"—(**B**) interrogating, and (**C**) alerting on Xcc-infected trees. Detector canine "Bady"—(**D**) interrogating, and (**E**) alerting on Xcc-infected trees. (**F**) "Juice" alerting on Xcc-infected grapefruit tree in commercial orchard. (**G**) Sample of three Xcc-infected grapefruit leaves from commercial orchard identified by "Juice"—note multiple small brown Xcc lesions surrounded by chlorotic halos.

2.5. Detection of Xcc Lesions of Various Ages

Ruby Red Grapefruit seedlings were inoculated using the methods described in Section 2.4 above. Inoculations were conducted over time, such that on the day of canine interrogation, a temporal array of Xcc-infected seedlings were 1, 3, 6, 9, and 12 weeks post-inoculation. Two canines, Bady and Maxi, each assessed a 50-tree grid (5 rows of 10 seedlings per row) five times (replications). For each replication, two seedlings with Xcc infections of the same age were randomly placed within a population of 48 additional non-infected seedlings of the same age. The experiment was repeated 7 days later using the same temporal array of seedlings, whose Xcc infections were now 1 week older. All tests were blinded as described in Section 2.4 above. Since there was no significant difference between the repeated experimental results, performance metrics were calculated for the combined data such that Xcc-infection age ranged 1–2, 3–4, 6–7, 9–10, and 12–13 weeks post-infection.

2.6. The Effect of Incidence of Xcc Lesions on Detection

Early observations with the canine "Juice" indicated the potential that he alerted differentially to trees infected with few versus prevalent lesions. Subsequently, canines were trained to alert to Xcc infections regardless of infection prevalence. To assess the effect of lesion incidence, Duncan grapefruit seedling trees were inoculated by pin-prick inoculation as described above to establish trees with 1, 5, 50 and 500 lesions each. Two canines, Bady and Maxi, each interrogated grids of 50 trees with one Xcc-infected tree of each lesion incidence and 46 non-infected trees. Each Xcc incidence level was interrogated twice by each canine with the location of the Xcc-infected trees re-randomized (using a random number generator) between replicates. All tests were blinded as described in Section 2.4 above. The experiment was repeated two months later with the same set of trees. Canine alerts were recorded, and latent class metrics were calculated to assess canine performance. For each lesion incidence, performance metrics were calculated considering one infected tree of the specific lesion incidence being evaluated in a population of 46 non-infected trees, ignoring the canine response of the other three Xcc-infected trees of other lesion incidence.

2.7. Detection of Xcc Infections in Decaying Foliage

Barring adverse environmental conditions or disease, the lifespan of citrus leaves is 1–3 years prior to abscission, after which they senesce and decay. However, Xcc-infected leaves experience foliar accumulation of elevated ethylene and often abscise early. To determine the duration of canine detection of Xcc-infections in abscised leaves, Duncan grapefruit leaves, *in planta*, were infected via a needless syringe inoculation described above. Non-infected leaves were physically abscised from the trees, and 20–30 leaves were placed in wire cages as non-infected controls. True positive targets were composed of cages with 20–30 non-infected leaves, with the addition of 2–6 leaves with 30-day Xcc infections. True positive (TP) and true negative (TN) cages were randomized in an open grassy field (Figure 3A,B). Two canines, Bady and Maxi, interrogated the leaf cages on 0, 1, 2, 5, 13, and 27 days post-abscission as the leaves decayed. Each canine interrogated the leaf cages in turn and then repeated the interrogation for two to five replications on each assay date; the number of replications/day/canine depended upon weather and handler availability (Figure 3C,D). All tests were blinded as described in Section 2.4 above. Canine alerts were recorded, data were combined across replications and canines, and latent class metrics were calculated to assess canine performance.

2.8. Detection of Xcc in Citrus Packinghouse Environments

To examine canine performance for the detection of Xcc in a commercial packinghouse, commercially packed cardboard boxes, each containing approximately 50 red grapefruit fruits were arrayed in a 10 × 10 grid, with ≈1.5 m between rows and boxes within a row (Figure 3E,F). Within target boxes, 2 Xcc-infected fruits, each with 2–20 lesions were randomly placed in the center within each TP target carton. The target boxes containing Xcc-positive fruit were randomly arrayed on the concrete packinghouse floor and re-randomized between replications. A single canine, "Juice", interrogated each box in the grid in sequence via a serpentine search pattern up and down the rows. The experiment was repeated twice 5 days apart. On the first day, the canine interrogated nine randomized arrays (900 cartons) with Xcc-incidence ranging from 1–4%, over a ≈3-h period, whereupon the study was halted due to excessive temperatures in an effort to preserve canine health. On the second day, the ambient conditions were more favorable, and the canine interrogated 19 randomized arrays (1900 cartons) with Xcc-incidence ranging from 1–10% over a ≈5-h period. All tests were blinded as described in Section 2.4 above. Human inefficiency of detection of packing-house inspected fruit was indicated by 100 commercial cartons selected at random post packinghouse processing, 2 of which contained 1 and 2 infected fruit with small lesions after passing multiple packing line inspections stations with trained fruit grader/inspectors and post packinghouse inspection by pathology technicians trained to detect Xcc symptoms. Canine detection of unknown true positives elucidated the human error, i.e., false negatives. These true positives were subsequently added into the grid data and considered true positives incorporated into the grid designs. Canine alerts were recorded, the results from the two days were combined, and latent class metrics were calculated to assess canine performance.

Figure 3. Canine detection of abscised Xcc-infected grapefruit leaves over time. (**A**) Mixed Xcc-infected and non-infected leaves in wire mesh cage, (**B**) close up of leaves in wire mesh decaying. Canine "Bady"—(**C**) interrogating, and (**D**) alerting on wire mesh cages with decaying Xcc-infected leaves. (**E**) Commercially packed grapefruit in cardboard carton with top layer of fruit removed to show Xcc-infected fruits—red arrows indicate infected fruit with Xcc lesions. (**F**) Grid of 100 cartons of commercial packed red grapefruit arrayed on packinghouse floor for canine interrogation; 1–6 cartons contain Xcc-infected fruits—positions of Xcc-infected cartons randomized between trials.

2.9. Assessment of Xcc Detection in Commercial Citrus Orchards

One canine, "Juice", surveyed two commercial red grapefruit orchard blocks in Indian River County, Florida with endemic low incidence Xcc infection (Figure 2F,G). Prior to canine assessment, the blocks were surveyed visually by human assessors to determine and map the location of Xcc-infected trees. Two assessors independently examined trees visually requiring ≈5 min/tree. If a tree had unusual symptoms or was difficult to assess, one to two additional assessors examined it as well. Subsequent to canine assessment, trees on which canines alerted were visually reassessed in an attempt to determine if human assessors could confirm the canine detections. However due to the previously documented inefficiency of human visual assessment ([11,36]; T. Riley, unpublished results), it is likely that many infected trees were missed and/or could not be confirmed. Canine alerts were recorded, and the results and latent class metrics were calculated to assess canine performance against putative human visual assessment.

2.10. Spatial Heterogeneity of Xcc Detection Errors

It is not uncommon for detector canines to acquire a target scent at some distance from the true target, occasionally alerting on a negative target within the scent plume [37,38]. To address the concern of FN and FP canine alerts on grids of Xcc-infected and non-infected targets, respectively, we analyzed the cumulative randomized placement of Xcc-infected trees and compared it to correct Xcc true positive (TP) and true negative (TN) tree positions, calculating the distances between TP and FP and locations. We conducted this spatial assessment for (1) disease incidence of lesions in simulated orchards, (2) lesion age in simulated orchards, and (3) incidence of cartons with Xcc-infected fruit in the packinghouse.

2.11. Direct Detection of Xcc Bacteria

Initially, it was assumed that canines trained to detect Xcc-infected trees were alerting to a complex scent signature composed of VOCs from the bacteria plus unique plant-based VOCs produced in response to Xcc infection. We questioned if trained canines would alert directly on VOCs from Xcc bacteria without the background citrus host odor and/or unique VOCs produced by the bacteria/host interaction. To answer this question, we grew Xcc in axenic culture on nutrient agar for 1 week, harvested the bacteria, suspended them in sterile phosphate-buffered saline (PBS; 0.14 M NaCl, 1.5 mM KH_2PO_4, 6.5 mM Na_2HPO_4, 2.6 mM KCl (pH7.4)), and adjusted the suspension spectrophotometrically by diluting with sterile PBS to approximately 10^6 cfu/mL. Subsequently, 400 µL of the suspension (containing $\approx 2 \times 10^5$ cfu/mL) or 400 µL of sterile PBS was pipetted onto sterile cotton filter discs.

Prior to this experiment, one canine (Mi) was trained exclusively to detect Xcc *in planta* from Xcc-infected plants and a second canine (Ti) was trained to detect Xcc in vitro from axenic culture. Both canines interrogated a row of 10 metal paint cans ≈ 2 m apart. Into one can, a sterile cotton filter disc infused with 400 µL of Xcc dilution in PBS was placed, and into the nine remaining cans, a sterile cotton filter disc infused with 400 µL sterile PBS was placed. The canines repeatedly interrogated the line of 10 cans 10 times, the cans were re-randomized between each replication, and the study was repeated once. All tests were blinded as described in Section 2.4 above. Canine alerts were recorded, and latent class metrics were calculated to assess canine performance.

The reciprocal experiment was also conducted, wherein we questioned if canines trained exclusively on cultured bacteria could detect the bacteria *in planta*, i.e., in Xcc-infected plants. Both canines repeatedly interrogated a line of 10 trees. The line was composed of one Xcc-infected and 9 non-infected Duncan grapefruit seedlings. The plants were re-randomized between each replication, and the study was repeated once. Canine alerts were recorded and latent class metrics were calculated to assess canine performance.

2.12. Estimation of Bacterial Detection Threshold

Having determined that canines trained on either Xcc-infected plants or on Xcc cultures were able to detect the bacteria directly in vitro as well as *in planta*, we wanted to determine the sensitivity, i.e., lower limit of bacteria needed for canine detection. To answer this question, we grew Xcc in axenic culture on nutrient agar for 1 week, harvested the bacteria, suspended them in sterile PBS, and adjusted the suspension spectrophotometrically by diluting with sterile PBS to approximately 10^4 cfu/mL as described above. Then, a dilution series (4×10^4, 4×10^2, 0×10^0, 0×10^{-1}, and 0×10^{-2}) was prepared using sterile PBS as the diluent, and 400 µL of each bacterial dilution was pipetted onto individual sterile cotton filter discs. Immediately following the canine detection trials (described below), the individual dilutions were plated on nutrient agar, incubated for 2 days, enumerated, and the results were used to calculate the number of bacteria on the cotton filter discs (approximately 26.4, 3.60, 0.27, 0, or 0 cfu) as presented to the canines at the time of the test.

The canine (Mi) previously trained to detect Xcc-infected plants and the canine (Ti) trained to detect Xcc from culture, each interrogated rows of 10 paint cans $\approx$2 m apart. A sterile cotton filter disc infused with a specified dilution of Xcc in PBS was placed into one can, and into the nine remaining cans a sterile cotton filter disc infused with PBS was placed. All dilutions were interrogated by both canines and the performance was assessed. A known TP consisting of a filter pad with 400 µL of $\approx 4 \times 10^2 = 25.2$ cfu Xcc culture was randomly placed in each line of cans as a TP control. Additionally, a TN control was assessed consisting of a single location of *Bacillus megaterium* to ensure that the canines were performing as expected. The original dilution was determined by subsequent culture to be 143 cfu/mL with $\approx$57 cfu/400 µL pipetted onto the sterile cotton filter disc.

To examine if canines were able to detect Xcc subcellular components, Xcc bacteria were collected from petri plate culture, suspended in PBS, and adjusted spectrophotmetrically to $\approx 10^2$ cfu/mL (25.2 cfu) by dilution with PBS. Then, the resulting suspension was passed through a 0.2 µm microbiological filter to remove bacterial cells and 400 µL of the filtrate was pipetted onto a sterile cotton filter disc. Filtrate was subsequently cultured and resulted in 0 cfu growth.

This "cell-free" suspension was placed in a single can and interrogated by both canines Mi and Ti in a line of 9 other cans into which were placed sterile cotton filter discs with 400 µL of PBS and one can with a sterile cotton filter disc infused with 400 µL of Xcc culture adjusted to $\approx 10^2$ cfu/mL = 25.2 cfu as a positive control. All tests were blinded as described in Section 2.4 above. Canine alerts were recorded and performance was assessed.

3. Results

3.1. Initial Training of Canines for Xcc Scent Signature Recognition

The Florida researchers were not involved in the initial training and evaluation of the first Xcc-detector canines in California. However, the collaborating trainer/handlers indicated that after a few days of repeated training using STU collected scent pads, the experienced criminology detector canines quickly learned and imprinted on the Xcc scent signature, alerting on scent pads from Xcc-infected trees with 95–98% accuracy (data not shown). This was a similar level of accuracy to that which trainers normally expect and achieve with criminology target scent signatures. This also led us to believe that training canines for Xcc detection was possible and thus could be viably explored further.

We established that longer sample collection times were better for recognition/training, presumably because the scent deposition on the pads was stronger, i.e., the concentration of the unique Xcc VOCs was higher. However, once trained, dogs were capable of detecting the Xcc signature on samples collected over exposure durations as low as 1 min, i.e., lower scent concentration. We also determined that the Xcc scent signature was not long lived. Once the heat-sealed, vapor-proof plastic bags were removed from refrigeration and opened to the ambient environment, the trained canines would alert on them for 2–3 weeks before the Xcc scent signature diminished to unreliable levels. Scent pad viability could be lengthened to some extent by reheat-sealing and refrigerating the pads between uses, but due to scent degradation concerns, we limited scent pad use to 2 weeks from time of collection. Therefore, the VOC composition of Xcc scent does not appear to be stable over time, unlike some scents such as human or animal scents that are very stable and can remain on clothing and other objects for long periods up to years. However, repeated experiments with a canine trained to recognize Xcc resulted in >95% correct recognition and differentiation from non-infected citrus scents, from properly handled scent pads.

3.2. Training Canines for Detection of Xcc-Infected Plant Material

During initial proof-of-concept studies, the 11 September 2001 terrorist attacks occurred, and many of the detector canines throughout the country were diverted to security-related tasks and became unavailable for research and our studies were curtailed. Over the next three years, Xcc had become more widely spread in Florida, reducing regulatory concerns for the movement of infected plant materials within state. Therefore, in 2004–2005, we were able to resume studies on canine detection of Xcc in collaboration with the NDDTC. However, the NDDTC's mission is to train and deploy agricultural contraband detector canines to US ports of entry and overseas ports of departure for US travelers and freight. Our research project utilized these same canines, who upon completion of their training were shipped out to their intended assignments, as were some of the staff. Thus, we did not have consistency of canines or staff, and we rotated through canines and trainers. Even so, trainers reported >95% detection accuracy after a few days training with plants in boxes, indicating good imprinting on Xcc-infected plant materials. When deployed on a single occasion to an infected orchard, the box-trained Labrador retriever also accurately detected Xcc-infected versus non-infected leaves from the orchard. Thus, we were able to discern that canines could be trained to detect Xcc using the differential of infected versus non-infected plant materials. Due to the high turnover rate of canines and handlers, the project was again curtailed.

3.3. Detection of Xcc in Simulated New Plantings

Canine detection studies resumed in 2006 and continued through 2020 with three collaborative commercial canine detection companies. The performance of the first canine "Juice" indicated excellent scent signature recognition (Video S1, see Supplementary Materials). Sensitivity, specificity, precision, and accuracy ranged from 0.7333–0.9167, 0.9943–1.0, 0.8759–1.0, and 0.9733–0.9967, respectively, indicating that canine false negative (FN) alerts were slightly more prolific than false positive (FP) alerts, especially at higher incidence levels. Performance metrics indicated detection was superior for lower (<5%) Xcc-incidence, eroding slightly when incidence was ≥7%. Even so, overall accuracy was 0.9842 (Table 2).

Table 2. Latent class metrics for canine Xcc-infected tree detection in simulated 100-tree Duncan grapefruit citrus orchard by tree incidence accumulated over all the months tested.

Metric [a]	Xcc-Infected Plant Incidence					
	2%	4%	5%	7%	10%	Overall
n	600	600	800	700	600	3300
TP	11	21	30	37	44	143
TN	587	573	758	647	540	3105
FP	1	3	3	4	0	11
FN	1	3	9	12	16	41
SEN	0.9167	0.8750	0.7692	0.7551	0.7333	0.7772
SPE	0.9983	0.9948	0.9961	0.9939	1.0000	0.9965
PPV	0.9167	0.8750	0.9091	0.9024	1.0000	0.9286
NPV	0.9983	0.9948	0.9883	0.9818	0.9712	0.9870
FPR	0.0017	0.0052	0.0039	0.0061	0.0000	0.0035
FDR	0.0833	0.1250	0.0909	0.0976	0.0000	0.0714
ACC	0.9967	0.9900	0.9850	0.9771	0.9733	0.9842
Month(s)	J,S,D	J,S,D	J,S,D,M	J,S,D,M	J,S,D	

[a] Performance metrics as described in Section 2.3 above. Orchard grid consisted of 100 trees with the indicated Xcc-infected tree incidence. The grid was re-randomized between each replication, at each incidence level and interrogated by a single canine—"Juice". Months tested: J = July, S = September, D = December, M = May.

Seasonality (i.e., July, summer; September, fall; December, winter; and May, spring) did not have a perceptible effect on canine detection performance when assay results were accumulated across incidence levels (Table 3). Canines were proficient at Xcc detection prior to assessing performance metrics; however, a training effect was noted. The canines improved notably in sensitivity and slightly in overall accuracy as they became more comfortable with the "game" of detection over time (Figure 4).

Table 3. Latent class metrics for canine Xcc-infected tree detection in simulated 100-tree Duncan grapefruit citrus orchard for each month tested accumulated over all tree incidence levels tested.

Metric [a]	Month of Test				
	July	September	December	May	Overall
n	900	1000	1000	400	3300
TP	42	40	44	17	143
TN	846	942	943	374	3105
FP	5	2	2	2	11
FN	7	16	11	7	41
SEN	0.8571	0.7143	0.8000	0.7083	0.7772
SPE	0.9941	0.9979	0.9979	0.9947	0.9965
PPV	0.8936	0.9524	0.9565	0.8947	0.9286
NPV	0.9918	0.9833	0.9885	0.9816	0.9870
FPR	0.0059	0.0021	0.0021	0.0053	0.0035
FDR	0.1064	0.0476	0.0435	0.1053	0.0714
ACC	0.9867	0.9820	0.9870	0.9775	0.9842

[a] Performance metrics as described in Section 2.3 above. Orchard grid consisted of 100 trees with the indicated Xcc-infected tree incidence. The grid was re-randomized between each replication, at each incidence level and interrogated by a single canine—"Juice".

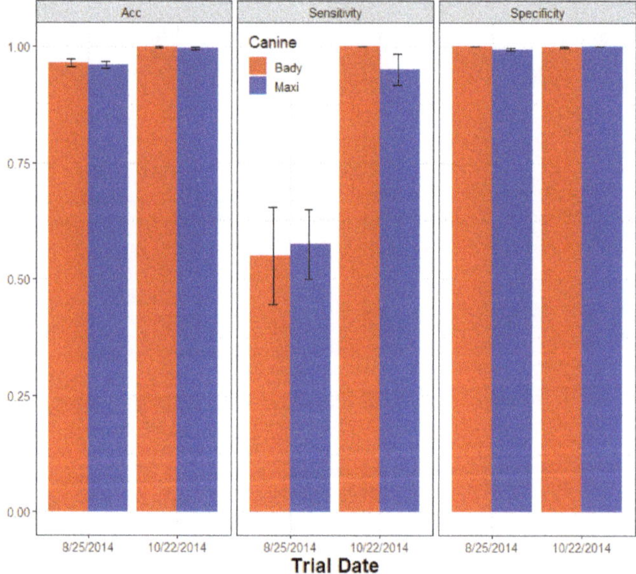

Figure 4. Latent class metrics for the effect of incidence of Xcc lesions on canine detection. The data demonstrate a training effect where canine detection of Xcc-infected trees (sensitivity) significantly improves between the first and second tests, which also improves slightly the overall accuracy metric. In essence, the canines learn the "game" of detecting Xcc-infected trees when presented with a grid imposed by the experimental design and become more proficient at detection over time.

Since seasonality effects were not significant in the first study, performance assessment of the second two canines (Bady and Maxi) was compressed to two separate days separated by 31 days during the second study. During this second study, randomized Xcc-infected tree incidence ranged from 1 to 6% to reduce the probability of scent acquisition from nearby Xcc-infected trees when using higher incidence levels. The sensitivity, specificity, precision, and accuracy performance metrics for canine Bady ranges were 1.0–1.0, 0.9894–1.0, 0.6667–1.0 and 0.9900–1.0, respectively; whereas the same performance metrics for canine Maxi ranged slightly higher: 0.9167–1.0, 1.0–1.0, 1.0–1.0, and 0.995–1.0, respectively. The overall combined performance for the same metrics for the two canines were 0.9856, 0.9974, 0.9257, and 0.9970, respectively (Table 4). Canine Bady alerted to 11 FP and 0 FN over 2200 interrogations, whereas Maxi alerted on 0 FP and 2 FN over 2300 interrogations, suggesting that Maxi was slightly more accurate overall and both of these newer trained canines were slightly more accurate compared with Juice, the earlier trained canine. The 11 FP alerts by Bady were distributed across Xcc-infected tree incidence levels.

3.4. Detection of Xcc Infections of Increasing Age

Latent class performance metrics for detecting Xcc lesions indicated no effect of increasing lesion age. For canine Bady sensitivity, specificity, precision = positive predicted value, and the accuracy performance ranges were 0.9–1.0, 1.0–1.0, 1.0–1.0 and 0.9960–1.0, respectively, whereas the same performance metrics for canine Maxi ranged slightly higher: 0.95–1.0, 1.0–1.0, 1.0–1.0, and 0.9980–1.0, respectively. The overall combined performance for the same metrics for the two canines were 0.9450, 1.0, 1.0, and 0.9978, respectively (Table 5). Neither canine had any FP alerts, and Bady and Maxi had eight and three FN alerts, respectively, which were distributed across Xcc-infection age groups with no apparent effect of age (Figure 5).

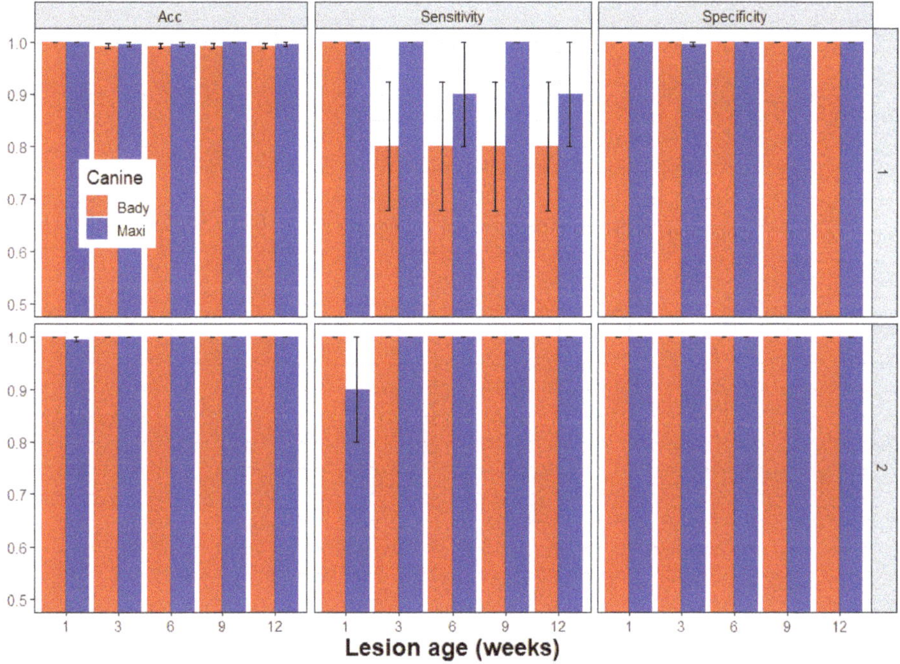

Figure 5. Latent class metrics for canine detection of Xcc-infections of increasing age. There was no relationship of lesion age on canine detection of Xcc-infected trees. However, the data demonstrate a training effect for canine detection of Xcc-infected trees (sensitivity) which significantly improves between the first (1) and second (2) tests as canines learn the "game" imposed by the experimental design and become more proficient at detection. Acc= Accuracy.

Table 4. Latent class metrics for canine Xcc-infected tree detection in simulated 100-tree Duncan grapefruit citrus orchard by tree incidence accumulated for each canine tested.

| | Bady | | | | | | | Maxi | | | | | | | Overall |
| | Xcc-Infected Plant Incidence | | | | | | | Xcc-Infected Plant Incidence | | | | | | | Both |
Metrics [a]	1%	2%	3%	4%	5%	6%	Overall	1%	2%	3%	4%	5%	6%	Overall	Canines
n	400	400	400	400	400	200	2200	400	400	400	400	300	200	2100	4300
TP	4	8	12	16	20	12	72	4	8	12	16	14	11	65	137
TN	396	388	387	380	380	186	2117	396	392	388	384	285	188	2033	4150
FP	0	4	1	4	0	2	11	0	0	0	0	0	0	0	11
FN	0	0	0	0	0	0	0	0	0	0	0	1	1	2	2
SEN	1.0000	1.0000	1.0000	1.0000	1.0000	1.0000	1.0000	1.0000	1.0000	1.0000	1.0000	0.9333	0.9167	0.9701	0.9856
SPE	1.0000	0.9898	0.9974	0.9896	1.0000	0.9894	0.9948	1.0000	1.0000	1.0000	1.0000	1.0000	1.0000	1.0000	0.9974
PPV	1.0000	0.6667	0.9231	0.8000	1.0000	0.857	0.8675	1.0000	1.0000	1.0000	1.0000	1.0000	1.0000	1.0000	0.9257
NPV	1.0000	1.0000	1.0000	1.0000	1.0000	1.000	1.0000	1.0000	1.0000	1.0000	1.0000	0.9965	0.9947	0.9990	0.9995
FPR	0.0000	0.0102	0.0026	0.0104	0.0000	0.011	0.0052	0.0000	0.0000	0.0000	0.0000	0.0000	0.0000	0.0000	0.0026
FDR	0.0000	0.3333	0.0769	0.2000	0.0000	0.143	0.1325	0.0000	0.0000	0.0000	0.0000	0.0000	0.0000	0.0000	0.0743
ACC	1.0000	0.9900	0.9975	0.9900	1.0000	0.9900	0.9950	1.0000	1.0000	1.0000	1.0000	0.9967	0.9950	0.9990	0.9970

[a] Performance metrics as described in Section 2.3 above. Orchard grid consisted of 100 trees with the indicated Xcc-infected tree incidence. The grid was re-randomized between each replication, at each incidence level and interrogated by two canines—"Bady" and "Maxi".

Table 5. Latent class metrics for canine detection of Xcc-infections of increasing age.

| | Canine "Bady" | | | | | | Canine "Maxi" | | | | | | Canines |
| | Xcc-Lesion Age (wks) | | | | | | Xcc-Lesion Age (wks) | | | | | | Combined |
Metric [a]	1 + 2	3 + 4	6 + 7	9 + 10	12 + 13	Overall	1 + 2	3 + 4	6 + 7	9 + 10	12 + 13	Overall	Overall
n	500	500	500	500	500	2500	500	500	500	500	500	2500	5000
TP	20	18	18	18	18	92	19	20	19	20	19	97	189
TN	480	480	480	480	480	2400	480	480	480	480	480	2400	4800
FP	0	0	0	0	0	0	0	0	0	0	0	0	0
FN	0	2	2	2	2	8	1	0	1	0	1	3	11
SEN	1.0000	0.9000	0.9000	0.9000	0.9000	0.9200	0.9500	1.0000	0.9500	1.0000	0.9500	0.9700	0.9450
SPE	1.0000	1.0000	1.0000	1.0000	1.0000	1.0000	1.0000	1.0000	1.0000	1.0000	1.0000	1.0000	1.0000
PPV	1.0000	1.0000	1.0000	1.0000	1.0000	1.0000	1.0000	1.0000	1.0000	1.0000	1.0000	1.0000	1.0000
NPV	1.0000	0.9959	0.9959	0.9959	0.9959	0.9967	0.9979	1.0000	0.9979	1.0000	0.9979	0.9988	0.9977
FPR	0.0000	0.0000	0.0000	0.0000	0.0000	0.0000	0.0000	0.0000	0.0000	0.0000	0.0000	0.0000	0.0000
FDR	0.0000	0.0000	0.0000	0.0000	0.0000	0.0000	0.0000	0.0000	0.0000	0.0000	0.0000	0.0000	0.0000
ACC	1.0000	0.9960	0.9960	0.9960	0.9960	0.9968	0.9980	1.0000	0.9980	1.0000	0.9980	0.9988	0.9978

[a] Performance metrics as described in Section 2.3 above. The simulated orchard grid consisted of 50 trees with two Xcc-infected trees of the indicated age group randomly placed with a population of 48 non-infected trees. The grid was re-randomized between each replication, at each incidence level and interrogated by two canines—"Bady" and "Maxi".

3.5. The Effect of Incidence of Xcc Lesions on Detection

Across all lesion incidence levels, both canines had a higher prevalence of FN than FP alerts (FDR = 0.1159, FPR = 0.0022, overall). Canine Bady had fewer FP than Maxi, but this effect was not consistent across the range of lesion incidence (Figure 6). The sensitivity of both canines increased with lesion incidence (0.6750 to 0.8750), whereas specificity remained high (0.9978) as did overall accuracy (0.9910–0.9952) across all incidence levels (Table 6).

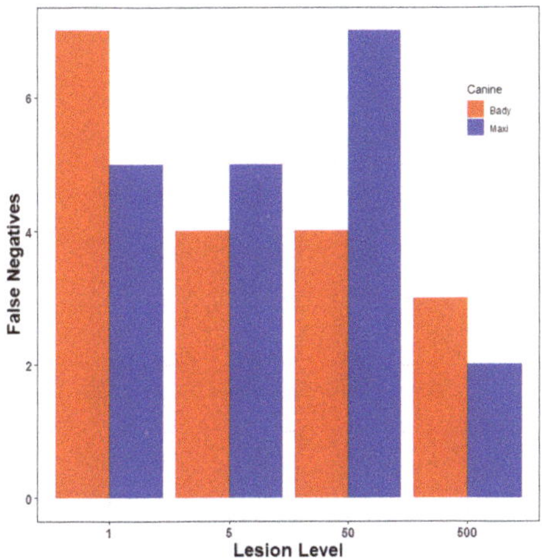

Figure 6. Effect of lesion incidence on false negative canine detections. The data demonstrate a general erosion of canine detection of Xcc-infected trees (sensitivity) as the incidence of infection within individual trees increases. As the scent signature becomes stronger due to heavy infection in some trees, canines begin to false alert on nearby trees because they acquire the scent farther away from the true source.

Table 6. Latent class metrics for the effect of incidence of Xcc lesions on canine detection.

							Canines
	Lesions/Tree				Bady	Maxi	Combined
Metric [a]	1	5	50	500	Overall	Overall	Overall
n	1880	1880	1880	1880	3760	3760	7520
TP	27	31	29	35	61	61	122
TN	1836	1836	1836	1836	3676	3668	7344
FP	4	4	4	4	4	12	16
FN	13	9	11	5	19	19	38
SEN	0.6750	0.7750	0.7250	0.8750	0.7625	0.7625	0.7625
SPE	0.9978	0.9978	0.9978	0.9978	0.9989	0.9967	0.9978
PPV	0.8710	0.8857	0.8788	0.8974	0.9385	0.8356	0.8841
NPV	0.9930	0.9951	0.9940	0.9973	0.9949	0.9948	0.9949
FPR	0.0022	0.0022	0.0022	0.0022	0.0011	0.0033	0.0022
FDR	0.1290	0.1143	0.1212	0.1026	0.0615	0.1644	0.1159
ACC	0.9910	0.9931	0.9920	0.9952	0.9939	0.9918	0.9928

[a] Performance metrics as described in Section 2.3 above. The simulated orchard grid consisted of 100 trees with one Xcc-infected tree of each incidence level (1, 5, 50, and 500 lesions/tree) randomly placed within a population of 96 non-infected trees. The grid was re-randomized between each replication and interrogated by two canines—"Bady" and "Maxi".

Both PPV (0.8710–0.8974) and NPV (0.9930–0.9973) remained relatively constant across Xcc incidence, indicating that canines were superior in predicting actual Xcc-non-infected trees (NPV) and slightly less predictive of actual Xcc-infected trees (PPV), although both canines had low false positive rates (FPR). These data also demonstrated the advantage of repetitive training. This training effect was seen as a general improvement in the sensitivity of Xcc detection with accumulated experience over an increasing number of trials. Conversely, accuracy remained high and improved only slightly, and specificity remained high and stable throughout (Figure 7).

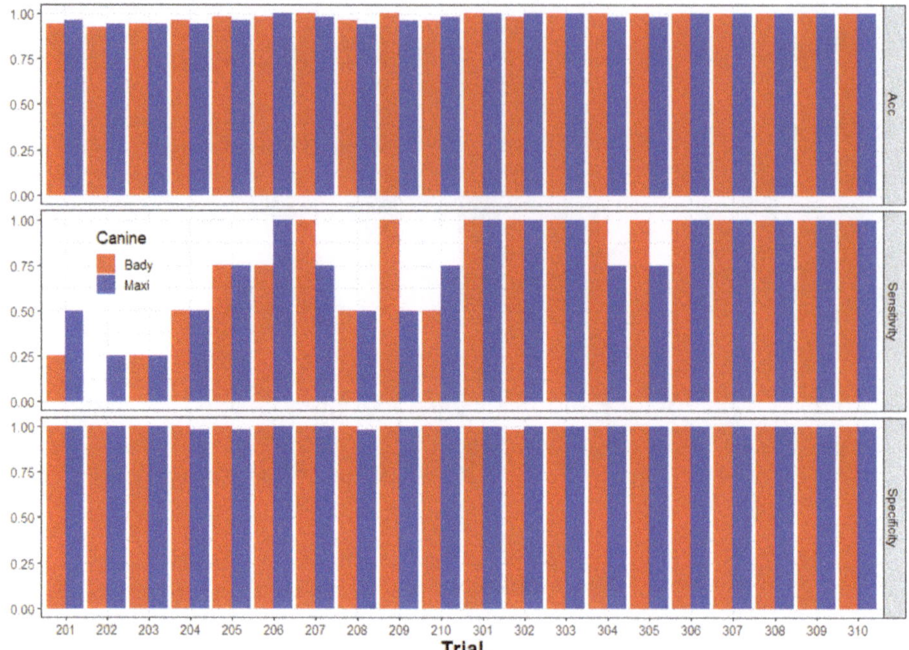

Figure 7. Effect of Xcc lesion incidence on overall accuracy, sensitivity, and specificity of canine detections. The data demonstrate a general improvement in the sensitivity of canine detection as training experience for both canines was accumulated over an increasing number of trials, whereas overall accuracy was high throughout and improved only slightly over accumulated trials and specificity remained high and stable. We use the overall results for each animal displayed in Table 6 to illustrate the diagnostic performance of the canines Bady and Maxi in an information theoretic framework. Figure 8 shows the results of this exercise.

The sequence of plots illustrates the diagnostic performance of the canine starting with the graphical summary provided by the likelihood ratio plot, which summarizes diagnostic performance using metrics that are independent of disease prevalence. The positive likelihood ratio (LR+ = TPP/FPP) is the slope of the solid line segment: for each animal, the values are Bady = 701.5, Maxi = 530.4. The slope of the dashed line segment is the negative likelihood ratio for each animal (LR− = 1 − TPP/(1 − FPP) = FNP/TNP). The values for each animal are Bady = 0.24, Maxi = 0.13. Generally, for effective diagnostic performance, LR+ >> 1 and LR− << 1 are required. Both canines achieved effective positive and negative diagnostic performance, with positive performance (i.e., confirmation of pathogen presence) being particularly effective.

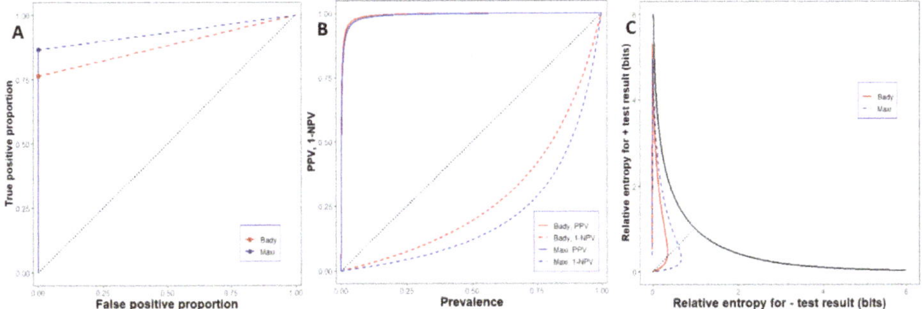

Figure 8. (**A**) Likelihood ratio plot for canines Bady and Maxi, based on average performance over a range of pathogen prevalence values in citrus canker detection trials. The axes are the same as those for a receiver operator characteristic (ROC) curve, each canine being represented by a single point. In general, the closer the point for the animals to the upper left corner with TPP = 1, FPP = 0, the better the overall diagnostic performance. The gradient of the solid line section is the positive likelihood ratio for cases for each canine. The gradient of the dashed section is the negative likelihood ratio. (**B**) A predictive value leaf plot for Bady and Maxi based on the likelihood ratio values displayed panel (**A**). The plot displays the relationship between possible disease prevalence (or prior in a Bayesian framework) and the possible post-diagnostic probability of disease given either positive or negative diagnostic outcomes. The canines have very similar positive alert performance, but they differ in the information they provide in negative alerts. In general, a negative alert by Maxi provides more information than one by Bady. For both canines, positive alerts result in a high post-test probability of disease even at low prior disease values. (**C**) A relative entropy "loop" plot for each animal based on the same likelihood ratios. For each animal, disease prevalence increases clockwise around the loop which shows the expected information supplied (in bits) for a positive vs. negative alert at each possible disease prevalence between 0 and 1 in steps of 0.0001. In effect, the loop plot shows the information gain from alerts corresponding to the change in probable disease prevalence following diagnosis displayed in the leaf plot in panel (**B**).

Likelihood ratios derived from TPP, FPP, TNP, and FNP allow Bayesian updating of disease prevalence to give disease prevalence conditional on diagnostic outcomes; i.e., they can be used to produce predictive values. Figure 8B shows a leaf plot for Bady and Maxi, based on the likelihood ratios displayed in Figure 8A. The leaf plot shows values for PPV and (1-NPV) as functions of initial disease prevalence over the range of prevalence (0,1). To interpret the plot, select a value for prior Xcc prevalence. Locate the value on the diagonal of the plot. To read off the post-diagnostic probability of Xcc presence following either positive or negative diagnosis, trace up or down (respectively) from the point on the diagonal until intersecting with the curves for either canine. The PPV and 1-NPV values can be read off the vertical scale from reading across from the upper and lower curves (respectively); the 1-NPV value being converted to NPV by simple arithmetic thereafter. We draw attention to the high posterior (i.e., PPV) probability of Xcc presence generated by both Bady and Maxi even at low prior disease prevalence. Figure 8B also illustrates that, in general, Bady and Maxi provided more information in positive than negative diagnostic outcomes.

Following Hughes et al. [34], we calculated the expected mutual information (i.e., relative entropy) for positive and negative diagnostic outcomes (I+ and I−, respectively) for Bady and Maxi corresponding to the likelihood ratios in Figure 8A and the predictive values in Figure 8B. The resulting "loop" plot is shown in Figure 8C. The black curve, descending from left to right across the plot, shows the theoretical maximum information (in bits) that could be obtained from an error-free definitive diagnosis of disease status for an unknown sample. Disease prevalence increases from 0 to 1 sequentially along the curve from left to right, indicating that at low disease prevalence, a positive diagnostic outcome contains more information than a negative one, with the opposite being true when disease prevalence is high. The short diagonal line intersecting the curve shows values where I+ = I−. Note that this diagonal intersects the curve at a value of I+ = I− = 1 bit, which occurs at a disease prevalence of 0.5; in effect, this corresponds to seeing the result of a coin flip.

Corresponding to the results in Figure 8B, the information loop plot shows clearly that Bady and Maxi provided more information about Xcc disease status in positive than in negative diagnostic outcomes. The maximum information supplied in a positive diagnosis for either canine was in the order of 4.5 bits, whereas the maximum value for negative diagnoses was less than 1 bit.

3.6. Detection of Xcc Infections in Decaying Foliage

On the day the leaves were abscised from infected trees (day 0) canines correctly identified all positive and negative leaf piles without error. There were very few FP errors throughout the study (3/448). However, as time post-abscission increased FN errors increased from 1 to 20 for days 1 and 27 post-abscission, respectively. On day 27, neither canine detected any of the decaying 20 Xcc-infected leaf piles after repeated attempts, i.e., FN error = 100% (Table 7).

Table 7. Latent class metrics for canine detection of Xcc-infections in leaves decaying over time post abscission.

	Days Post-Abscission						**Totals**
Metric [a]	**0**	**1**	**2**	**5**	**13**	**27**	**Over Time**
n	32	64	80	80	112	80	448
TP	8	15	8	9	16	0	56
TN	24	48	59	60	82	60	333
FP	0	0	1	0	2	0	3
FN	0	1	12	11	12	20	56
SEN	1.0000	0.9375	0.4000	0.4500	0.5714	0.0000	0.5000
SPE	1.0000	1.0000	0.9833	1.0000	0.9762	1.0000	0.9911
PPV	1.0000	1.0000	0.8889	1.0000	0.8889	NA	0.9492
NPV	1.0000	0.9796	0.8310	0.8451	0.8723	0.7500	0.8560
FPR	0.0000	0.0000	0.0167	0.0000	0.0238	0.0000	0.0089
FDR	0.0000	0.0000	0.1111	0.0000	0.1111	NA	0.0508
ACC	1.0000	0.9844	0.8375	0.8625	0.8750	0.7500	0.8683

[a] Performance metrics as described in Section 2.3 above. Two canines "Bady" and "Maxi", interrogated piles of Xcc-infected and non-infected decaying leaves at various assessment times post-leaf abscission. Leaf piles were continuously exposed to ambient conditions over time. Data were accumulated and combined over canines and replications. NA—metric could not be calculated due to a complete lack of canine detection by both canines, resulting in a calculated TP + FP = 0 value in the denominator.

3.7. Detection of Xcc in Citrus Packinghouse Environments

Canine Juice detected just a few Xcc lesions on 1–2 grapefruit fruits when commercially packed into cartons of ≈50 fruit per carton, especially at low incidence. FN errors generally exceeded FP errors throughout the packinghouse study. However, the total number of FN + FP errors began to increase when ≥4% Xcc-incidence of cartons were arrayed in the grid of 100 cartons on the packinghouse floor (Table 8, Figure 9 and Video S2, see Supplementary Materials).

We conducted higher Xcc-incidence interrogations later in the day, which corresponded to a general rise in temperature in the packinghouse environment, resulting in fatigue of the canine. It was observed that the canine often acquired the Xcc scent signature from cartons farther away depending on the direction of the airflow. This was evidenced by the canine's desire to disperse with the required serpentine search pattern through the grid in favor of going directly to a detected target at some distance. Additionally, the spatial proximity to other Xcc-infected cartons as well as the prior locations of Xcc-infected cartons with residual odor led to some FP results (see spatial analyses below).

Table 8. Latent class metrics for canine detection of Xcc-infections of various incidence in commercially packed boxes of red grapefruit.

	Incidence of Cartons Containing Xcc-Infected Fruit										
Metric [a]	1	2	3	4	5	6	7	8	9	10	Total
n	100	700	400	400	100	200	200	300	200	200	2800
TP	1	12	11	11	3	9	10	15	14	12	98
TN	99	685	387	385	95	186	183	273	180	179	2652
FP	0	1	1	0	1	2	3	3	2	1	14
FN	0	2	1	4	1	3	4	9	4	8	36
SEN	1.0000	0.8571	0.9167	0.7333	0.7500	0.7500	0.7143	0.6250	0.7778	0.6000	0.7313
SPE	1.0000	0.9985	0.9974	1.0000	0.9896	0.9894	0.9839	0.9891	0.9890	0.9944	0.9947
PPV	1.0000	0.9231	0.9167	1.0000	0.7500	0.8182	0.7692	0.8333	0.8750	0.9231	0.8750
NPV	1.0000	0.9971	0.9974	0.9897	0.9896	0.9841	0.9786	0.9681	0.9783	0.9572	0.9866
FPR	0.0000	0.0015	0.0026	0.0000	0.0104	0.0106	0.0161	0.0109	0.0110	0.0056	0.0053
FDR	0.0000	0.0769	0.0833	0.0000	0.2500	0.1818	0.2308	0.1667	0.1250	0.0769	0.1250
ACC	1.0000	0.9957	0.9950	0.9900	0.9800	0.9750	0.9650	0.9600	0.9700	0.9550	0.9821

[a] Performance metrics as described in Section 2.3 above. Incidence (1–10) indicates the number cartons containing Xcc-infected fruit placed in a grid of 100 commercially packed boxes each containing ≈50 red grapefruit.

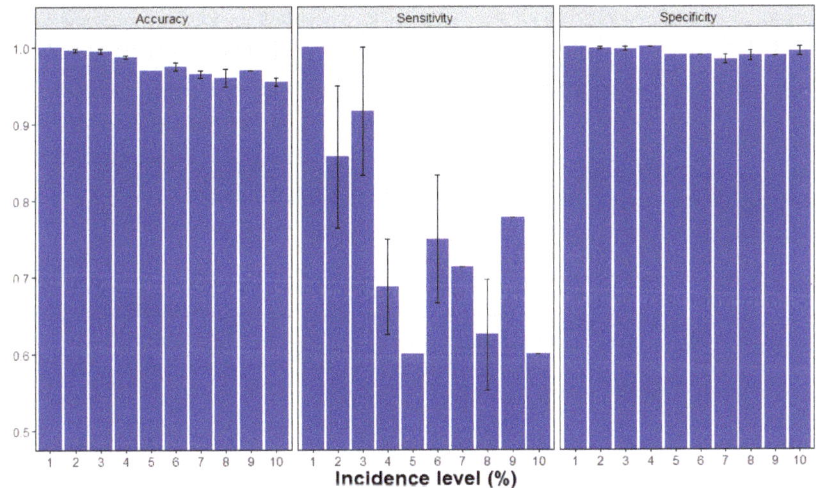

Figure 9. Effect of incidence (proportion) of commercial fruit boxes containing Xcc-infected grapefruit on false negative canine detections. The data demonstrate a general erosion of canine detection of Xcc-positive boxes (sensitivity) as the incidence of infected boxes increases within the grid in the packinghouse. As the scent signature becomes more prevalent within the test grid commensurate with the number of boxes containing Xcc-infected fruit, canines begin to false alert on nearby boxes in close proximity to boxes with actual infected fruit because the canines acquire the scent farther away for the true source.

3.8. Assessment of Xcc Detection in Commercial Citrus Orchards

For the canine Juice, sensitivity, specificity, precision = positive predicted value, negative predictive value and accuracy performance metrics were 1.0, 0.9842, 0.2222, 1.0, and 0.9843 for Orchard 1, and 0.8667, 0.9848, 0.7222, 0.9939 and 0.9797 for Orchard 2, respectively (Table 9). When we progress from a known infection status of potted trees in a simulated orchard (composed of confirmed Xcc-infected and non-infected trees, held in an isolation greenhouse to ensure no additional infections — see simulated new planting results above) to a commercial orchard (Figure 10) (in which disease status is unknown and subject to human visual confirmation and errors), detection precision metrics declined. This decline relates to our inability to determine the infection status unequivocally via human visual inspection. As previously demonstrated, in mature trees it is difficult-to-impossible to detect all incipient and low incidence infections by human visual inspection ([11,39]; T. Riley, unpublished results).

Table 9. Latent class metrics for canine detection of Xcc-infections in commercial red grapefruit orchards in Indian River County, Florida.

			Orchard 2
Metric [a]	Orchard 1	Orchard 2	Theoretical
n	445	345	345
TP	2	13	18
TN	436	325	325
FP	7	5	0
FN	0	2	2
SEN	1.0000	0.8667	0.9000
SPE	0.9842	0.9848	1.0000
PPV	0.2222	0.7222	1.0000
NPV	1.0000	0.9939	0.9939
FPR	0.0158	0.0152	0.0000
FDR	0.7778	0.2778	0.0000
ACC	0.9843	0.9797	0.9942

[a] Performance metrics as described in Section 2.3 above. Orchard 1—10 rows of trees, 38 trees per row, 8 missing trees. Orchard 2—17 rows of trees, 29 trees per row, 47 missing trees. Orchard 2 (theoretical results for discussion) considers that the 5 FP detections were Xcc-infected but not detected by human visual survey.

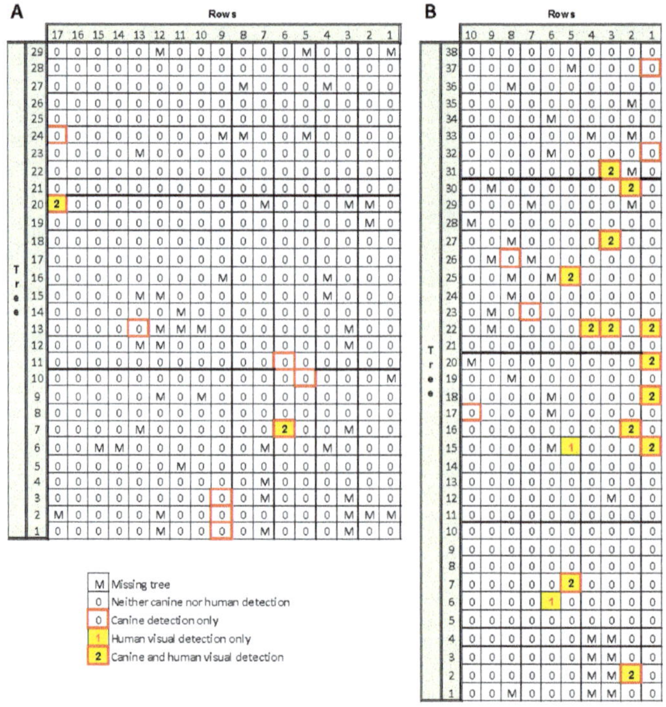

Figure 10. Canine versus human visual detection of Xcc-infection in commercial citrus orchards in Indian River County, Florida. (**A**) Orchard 1—Mature 42-year-old red grapefruit on sour orange rootstock, (**B**) Orchard 2—7-year-old red grapefruit.

3.9. Analyses for Spatial Heterogeneity of Xcc Detection Errors

Sufficient data were available for the spatial analysis of detection errors for three portions of the overall Xcc detection study for the following trials: (1) Disease incidence of lesions in simulated orchards, (2) Lesion age in simulated orchards, and (3) Incidence of cartons with Xcc-infected fruit in the packinghouse. In all three trials, there were more FN than FP errors. Errors were omnidirectional and more prevalent at shorter distances, especially <6.5 m from a true positive (TP) (Figure 11).

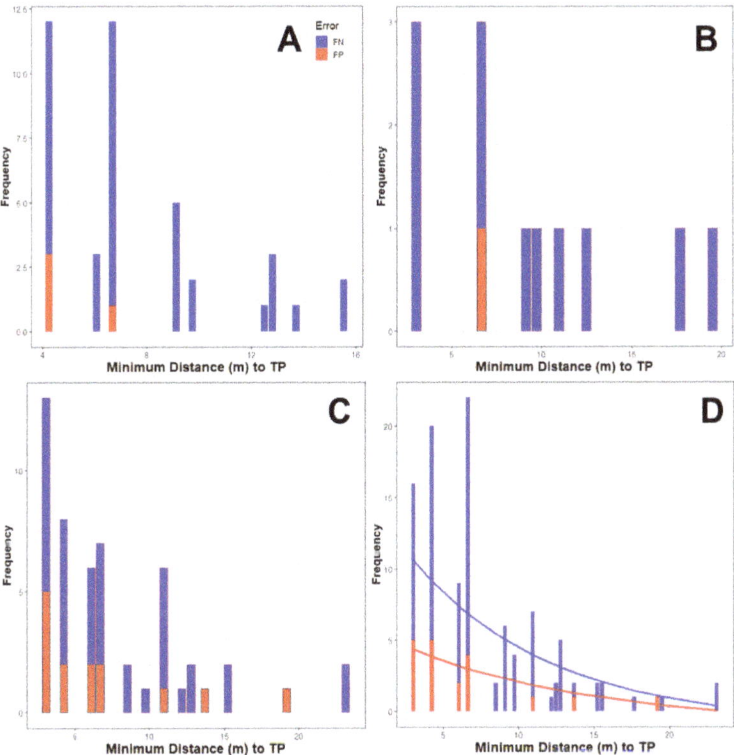

Figure 11. Spatial heterogeneity analysis of canine detection errors for trials of (**A**) lesion incidence, (**B**) lesion age, (**C**) packinghouse, and (**D**) combined data A through C. In all three trials there was a greater number of false negative (FN) than false positive (FP) errors. All trials had a greater prevalence of FN errors and errors were more prevalent at shorter distances from a true positive (TP). Distance is presented as multiples of the distance between plants or cartons (packinghouse) in the grid, i.e., 3.048 m (10 ft) within and between rows.

3.10. Direct Detection of Xcc Bacteria

The canine Mi trained on *in planta* Xcc-infected trees was competent at detecting Xcc bacteria harvested from in vitro axenic cultures (Videos S3 and S4, see Supplementary Materials). The reciprocal was also true, in that the canine Ti trained to detect Xcc from in vitro axenic cultures was able to detect the pathogen *in planta* from Xcc-infected trees (Videos S5 and S6, see Supplementary Materials). Both canines detected the reciprocal target without any additional training. There was a slight trend in that canine Mi was slightly superior at detecting Xcc-infected plants and canine Ti was slightly better at detecting Xcc from culture (Table 10).

Table 10. Latent class metrics for canine detection of Xcc-infected plants versus Xcc culture.

Metric [a]	Xcc-Infected Plants			Xcc Culture		
	Mi	Ti	Overall	Mi	Ti	Overall
n	120	230	250	140	100	240
TP	10	13	23	11	9	20
TN	107	207	314	123	89	212
FP	1	0	1	3	1	4
FN	2	10	12	3	1	4
SEN	0.8333	0.5652	0.6571	0.7857	0.9000	0.8333
SPE	0.9907	1.0000	0.9968	0.9762	0.9889	0.9815
PPV	0.9091	1.0000	0.9583	0.7857	0.9000	0.8333
NPV	0.9817	0.9539	0.9632	0.9762	0.9889	0.9815
FPR	0.0093	0.0000	0.0032	0.0238	0.0111	0.0185
FDR	0.0909	0.0000	0.0417	0.2143	0.1000	0.1667
ACC	0.9750	0.9565	0.9629	0.9571	0.9800	0.9667

[a] Performance metrics as described in Section 2.3 above. Two canines, Mi and Ti, were trained to detect Xcc-infected plants or Xcc cultures, respectively. Both canines interrogated both Xcc-infected plants or Xcc cultures to determine the effect of target used for training. Data represent the combination of two replications.

3.11. Estimation of Bacterial Detection Threshold

The result that canines trained on either Xcc-infected plants or on Xcc axenic cultures can detect the bacteria directly in vitro as well as *in planta*, with near equivalent accuracy, implied that canines may be detecting the bacteria directly in plants and not volatiles generated by the host plants in response to infection. Therefore, we sought to determine the detection limits, i.e., the lowest concentration of bacteria required by the canines for scent detection. Thus, both canines were challenged with the task of detecting Xcc harvested from axenic culture, suspended in PBS, over a range of dilutions from $\approx 4 \times 10^2$ to 0×10^{-2} cfu as described above, with final bacteria on sterile cotton filter discs estimated to be 26.4, 3.60, 0.27, 0, and 0 cfu. Both canines detected Xcc throughout the dilution range, indicating that a sufficient concentration of the Xcc scent signature was present throughout the range tested (Table 11 and Videos S7 and S8, see Supplementary Materials). Neither canine reacted to the *B. megaterium* isolate at the estimated 57 cfu on the scent pad, indicating that the scent signature was apparently specific to Xcc bacteria and not a ubiquitous bacteria scent. Neither canine reacted to a 1×10^2 cfu/mL concentration of 0.2 µm bacterial culture filtrate, indicating that the subcellular component(s) composing the scent signature was not diluted out at any concentration tested but instead was filtered out. This indicates that the canines are detecting Xcc at below the cellular population level, i.e., below a single bacterial cell.

Table 11. Latent class metrics for canine detection of Xcc culture dilutions, detection threshold.

Metric [a]	Canine Mi						
	Xcc Bacteria on Filter Pad Targets					Xcc Filtrate [c]	B. Meg [d]
	10^4 cfu/mL = 26.4 cfu [b]	10^2 cfu/mL = 3.60 cfu	10^0 cfu/mL = 0.27 cfu	10^{-1} cfu/mL = 0 cfu	10^{-2} cfu/mL = 0 cfu		
n	10	10	10	10	10	27	10
TP	1	1	1	1	1	1	0
TN	9	8	9	9	9	27	9
FP	0	1	0	0	0	0	0
FN	0	0	0	0	0	2	1
TPR	1.0000	1.0000	1.0000	1.0000	1.0000	0.3333	0.0000
SEN	1.0000	1.0000	1.0000	1.0000	1.0000	0.3333	0.0000
SPE	1.0000	0.8889	1.0000	1.0000	1.0000	1.0000	1.0000
PPV	1.0000	0.5000	1.0000	1.0000	1.0000	1.0000	0.0000
NPV	1.0000	1.0000	1.0000	1.0000	1.0000	0.9310	0.9000
FPR	0.0000	0.1111	0.0000	0.0000	0.0000	0.0000	0.0000
FDR	0.0000	0.5000	0.0000	0.0000	0.0000	0.0000	0.0000
ACC	1.0000	0.9000	1.0000	1.0000	1.0000	0.9333	0.9000
	Canine Ti						
n	10	20	10	10	10	30	20
TP	1	1	1	1	1	0	0
TN	9	18	9	9	9	27	18
FP	0	0	0	0	0	0	0
FN	0	1	0	0	0	3	2
TPR	1.0000	0.5000	1.0000	1.0000	1.0000	0.0000	0.0000
SEN	1.0000	0.5000	1.0000	1.0000	1.0000	0.0000	0.0000
SPE	1.0000	1.0000	1.0000	1.0000	1.0000	1.0000	1.0000
PPV	1.0000	1.0000	1.0000	1.0000	1.0000	0.0000	0.0000
NPV	1.0000	0.9474	1.0000	1.0000	1.0000	0.9000	0.9000
FPR	0.0000	0.0000	0.0000	0.0000	0.0000	0.0000	0.0000
FDR	0.0000	0.0000	0.0000	0.0000	0.0000	0.0000	0.0000
ACC	1.0000	0.9500	1.0000	1.0000	1.0000	0.9000	0.9000

Table 11. *Cont.*

Metric [a]	Canine Mi						
	Xcc Bacteria on Filter Pad Targets					Xcc Filtrate [c]	B. Meg [d]
	10^4 cfu/mL = 26.4 cfu [b]	10^2 cfu/mL = 3.60 cfu	10^0 cfu/mL = 0.27 cfu	10^{-1} cfu/mL = 0 cfu	10^{-2} cfu/mL = 0 cfu		
	Overall						
n	20	30	20	20	20	60	30
TP	2	2	2	2	2	1	0
TN	18	26	18	18	18	54	27
FP	0	1	0	0	0	0	0
FN	0	1	0	0	0	5	3
TPR	1.0000	0.6667	1.0000	1.0000	1.0000	0.1667	0.0000
SEN	1.0000	0.6667	1.0000	1.0000	1.0000	0.1667	0.0000
SPE	1.0000	0.9630	1.0000	1.0000	1.0000	1.0000	1.0000
PPV	1.0000	0.6667	1.0000	1.0000	1.0000	1.0000	0.0000
NPV	1.0000	0.9630	1.0000	1.0000	1.0000	0.9153	0.9000
FPR	0.0000	0.0370	0.0000	0.0000	0.0000	0.0000	0.0000
FDR	0.0000	0.3333	0.0000	0.0000	0.0000	0.0000	0.0000
ACC	1.0000	0.9333	1.0000	1.0000	1.0000	0.9167	0.9000

[a] Performance metrics as described in Section 2.3 above. Two canines, Mi and Ti were trained to detect Xcc-infected plants or Xcc cultures, respectively. Both dogs interrogated Xcc culture dilutions to determine the lower threshold of bacterial concentration that canines can detect. [b] Initial dilutions were prepared spectrophotometrically to 10^4, 10^2, 10^0, 10^{-1} and 10^{-2} cfu/mL. Dilutions were presented to the canines and cultured the same day. Three days later, cultures were quantified and used to calculate the number of bacteria that were pipetted onto filter pads and presented to canines as targets, i.e., 26.4, 3.60, 0.27, 0, and 0 cfu, respectively. [c] **Xcc Filtrate**—bacterial culture filtrate prepared by filtration of $\approx 10^2$ cfu/mL culture through 0.2 μm filter to eliminate all bacterial cells. Filtrate was cultured and resulted in 0 cfu growth. [d] **B. Meg** = *Bacillius megaterium*, additional bacterium used as a negative control to ensure that canines were not reacting to random bacteria. Original dilution was 143 cfu/mL with ≈ 57 cfu/400 μL pipetted onto a sterile cotton filter disc.

4. Discussion

4.1. Canine Detection of Xcc in Simulated Plantings and Information Theoretic Analyses

Through the multiple trials of this study, we were able to demonstrate unequivocally that canines can discriminate and detect Xcc *in planta* (in infected foliage and fruit) by the use of canine olfactory surveillance in simulated and commercial orchards and packinghouses.

We found that canines alerted to Xcc infections over a range of lesion ages from 1 to 12 weeks post-infection with equal accuracy. These results indicate that the scent signature of Xcc-infection as perceived by canine detectors does not change significantly, if at all, with age of infection. At one week post-infection, the lesions are not yet erumpent and only visible with 10× magnification. At the non-erumpent stage, the numbers of bacteria in such lesions are low and do not directly expose the bacteria to the environment (i.e., lesions have not broken through the intact epidermis and cuticle), yet a sufficient concentration of scent signature appears to emanate from leaves for canine recognition.

Canines were tested in simulated plantings with 2% to 10% (canine Juice initial test) and 1% to 6% (canines Bady and Maxi) incidence of infected trees. In the initial test with the canine Juice, sensitivity increased with lesion incidence, whereas the specificity and overall accuracy remained static across all incidence levels. In a more expansive test with two more extensively trained canines, sensitivity for canine Bady appeared to be unaffected by incidence of infected trees, whereas the sensitivity of canine Maxi seemed to erode slightly with increased incidence of infected trees. Bady had a number of FN alerts, eroding specificity and precision somewhat that were not related to infected tree incidence, whereas Maxi had none and experienced no such erosion. Overall, canines were superior at predicting Xcc-non-infected trees (NPV) and slightly less predictive of actual Xcc-infected trees (PPV). As a field deployable early detection tool, a slight trend toward false negatives is usually accepted by growers, as they are willing to tolerate an assay that misses a few infections rather than misidentifying disease-free trees as diseased, resulting in tree removal and commensurate loss of production.

Xcc lesion populations can range from a single lesion to thousands per tree depending on inoculum prevalence and the susceptibility of tissues following an inoculum dispersal event. Across the range of lesion incidence levels assayed, both canines had a higher prevalence of FN than FP alerts. The sensitivity of both canines increased with lesion incidence, whereas the specificity and overall accuracy was >99% across all incidence levels with low false positive rates (FPR).

Information theoretic analyses of the overall diagnostic performance of both canines showed very clearly that the information provided by positive diagnostic outcomes was far in excess of that provided by negative outcomes, which is in relation to the impact on the probable presence of Xcc. We used a graphical approach recently developed by Hughes et al. [34] to illustrate the diagnostic capacity of the animals. One noteworthy aspect of canine performance is their very high positive likelihood ratios for cases. It has been noted by several authors that disease diagnosis as a tool in decision making at low disease prevalence is problematic [40,41] because, with the diagnostic capability of many commonly used approaches, the post-test probability of disease is still relatively low, even after a positive test outcome, and the sampling error is pervasive (see Section 4.5 below). The positive likelihood ratios achieved by the canines assessed here are in the region of two orders of magnitude higher than values reported in the literature for many plant and human diseases [38,42,43]. As illustrated in Figure 8B this diagnostic capacity allows for effective disease screening even when the background disease prevalence is low.

4.2. Detection of Xcc Infections in Decaying Foliage

Lesions of Xcc cause the surrounding citrus tissues to produce ethylene, which can cause abscission if lesions occur near the junction where a leaf petiole joins a stem. When Xcc-infected leaves fall to the ground near citrus trees, rain or irrigation splash can cause inoculum dispersal and re-infection However, Xcc survival in lesions of decaying leaves decreases exponentially over time [44,45]. Canines were able to detect Xcc in lesions when leaves were newly abscised, but canine detection eroded over time. As time post-abscission increased, FN errors increased, and canines were unable to detect Xcc in leaves 27 days post-abscission. The results indicate a rapid decline in the emanation of Xcc scent signature over time as Xcc-infected foliage decays live Xcc population concurrently decline, resulting in an increase in FN errors through time post-abscission and that abscised infected leaves may not be a reliable indicator of a tree's infection status.

4.3. Detection of Xcc in Citrus Packinghouse Environments

Canine detection of Xcc lesions in boxed citrus fruit in the packinghouse was exceptional considering the noisy, hot, and highly distracting environment. Total FN + FP errors increased when the incidence of Xcc-incidence of cartons was ≥4%. We conducted higher Xcc-incidence interrogations later in the day, which corresponded to a general rise in temperature in the packinghouse environment, causing fatigue of the canine. Additionally, the spatial proximity to adjacent or proximal Xcc-infected cartons may have led to FP alerts. Furthermore, FP alerts occasionally occurred due to persistent residual odor when Xcc-free cartons were relocated (during re-randomization between replications) to where a prior Xcc-infected carton had previously resided. This issue has been noted often by canine trainers across an array of target odor types. Therefore, we re-examined the data for error effects due to spatial proximity to Xcc-infected cartons and residual scent signature odor temporarily permeating the concrete packinghouse floor (see Section 4.4 below).

In an initial study prior to the full experiment, we noted that the canine alerted on two cartons that had recently passed through the packing line. These cartons were visually inspected thoroughly at multiple points in the packing line by commercial packinghouse inspectors and were determined to be Xcc-negative. At first, we thought these were FP detections, but when the grid was re-randomized and the cartons were relocated to different positions, the canine continued to alert on these same two cartons. Upon careful re-inspection, we determined that these two cartons contained a single fruit each with one and two small lesions (≈1 mm dia.), respectively. These observations demonstrate the keen sensitivity of canines to detect even a trace amount of odor and that human visual inspection has limitations for the detection of low-incidence infections especially with small lesions. Canine detection in the packinghouse is superior at low incidence levels. Current detection and elimination/discard of infected fruit were conducted by multiple skilled inspectors observing fruit as it passes through the multiple designated visual examination stations of the packing line. As expected, small Xcc lesions are the most difficult for inspectors to detect and differentiate from other small blemishes. However, Xcc lesions that go undetected can cause a rejection of fruit shipments when inspected at ports of destination. Thus, a highly sensitive detection tool such as canines that does not depend on visual detection could greatly enhance the sanitation of packed fruit shipments and diminish the proportion of rejections at destination markets.

4.4. Spatial Heterogeneity of Xcc Detection Errors

A spatial assessment of the detection data allowed us to determine if the proximity of Xcc-positive trees positioned immediately adjacent to, at oblique angles to, or upwind from canine FNs or FPs were related to the cause of the error. In the trials of this study, the majority of errors, both FN and FP, were within 1 to 2 plant or carton (packinghouse) spaces (within row, across row, or at oblique angles) of a TP. Errors also became more prevalent when the incidence of TP targets increased within the trial grids. However, there was no prominent directionality to errors. False positive alerts can be caused by the dissemination of the odor plume around an odor source. Jezierski states, *"Ideally, dogs should alert as close as possible to the site where the odorous material is hidden by comparing the differences in odor concentration inside the odor plume. It is common for a dog to enter, then exit and reenter the scent cone during odor detection, which may account for the number of times a canine passed a hide as demonstrated in the data. The role of the distribution of the odor plume was evident in our experiment when comparing the percentage of false alerts in particular searching sites. When searching outdoors, the distribution of the odor plume may often enable a more easily directional scenting and localization of odor source, which thus takes less time with more correct and fewer false alerts"* [46]. Craven found that the odor plume of a drug moves and disperses, depending on air currents, humidity, temperature, or features in the terrain, which may also influence the detection performance [47]. The fluid dynamics of odor transport during canine scent detection is highly complex and has not been examined extensively. Angle stated that, *"Much more research needs to be conducted in order to understand the movement of biological VOC within the thermal plume (e.g., microcurrents) and in the aerodynamic wake/wind currents in order to develop search patterns to optimize biomedical detection"* [39].

Although errors were relatively few overall, FN alerts were more common with Xcc detector canines than FP alerts. In an orchard environment, airflow is highly channeled between hedge rows of trees, dynamic, and convoluted with ubiquitous eddies. The complexity of using canines as detectors in a highly heterogeneous open-air commercial orchard environment, or indoor in a harsh noise and distraction rich packinghouse, presents a myriad of potential causes for error, not all of which do we understand or recognize. Our data indicate that the relatively few FN errors increases slightly with an increase in the incidence of TP targets in the environment, especially when these TP targets are in close proximity. One explanation could be that when a canine correctly locates a TP (TP1), infrequently, it can be confused by another TP (TP2) in close proximity. In this example, the canine may misinterpret that the scent it acquired near TP2 is originating from TP1, disregarding TP2 as an additional scent source. Conversely, if the scent emanating from TP1 is of significantly greater concentration than the scent emanating from TP2, the canine might track to the "maximum scent concentration" of TP1, thus ignoring the "minor" scent signature of TP2.

It is well demonstrated that canine detectors can pick up a target scent at some distance. In our experience, training canines for Xcc detection, and additionally for CLas [26], plum pox virus, and vegetable virus detection (unpublished), we have often observed that canines acquire a scent signature at considerable distance from the known target. This is consistently experienced when training canines in spatial grids of predominantly non-infected plants with a low incidence of pathogen-infected plants randomly placed in the grid. Normally, we urge the canine to interrogate each plant in the grid in a serpentine pattern up and down each row so that we can ensure that all plants are assessed equally and collect performance data. However, if the canine is allowed to interrogate the grid off leash, it will often acquire a target scent plume and divert its trajectory obliquely across multiple rows directly to an infected plant: then, it will alert. In other words, the canine has already acquired a viable target at a considerable distance, and it is likely that by odor strength and gradient characteristics, the canine developed a mental picture of the target's estimated location that the canine refines as it hones in on the source.

4.5. Assessment of Xcc Detection in Commercial Citrus Orchards

As intimated in the results above, when we examined canine performance in grids for simulated orchards, lesion age, and lesion incidence, we had unequivocal knowledge of the Xcc infection status of all individual trees. This is because we performed the inoculations, enumerated the Xcc infections on each tree, and all trees were maintained in an isolation greenhouse prior to use where disease spread was highly unlikely. However, in a commercial orchard environment, precise mapping of disease is immensely more difficult. In orchards, we were hindered by the difficulty of Xcc detection and our reliance on human visual survey, especially when trying to accurately map low incidence infections. Tree canopies can be large, composed of >100,000 leaves per tree, and some leaves are in difficult-to-view positions — high in a tree or visually occluded by surrounding branches and foliage. Angle and quality of light, cloudy or bright sun can enhance or diminish visual acuity. For example, when visually surveying trees during the citrus canker epidemic for Xcc infections during the Citrus Canker Eradication Campaign in Miami, Florida, it required an average of 106 days (range 30–270 da) post-infection to visually detect Xcc infections in trees on residential properties. Thus, incipient and low incidence infections were rarely found [11,48]. In a study conducted by USDA, APHIS to determine the efficacy of visual detection of Xcc, 18 two-person teams of trained inspectors surveyed a 12.1 ha (30 ac) block of commercial citrus with known Xcc infections. No two teams detected the same Xcc-infected trees, no teams detected all Xcc-infected trees, and each team found Xcc infections and infected trees previously unrecorded. [Riley, unpublished results].

In the current study, in Orchard 2, the canine alerted on five trees previously unconfirmed by human visual assessment (Figure 10). In an attempt to confirm or refute the canine alert on one of these trees, two highly trained technicians required over two hours on the ground and climbing in the 7-year-old tree, ≈3 m tall canopy to find a single leaf lesion that was obscured by sooty mold. Other lesions may have existed in this tree as well but eluded visual inspection. Visual inspection was limited due to the large number of trees that needed to be visually scrutinized for low incidence lesions and because of the immense amount of time needed to fully assess each tree. Therefore, the infection statuses of these five trees were categorized as FP alerts by the canine (Table 9, Orchard 2). However, if we give the benefit of the doubt to the canine and assume that these trees escaped human visual detection, and were thus TP trees that were correctly identified by the canine, the precision metrics improve considerably, with sensitivity, specificity, precision, negative predictive value, and accuracy performance metrics rising to 0.9, 1.0, 1.0, 0.9939, and 0.9942 (Table 9, Orchard 2—Theoretical). Thus, the precision = positive predicted value (PPV) increases from 0.7222 to 1.0 and overall accuracy increases from 0.9797 to 0.9942, both theoretically.

Humans have great difficulty detecting Xcc in tree canopies with low Xcc-infected leaf incidence. Whereas the canine was a highly sensitive detector (i.e., discovered one lesion within a large canopy) that very rapidly detected the infection and alerted within 1–2 s. The differential between canine and human sensitivity and speed, canine-2 s vs. human team-2 h, becomes apparent and exemplifies the significant differences in probability of detection. These differences translate directly into cost of survey, leaning heavily toward the superiority of canine detection. Additionally, from our experience, it was inappropriate to use a less sensitive detection method (human visual) to validate a more sensitive method (canines). This is demonstrated by the example of commercial Orchard 2 (Table 9, Figure 10), for which five TN trees (as determined by human visual survey) were probably incorrect although identified by the trained canine detector as alerts, and therefore diminished the true estimates of canine detection performance.

4.6. PPV Versus NPV, the Choice between Risk-Aversion and Risk-Acceptance

Growers and regulatory agencies need to ask themselves what is more important depending upon their detection requirements. Is it important to detect all Xcc-positive plants (PPV = 1.0), even if it means a few FP plants (NPV < 1.0) will be indicated as well (i.e., risk averse: willing to cull or treat some Xcc-negative plants in an attempt to best control/mitigate a disease epidemic)? Conversely, does a grower prefer a diagnostic that never falsely implicates a plant as Xcc-positive (NPV = 1.0), and is willing to accept a few FN indications (PPV < 1.0) (i.e., risk accepting: does not want to cull or treat any non-Xcc-positive plants, for example attempting to avoid an adversely harsh regulatory response during an eradication campaign)? Ideally, we want a diagnostic with the highest PPV and NPV possible. However, PPV and NPV are influenced by disease prevalence [30,31]. For example, if we hold sensitivity and specificity constant, the lower the disease prevalence, the higher the PPV. In contrast, as the incidence of the disease increases in a population, PPV improves. Therefore, when evaluating the PPV and NPV metrics for canines or any other diagnostics, we need to consider both the disease incidence within the population and the population size tested.

In general, near-perfect diagnostics are rare in both medicine and plant pathology for a wide array of reasons. Even near-perfect diagnostics are often plagued by sampling error. For example, PCR detection of another citrus bacterial pathogen, CLas, the causal agent of citrus Huanglongbing, is near 100% accurate when testing infected tissue, but when used to assay field trees, it often gives FN results with accuracy of ≈20% due to sampling error, due to the scarcity of infected cells even in systemically infected trees [26]. In mature trees with >100,000 leaves, selecting a leaf with CLas even from a systemically infected tree or even selecting tissue with CLas from an incompletely infected leaf can be very improbable. In contrast, the canines are interrogating the tree holistically; that is, the CLas scent signature can be acquired regardless of where the bacteria are located in the tree, which circumvents both the potential paucity of CLas-infected tissue and the sampling problem. Canine detection of Xcc exhibited many of these same CLas hallmarks, although Xcc and CLas are very different types of phytobacteria. As noted above, in one case, the canine detected Xcc in a 7-year-old field tree with a single lesion obscured by sooty mold that required two trained technicians over 2 h to locate, and in another case, the canine detected single small lesions in fruit cartons missed by trained inspectors. Whereas human inspectors must spend several minutes examining each tree, canines trot along a row of orchard trees interrogating at a rate of ≈1 tree/2 s, continuously drawing in air parcels multiple times per second, thus efficiently surveilling large orchard areas quickly. Molecular or serological detection methods require collecting multiple samples per tree, returning to the lab to process and assay, and the use of moderate to extensive consumables; in addition, assay results are delayed depending upon laboratory backlog and are expensive (sampling and assay costs can exceed $40 US per sample at the time of this writing). Conversely, canines are rapid (1–2 s/tree), results are essentially instantaneous, more cost effective (≈$4.50/tree depending on orchard size, conditions, access, etc.), and more accurate due to the lack of sampling issues when using canines versus other detection methodologies [26].

Additionally, Xcc detector canines were successfully deployed by a commercial canine detection company to assess a commercial citrus nursery and successfully detected multiple Xcc-infected nursery plants unknown to the nurseryman, although no quantification of detection was documented. Such early detection can ensure that Xcc-infected plants are identified, eliminated in the nursery, and thus not transplanted to orchards. Avoiding the introduction of initial inoculum when establishing new plantings is an obvious advantage to mitigating an otherwise potential epidemic. Our collective results from this study imply that canines can be trained as viable early detectors of the pathogen in various agricultural environments, including citrus orchards, packinghouses, and nurseries.

4.7. Direct Detection of Xcc Bacteria and Estimation of Bacterial Detection Threshold

We also trained canines exclusively to two targets: Xcc-infected trees and Xcc axenic cultures. We found that canines trained to either target inherently detected the heterologous targets, although both canines were superior at detecting the homologous target they were trained upon as opposed to the heterologous target. This implies that the scent signature does not need to be augmented with background citrus host VOCs and/or unique VOCs produced by the bacteria/host interaction. Therefore, Xcc cultures might be a sufficient training target if Xcc-infected plants are not available such as in a quarantine situation. In a recent study of five and four respiratory viruses and bacteria, respectively, researchers detected 12 and six VOC that were associated with bacterial and viral growth, respectively, and they identified two VOCs that differentiated bacterial and viral infection [49]. Angle and colleagues discuss the opportunity and complexity of discriminating VOC biomarker detection from diseased individuals [50]. Thus, the use of canines to detect and discriminate a phytobacterial pathogen is not unfounded. The precise and discriminating biomarker VOCs detected from Xcc bacterial cultures needs further examination. Such examination should include multiple canines and perhaps determination of the optimum Xcc-culture concentration on which to train canines to achieve optimized Xcc-infected plant detection, which is beyond the scope of this study.

In a deeper examination of direct in vitro detection, canines trained to either target were able to detect highly dilute Xcc culture solutions as low as 0×10^{-2} cfu, i.e., 100-fold less than a single bacterial cell. This implies that the scent signature is composed of bacterial cell VOC constituents or exudates that occur at concentrations many-fold that of the bacterial cells. To explore this further, we had the canines interrogate a bacterial culture filtrate and found that the canines did not react to the filtrate without bacterial cells present. When considered concomitantly with the canines reacting to culture dilutions as low as 10^{-2} cfu/mL, this led us to suspect that the canines are reacting to a subcellular component that is larger than the 0.2 μm filter pore size. Axenic bacterial cultures are composed of both intact cells and fragments of older dead and decomposing cells. One possibility is that bacterial cell fragments may play a role in the Xcc scent signature. The chemical fractionation and canine testing of such of filtrates and eventual identification of the single or multiple VOCs that compose the Xcc scent signature is beyond this study, as is their individual concentrations necessary for canine detection. However, the realization of a scent signature composition that is at least in part sub-cellular opens a clear and exciting path for future explorations.

5. Conclusions

For millennia, humans have benefitted from application of the acute canine sense of smell to hunt, track, and find targets of importance. In this study, we demonstrated that canines can detect the Xcc phytobacterial pathogen of Asiatic citrus canker in simulated orchards, commercial orchards, and in a commercial packinghouse with high sensitivity, specificity, accuracy, and precision. Canines detected Xcc within 1–2 s of target interrogation time. Canines also alerted across a range of 1–12-week-old infections as well as across a range of pathogen prevalence with equal accuracy. Information theoretic analyses illustrate the diagnostic capacity of canines via their very high positive likelihood ratios for cases across pathogen prevalence at two orders of magnitude higher than values reported for other plant and human diseases. When trained to either Xcc-infected trees or Xcc axenic cultures, canines inherently detected the homologous and heterologous targets, suggesting they can detect Xcc directly rather than only volatiles produced by the host following infection. Canines were also able to detect the Xcc scent signature across a range of axenically cultured Xcc concentrations (10^4 to 10^0 = single cell) and even <1 bacterial cell, which implies that the scent signature is composed of bacterial cell volatile organic compound constituents or exudates that occur at concentrations many fold that of the bacterial cells. These findings indicate that Xcc cultures are a valuable surrogate targeting tool in the absence of infected plants. Results imply that canines can be trained and deployed as viable early detectors of Xcc across a diversity of environments and outperform the prevailing detection method, i.e., human visual detection.

Supplementary Materials: The following are available online at http://www.mdpi.com/1099-4300/22/11/1269/s1, Video S1: Canine "Juice" surveilling simulated new planting of 100 young grapefruit trees. Trees with Xcc infection were randomized within rows. Note that the canine hesitates when it acquires the Xcc scent signature and the handler continues to pull on the leash in an attempt to dissuade the canine and thereby confirm the detection. Canine cannot be dissuaded to leave the TP Xcc-infected tree and eventually alerts by sitting next to the infected tree. Canines utilized in future trials were trained to alert (sit) immediately upon detection. Video S2: Canine "Juice" surveilling for 1–2 fruit with Xcc infections hidden within cardboard cartons of ≈50 commercially packed red grapefruit in a Florida packinghouse. Note canine hesitates when it acquires the Xcc scent signature and handler continues to pull on leash in an attempt to dissuade the canine and thereby confirm the detection. Canine cannot be dissuaded to leave the carton and eventually alerts by sitting next to a carton containing TP fruit with Xcc infections. Canines utilized in future trials were trained to alert (sit) immediately upon detection. Video S3: Canine "Mi" training to detect TP Xcc-infected tree randomized within a line of nine other TN trees. Canine alerts by sitting next to the TP Xcc-infected tree. Video S4: Canine "Mi" previously trained to detect Xcc-infected plants, correctly detects (alerts by sitting) on a metal can containing a TP cotton pad infused with axenic Xcc-culture suspension in PBS buffer in a line of cans with one TP and nine TN cans (with cotton pads infused with PBS buffer only) without prior training on Xcc culture. Video S5: Canine 'Ti' previously trained to detect TP cotton pads infused with axenic Xcc-culture suspension in PBS buffer, correctly detects (alerts by sitting) TP Xcc-infected tree in a line with one TP and nine TN trees without prior training on trees. Video S6: Canine "Ti" previously trained to detect TP cotton pads infused with axenic Xcc-culture suspension in PBS buffer, correctly detects (alerts by sitting) on a metal can containing a TP cotton pad infused with a 10^0 cfu/mL concentration of Xcc bacteria (≅0.27 bacteria/pad), in a line of nine TN pads in cans. Indicates that canine may be acquiring the Xcc-scent signature and alerting on subcellular bacterial components. Note canine surveilled line of cans off leash. Video S7: Canine "Mi" interrogates a row of 11 suspect cans. The canine alerts by sitting next to TP can #6 containing a cotton pad infused with a 400 µL Xcc bacterial suspension in PBS buffer of 10^{-2} cfu/mL (≅0 bacteria/pad, subsequent determined via culturing). Remaining 10 cans contained TN cotton pads infused with 400 µL PBS buffer only. Handler rewards canine for correct detection with a few moments of play with a hard rubber 'kong' toy. Detection of bacterial concentration containing less than one bacteria indicates canine recognizes a scent signature composed of subcellular bacterial components. Video S8: Canine 'Ti' interrogates a row of 11 suspect cans. The canine alerts by sitting next to TP can #6 containing a cotton pad infused with a 400 µL Xcc bacterial suspension in PBS buffer of 10^{-2} cfu/mL (≅0 bacteria/pad, subsequent determined via culturing). Remaining 10 cans contained TN cotton pads infused with 400 µL PBS buffer only. Handler rewards canine for correct detection with a few moments of play with a hard rubber "kong" toy. Detection of bacterial concentration containing less than one bacteria indicates that the canine recognizes a scent signature composed of subcellular bacterial components.

Author Contributions: Conceptualization, T.G. and G.P.; methodology, T.G., and G.P.; software, W.L, D.P., N.M.; validation, T.G., G.P., W.S., E.T. and N.M.; formal analysis, T.G., W.L., D.P., and N.M.; investigation, T.G., G.P., E.T., W.S., and N.M.; resources, T.G. and W.S.; data curation, T.G., G.P.; writing—original draft preparation, T.G.; writing—review and editing, G.P., E.T., W.L., D.P., S.A., W.S., and N.M.; visualization, T.G., W.L., D.P. and N.M.; supervision, T.G., W.S. and G.P.; project administration, T.G.; funding acquisition, T.G. All authors have read and agreed to the published version of the manuscript.

Funding: This work was funded by USDA, APHIS Farm Bill grant 13-8130-0313-CA, and HLB MAC Grant 15-8130-0475. Mention of a trademark, warranty, proprietary product, or vendor does not constitute a guarantee by the USDA and does not imply its approval to the exclusion of other products or vendors that may also be suitable.

Acknowledgments: We thank Alan Hardison (deceased) for bringing canine detection to our attention many years ago. We thank Jerry Bishop, Bryan Brice, Tyler Meck, and William Moraitis F1-K9, Palm Coast FL for their deep collaboration and Peggy and Bill Heiser Coast to Coast K9 Teams, New Smyrna, FL specifically, on this project as well as K-9 Security and Detection International, Orange Co., CA, and Pepe Peruyero J&K Canine Academy, Inc., Alachua, FL, and the USDA, Animal and Plant Health Inspection Service (APHIS), National Detector Dog Training Center (NDDTC), Orlando, FL for their collaborations and assistance. We also thank Daniel Scott Citrus, Vero Beach, Florida, Dan Richie Riverfront Citrus Inc., Vero Beach, Florida and Michel Sallin, IMG Citrus, Vero Beach, Florida for research access to commercial orchards and packinghouse. We also thank Phil Berger and Laurene Levy (deceased), USDA, APHIS for valuable consultation and support. We thank Greg Para, USDA, APHIS, for administrative oversight and support. Finally we thank Len Therrien, Greg Brock, Joann Hodge, and Leigh Sitler for technical assistance.

Conflicts of Interest: The authors declare no conflict of interest. In addition, the funders had no role in the design of the study; in the collection, analyses, or interpretation of data; in the writing of the manuscript, or in the decision to publish the results.

References

1. Graham, J.H.; Brooks, C.; Yonce, H. Importance of early season copper sprays for protection of hamlin orange fruit against citrus canker infection and premature fruit drop. In *Proceedings of the Florida State Horticultural Society*; American Society for Horticultural Sciences: Alexandria, VA, USA, 2016; Volume 129, pp. 74–78.

2. Irey, M.; Gottwald, T.R.; Graham, J.H.; Riley, T.D.; Carlton, G. Post-hurricane analysis of citrus canker spread and progress towards the development of a predictive model to estimate disease spread due to catastrophic weather events. *Plant Health Prog.* **2006**. [CrossRef]

3. Gottwald, T.R.; Irey, M. Post-hurricane analysis of citrus canker II: Predictive model estimation of disease spread and area potentially impacted by various eradication protocols following catastrophic weather events. *Plant Health Prog.* **2007**. [CrossRef]

4. Bock, C.H.; Parker, P.E.; Gottwald, T.R. Effect of simulated wind-driven rain on duration and distance of dispersal of *Xanthomonas axonopodis* pv. *citri* from canker infected citrus trees. *Plant Dis.* **2005**, *89*, 71–80.

5. Gottwald, T.R.; Graham, J.H.; Schubert, T.S. An epidemiological analysis of the spread of citrus canker in urban Miami, Florida, and synergistic interaction with the Asian citrus leaf miner. *Fruits* **1997**, *52*, 383–390.

6. Gottwald, T.R.; Graham, J.H.; Schubert, T.S. Citrus canker in urban Miami: An analysis of spread and prognosis for the future. *Citrus Ind.* **1997**, *78*, 72–78.

7. Schubert, T.S.; Rizvi, S.A.; Sun, X.; Gottwald, T.R.; Graham, J.H.; Dixon, W.N. Meeting the challenge of eradicating citrus canker in Florida-Again. *Plant Dis.* **2001**, *85*, 340–356.

8. Timmer, L.W.; Gottwald, T.R.; Zitco, S.E. Bacterial exudation from lesions of Asiatic citrus canker and citrus bacterial spot. *Plant Dis.* **1991**, *75*, 192–195.

9. Achor, D.S.; Browning, H.W.; Albrigo, L.G. Anatomical and histological modification in citrus leaves caused by larval feeding of citrus leaf miner (*Phyllocnistis citrella* Staint). In Proceedings of the International Conference Citrus Leafminer, Orlando, FL, USA, 23–25 April 1996; p. 69.

10. Graham, J.H.; Gottwald, T.R.; Browning, H.S.; Achor, D.S. Citrus leafminer exacerbated the outbreak of Asiatic citrus canker in South Florida. In Proceedings of the International Conference Citrus Leafminer, Orlando, FL, USA, 23–25 April 1996; p. 83.

11. Gottwald, T.R.; Sun, X.; Riley, T.; Graham, J.H.; Ferrandino, F.; Taylor, E.L. Geo-referenced spatiotemporal analysis of the urban citrus canker epidemic in Florida. *Phytopathology* **2002**, *92*, 361–377.

12. Frantz, L.A.; Mullin, V.E.; Pionnier-Capitan, M.; Lebrasseur, O.; Ollivier, M.; Perri, A.; Linderholm, A.; Mattiangeli, V.; Teasdale, M.D.; Dimopoulos, E.A.; et al. Genomic and archaeological evidence suggest a dual origin of domestic dogs. *Science* **2016**, *352*, 1228–1231. [CrossRef]

13. Savolainen, P.; Zhang, Y.-P.; Luo, J.; Lundeberg, J.; Leitner, T. Genetic Evidence for an East Asian Origin of Domestic Dogs. *Science* **2002**, *298*, 1610–1613.

14. Vilà, C.; Savolainen, P.; Maldonado, J.E.; Amorim, I.R.; Rice, J.E.; Honeycutt, R.L.; Crandall, K.A.; Lundeberg, J.; Wayne, R.K. Multiple and Ancient Origins of the Domestic Dog. *Science* **1997**, *276*, 1687–1689.

15. Leonard, J.A.; Wayne, R.K.; Wheeler, J.; Valdez, R.; Guillén, S.; Vilà, C. Ancient DNA evidence for old world origin of new world dogs. *Science* **2002**, *298*, 1613–1615.

16. Lorenzo, N.; Wan, T.L.; Harper, R.J.; Hsu, Y.L.; Chow, M.; Rose, S.; Furton, K.G. Laboratory and field experiments used to identify *Canis lupus* var. *familiaris* active odor signature chemicals from drugs, explosives, and humans. *Anal. Bioanal. Chem.* **2003**, *376*, 1212–1224.

17. Galibert, F.; Quignon, P.; Hitte, C.; André, C. Toward understanding dog evolutionary and domestication history. *C. R. Biol.* **2011**, *334*, 190–196.

18. Jezierski, T.; Ensminger, J.; Paper, L.E. *Canine Olfaction Science and Law: Advances inn Forensic Science, Mewdicine, Conservation, and Environmental Remediation*; CRC Press Taylor and Francis Group: Boca Raton, FL, USA; London, UK; New York, NY, USA, 2016; p. 482.

19. Peterson, R.O.; Ciucci, P. The wolf is a carnivore. In *Wolves, Behavior, Ecology and Conservation*; Mech, L.D., Boitani, L., Eds.; University of Chicago Press: Chicago, IL, USA, 2003; pp. 104–130.

20. Hall, N.J.; Protopopova, A.; Wynne, C.D.L. The role of environmental and owner-provided consequences in canine stereotypy and compulsive behavior. *J. Vet. Behav. Clin. Appl. Res.* **2015**, *10*, 24–35. [CrossRef]

21. Leitch, O.; Anderson, A.; Kirkbride, K.P.; Lennard, C. Biological organisms as volatile compound detectors: A review. *Forensic Sci. Int.* **2013**, *232*, 92–103.

22. Browne, C.; Stafford, K.; Fordham, R. The use of scent-detection dogs. *Irish Vet. J.* **2006**, *59*, 97–104.

23. Moser, E.; McCulloch, M. Canine scent detection of human cancers: A review of methods and accuracy. *J. Vet. Behav.* **2010**, *5*, 145–152.

24. Willis, C.M.; Church, S.M.; Guest, C.M.; Cook, W.A.; McCarthy, N.; Bransbury, A.J.; Church, M.R.T.; Church, J.C.T. Olfactory detection of human bladder cancer by dogs: Proof of principle study. *BMJ* **2004**, *329*, 712–714.

25. Gordon, R.T.; Schatz, C.B.; Myers, L.J.; Kosty, M.; Gonczy, C.; Kroener, J.; Tran, M.; Kurtzhals, P.; Heath, S.; Koziol, J.A.; et al. The Use of Canines in the Detection of Human Cancers. *J. Altern. Complement. Med.* **2008**, *14*, 61–67.

26. Gottwald, T.; Poole, G.; McCollum, T.; Hall, D.; Hartung, J.; Bai, J.; Luo, W.; Posny, D.; Duan, Y.P.; Taylor, E.; et al. Canine Olfactory Detection of a Vectored Phytobacterial Pathogen. *Liberibacter asiaticus*, and Integration with Disease Control. *Proc. Natl. Acad. Sci. USA* **2020**. Available online: https://www.pnas.org/content/early/2020/01/28/1914296117 (accessed on 7 September 2020). [CrossRef]

27. Mendel, J.; Furton, K.G.; Mills, D. An Evaluation of Scent-discriminating Canines for Rapid Response to Agricultural Diseases. *Hortic. Technol.* **2018**, *28*, 102–108. [CrossRef]

28. Eckenrode, B.A.; Ramsey, S.A.; Stockham, R.A.; Van Berkel, G.J.; Asano, K.G.; Wolf, D.A. Performance evaluation of the Scent Transfer Unit (STU-100) for organic compound collection and release. *J. Forensic Sci.* **2006**, *51*, 780–789. [CrossRef] [PubMed]

29. Formann, A.K.; Kohlmann, T. Latent class analysis in medical research. *Stat. Methods Med. Res.* **1996**, *5*, 179–211. [CrossRef] [PubMed]

30. Akobeng, A.K. Understanding diagnostic tests 1: Sensitivity, specificity and predictive Values. *Acta Paediatr.* **2006**, *96*, 338–341. [CrossRef]

31. Lalkhen, A.G.; McCluskey, A. Clinical tests: Sensitivity and specificity. *Contin. Educ. Anaesth. Crit. Care Pain* **2008**, *8*, 221–223. [CrossRef]

32. Yerushalmy, J. Statistical Problems in Assessing Methods of Medical Diagnosis, with Special Reference to X-Ray Techniques. *Public Health Rep.* **1947**, *62*, 1432–1449. [CrossRef] [PubMed]

33. Biggerstaff, B.J. Comparing diagnostic tests: A simple graphic using likelihood ratios. *Stat. Med.* **2000**, *19*, 649–663. [CrossRef]

34. Hughes, G.; Reed, J.; McRoberts, N. Information graphs incorporating predictive values of disease forecasts. *Entropy* **2020**, *22*, 361. [CrossRef]

35. Bock, C.H.; Gottwald, T.R.; Graham, J.H. A Comparison of Pathogen Isolation in Culture and Injection–infiltration Bioassay of Citrus Leaves for Detecting *Xanthomonas citri* subsp. *citri*. *J. Phytopathol.* **2014**. [CrossRef]

36. Gottwald, T.R.; Graham, J.H.; Schubert, T.S. Citrus canker: The pathogen and its impact. Online. *Plant Health Prog.* **2002**. [CrossRef]

37. Cablik, M.E.; Sagebiel, J.C.; Heaton, J.S.; Valentin, C. Olfaction-based Detection Distance: A quantitative analysis of how far away dogs recognize tortoise odor and follow it to source. *Sensors* **2008**, *8*, 2208–2222. [CrossRef]

38. Wells, D.L.; Hepper, P.G. Directional tracking in the domestic dog, *Canis familiaris. Appl. Anim. Behav. Sci.* **2003**, *84*, 297–305. [CrossRef]

39. Angle, T.C.; Passler, T.; Waggoner, P.L.; Fischer, T.D.; Rogers, B.; Galik, P.K. Real-time detection of a virus using detection dogs. *Front. Vet. Sci.* **2016**, *2*, 79. [CrossRef]

40. Yuen, J.; Hughes, G. Bayesian analysis of plant disease prediction. *Plant Pathol.* **2002**, *51*, 407–412. [CrossRef]

41. Madden, L.V. Botanical epidemiology: Some key advances and its continuing role in disease management. *Eur. J. Plant Pathol.* **2006**, *115*, 3–23. [CrossRef]

42. Fabre, F.; Plantegenest, M.; Yuen, J. Financial benefit of using crop protection decision rules over systematic spraying strategies. *Phytopathology* **2007**, *97*, 1484–1490. [CrossRef]

43. United States Preventative Services Task Force. Available online: https://www.uspreventiveservicestaskforce.org/uspstf/topic_search_results?topic_status=P (accessed on 6 July 2020).

44. Graham, J.H.; McGuire, R.G.; Miller, J.W. Survival of *Xanthomonas campestris* pv. *citri* in citrus debris and soil in Florida and Argentina. *Plant Dis.* **1987**, *71*, 1094–1098. [CrossRef]

45. Gottwald, T.; Graham, J.; Bock, C.; Bonn, G.; Civerolo, E.; Irey, M.; Leite, R.; McCollum, G.; Parker, P.; Ramallo, J.; et al. The epidemiological significance of post-packinghouse survival of *Xanthomonas citri* subsp. *citri* for dissemination of Asiatic citrus canker via infected fruit. *Crop Prot.* **2009**, *29*, 508–524.

46. Jezierski, T.; Adamkiewicz, E.; Walczak, M.; Sobczyńska, M.; Górecka-Bruzda, A.; Ensminger, J.; Papet, E. Efficacy of drug detection by fully-trained police dogs varies by breed, training level, type of drug and search environment. *Forensic Sci. Int.* **2014**, *237*, 112–118. [CrossRef] [PubMed]

47. Craven, B.A.; Paterson, E.G.; Settles, G.S. The fluid dynamics of canine olfaction: Unique nasal airflow patterns as an explanation of macrosmia. *J. R. Soc. Interface* **2010**, *7*, 933–943. [CrossRef] [PubMed]

48. Gottwald, T.R.; McCollum, T.G. Huanglongbing solutions and the need for anti-conventional thought. *J. Citrus Pathol.* **2017**. Available online: https://escholarship.org/uc/item/2fp8n0g1 (accessed on 6 December 2019).

49. Abd El Qader, A.; Lieberman, D.; Avni, Y.S.; Lazarovitch, T.; Sagi, O.; Zeiri, Y. Volatile organic compounds generated by cultures of bacteria and viruses associated with respiratory infections. *Biomed. Chromatogr.* **2015**, *29*, 1783–1790. [CrossRef]

50. Angle, C.; Waggoner, L.P.; Ferrando, A.; Haney, P.; Passler, T. Canine Detection of the Volatilome: A Review of Implications for Pathogen and Disease Detection. *Front. Vet. Sci.* **2016**, *3*, 47. [CrossRef]

Publisher's Note: MDPI stays neutral with regard to jurisdictional claims in published maps and institutional affiliations.

Article

Dynamics of Ebola Disease in the Framework of Different Fractional Derivatives

Khan Muhammad Altaf [1,†] and **Abdon Atangana** [2,*,†]

[1] Department of Mathematics, City University of Science and Information Technology, Peshawar 25000, Pakistan; altafdir@gmail.com
[2] Institute for Groundwater Studies, Faculty of Natural and Agricultural Sciences, University of the Free State, Bloemfontein 9300, South Africa
* Correspondence: AtanganaA@ufs.ac.za
† These authors contributed equally to this work.

Received: 18 February 2019; Accepted: 7 March 2019; Published: 21 March 2019

Abstract: In recent years the world has witnessed the arrival of deadly infectious diseases that have taken many lives across the globe. To fight back these diseases or control their spread, mankind relies on modeling and medicine to control, cure, and predict the behavior of such problems. In the case of Ebola, we observe spread that follows a fading memory process and also shows crossover behavior. Therefore, to capture this kind of spread one needs to use differential operators that posses crossover properties and fading memory. We analyze the Ebola disease model by considering three differential operators, that is the Caputo, Caputo–Fabrizio, and the Atangana–Baleanu operators. We present brief detail and some mathematical analysis for each operator applied to the Ebola model. We present a numerical approach for the solution of each operator. Further, numerical results for each operator with various values of the fractional order parameter α are presented. A comparison of the suggested operators on the Ebola disease model in the form of graphics is presented. We show that by decreasing the value of the fractional order parameter α, the number of individuals infected by Ebola decreases efficiently and conclude that for disease elimination, the Atangana–Baleanu operator is more useful than the other two.

Keywords: Ebola model; Caputo derivative; Caputo–Fabrizio derivative; Atangana–Baleanu derivative; numerical results

1. Introduction

Ebola caused many deaths in Western Africa, especially in the outbreak of 2014. It includes more than 16 thousand laboratory cases with 70% death cases, which is regarded the deadliest outbreak in history since 1976 with 20 Ebola threats. It is evident that in each outbreak, the first case of infection occurred due to contact with infected animals such as monkeys, fruit bats, etc., which shows the spread of the virus through indirect contact [1]. It is documented in [2] that some percentage of the Ebola-Zaire type survived after two weeks on glass at 4 °C and (10%) on plastic, and on surfaces (3%). Moreover, 0.1% to 1 % of the Ebola virus particle can remain up to 50 days at 4 °C [3]. The survival of the Ebola virus in the environment due to poor sanitary and hygienic conditions considerably become another source of Ebola infection in Africa. In Africa, regions were affected greatly by the Ebola virus outbreak due to their inhabitants being involved in hunting food, being close to the rain-forest, and harvesting forest fruits for food [4,5].

The Ebola disease outbreaks and their transmission have been documented in many articles (see [6–10] and the references therein) and the main focus was to study the human population and the direct transmission. Some models of the type SI, SIR, SEIR, and other types also considered the dynamics of the Ebola disease outbreaks [9,11,12]. Recently, in [13] studied an Ebola virus disease through a simple mathematical model of the type SIR with the inclusion of environment effect. Due to the fact that Ebola virus survives in the environment, this warrants that future epidemics can occur. Thus, the inclusion of the environment effect in Ebola disease spread should be studied more and some preventive and other measures should be used to protect people further from this deadly infection. Therefore, based on the model presented in [13], we aim to study the Ebola disease model in the framework of the fractional calculus. The reason for the use of the fractional calculus in Ebola disease is that it has many advantages. Some of them are the heredity and memory effects, the parameter estimations are better, the crossover behavior of the model, and effective strategies for the case of arbitrary order. Some other works used it to study the dynamics of complex networks [14–16]. In [14] the authors studied the dynamics of information and the uses in complex networks. Coupling dynamics of an epidemic spreading with information diffusion is analyzed in [15]. The events that determine spreading dynamics and the information transmission through internal and external influences are considered in [16].

Fractional calculus and its applications to real life problems is found extensively in the literature, for example [17–21]. In all these mentioned papers the focus is to eliminate the infection from the community and it is proven that the fractional models have the ability to model such epidemic disease efficiently and provide reasonable results for the case of non-integer. It is shown that the fractional models are useful for the data fitting [22]. The results suggest in [22] that fractional models are efficient to study disease dynamics well. Therefore, motivated with the above applications, we aim to study an Ebola disease model in the fractional order. We consider three different fractional operators, that is, the Caputo, Caputo–Fabrizio, and the Atangana–Baleanu derivatives. According to the authors' knowledge no one has applied the three operators to an epidemic model. So, this work is a useful study to analyze Ebola disease with different fractional operators. The rest of the work on Ebola disease is categorized as follows: The fractional background material are shown in Section 2. A mathematical model on Ebola disease is presented in Section 3 with basic mathematical results. In Section 4, a mathematical model in the frame of the Caputo derivative and their numerical results, the Caputo–Fabrizio derivative is used to formulate the model and their relevant results are presented, and we further consider the Atangana–Baleanu model for Ebola disease and discuss its existence and uniqueness and a useful numerical scheme for their solution, and lastly in this section, the comparison results for these operators with various fractional order parameters are shown. The Ebola disease models and their fractional results are summarized in Section 5.

2. Fundamental Concepts

Here, we recall the fundamental concepts regarding the Caputo, Caputo–Fabrizio, and the Atangana–Baleanu derivative.

Definition 1. *For a function $f : \mathbb{R}^+ \to \mathbb{R}$, then the fractional integral of order $\alpha > 0$ is given by*

$$I_t^\alpha(f(t)) \quad = \quad \frac{1}{\Gamma(\alpha)} \int_0^t (t-z)^{\alpha-1} f(z)dz.$$

where Γ shows the Gamma function and α is the fractional order parameter.

Definition 2. *For a function $f \in C^n$, then the Caputo derivative with order α is defined as*

$$^C D_t^\alpha(f(t)) = I^{n-\alpha} D^n f(t) = \frac{1}{\Gamma(n-\alpha)} \int_0^t \frac{f^n(z)}{(t-z)^{\alpha+n-1}} dz,$$

that is defined for the absolute continuous functions and $n - 1 < \alpha < n \in N$. Obviously, $^C D_t^\alpha(f(t))$ tends to $f'(t)$ as $\alpha \to 1$.

Definition 3. *[23]. Let $z \in H^1(a,b)$, with $b > a$, and $0 \leq \alpha \leq 1$, then the Caputo–Fabrizio derivative can be written as*

$$D_t^\alpha(z(t)) = \frac{\mathcal{K}(\alpha)}{1-\alpha} \int_a^t z'(x) \exp\left[-\alpha\frac{t-x}{1-\alpha}\right] dx, \tag{1}$$

the normalized function is shown by $\mathcal{K}(\alpha)$ and it holds $\mathcal{K}(0) = \mathcal{K}(1) = 1$. Consider the case for which $z \notin H^1(a,b)$ then, we have the following:

$$D_t^\alpha(z(t)) = \frac{\alpha\,\mathcal{K}(\alpha)}{1-\alpha} \int_a^t (z(t) - z(x)) \exp\left[-\alpha\frac{t-x}{1-\alpha}\right] dx. \tag{2}$$

Remark 1. *[24]. Let $v = \frac{1-\alpha}{\alpha} \in [0,\infty)$, $\alpha = \frac{1}{1+v} \in [0,1]$, then equation given by (2) can be expressed is as follows,*

$$D_t^v(z(t)) = \frac{\mathcal{K}(v)}{v} \int_a^t z'(x) \exp\left[-\frac{t-x}{v}\right] dx, \quad \mathcal{K}(0) = \mathcal{K}(\infty) = 1. \tag{3}$$

Further,

$$\lim_{v \to 0} \frac{1}{v} \exp\left[-\frac{t-x}{v}\right] = \varphi(x-t). \tag{4}$$

Definition 4. *Consider $\alpha \in (0,1)$, for a function $z(x)$ then we can write the integral of fractional order α is as follows,*

$$I_t^\alpha(z(t)) = \frac{2(1-\alpha)}{(2-\alpha)\mathcal{K}(\alpha)} g(t) + \frac{2\alpha}{(2-\alpha)\mathcal{K}(\alpha)} \int_0^t z(s)ds, t \geq 0. \tag{5}$$

Remark 2. *In Equation (4), the remainder of the Caputo type non-integer order integral of the function with order $\alpha \in (0,1)$ is a mean into z with integral of order 1. Thus, it requires,*

$$\frac{2}{2\mathcal{K}(\alpha) - \alpha\mathcal{K}(\alpha)} = 1, \tag{6}$$

implies that $\mathcal{K}(\alpha) = \frac{2}{2-\alpha}$, $\alpha \in (0,1)$. Based on Equation (6), a new Caputo derivative is suggested with $\alpha \in (0,1)$ and is given by

$$D_t^\alpha(z(t)) = \frac{1}{1-\alpha} \int_0^t z'(x) \exp\left[-\alpha\frac{t-x}{1-\alpha}\right] dx. \tag{7}$$

In the following we present the new derivative known as the Atangana–Baleanu derivatives having non-singular and non-local kernel [25].

Definition 5. *Let $f \in H^1(a,b)$, $b > a$, $\alpha \in [0,1]$, then in the Caputo sense the Atangana–Baleanu derivative is defined as:*

$$_a^{ABC}D_t^\alpha f(t) = \frac{\mathcal{K}(\alpha)}{1-\alpha} \int_a^t f'(z) E_\alpha\left[-\alpha\frac{(t-z)^\alpha}{1-\alpha}\right] dz. \tag{8}$$

Definition 6. *The fractional integral associated with the Atangana–Beleanu derivative is given by:*

$$_a^{ABC}I_t^\alpha f(t) = \frac{1-\alpha}{\mathcal{K}(\alpha)} f(t) + \frac{\alpha}{\mathcal{K}(\alpha)\Gamma(\alpha)} \int_a^t f(z)(t-z)^{\alpha-1}dz \tag{9}$$

when the fractional order turns to zero, we can obtain the original function.

Theorem 1. *Consider the function $f \in C[a,b]$, then the following holds [25]:*

$$\|_a^{ABC}D_t^\alpha(f(t))\| < \frac{\mathcal{K}(\alpha)}{1-\alpha}\|f(t)\|, \ \text{where} \ \|f(t)\| = max_{a \leq t \leq b}|f(t)|. \tag{10}$$

Further, for the newly derivative the Lipschitz condition can be easily satisfied [25]:

$$\|_a^{ABC}D_t^\alpha f_1(t) - {}_a^{ABC}D_t^\alpha f_2(t)\| < \leftrightarrow_1\|f_1(t) - f_2(t)\|. \tag{11}$$

Theorem 2. *A given fractional differential equation:*

$$_a^{ABC}D_t^\alpha f(t) = s(t), \tag{12}$$

has the unique solution given by [25]:

$$f(t) = \frac{1-\alpha}{\mathcal{K}(\alpha)}s(t) + \frac{\alpha}{\mathcal{K}(\alpha)\Gamma(\alpha)}\int_a^t s(z)(t-z)^{\alpha-1}dz. \tag{13}$$

3. Model Formulation

We begin to formulate the Ebola epidemic disease by considering the human population in three compartments, that is, the susceptible individuals, $S(t)$, individuals infected with Ebola virus, $I(t)$, and the individuals recovered from the Ebola virus, $R(t)$. The individuals infected with Ebola and the deceased is $D(t)$ and $P(t)$ is the class for the Ebola virus pathogen in the environment. The model that describes the dynamics of Ebola disease modeled through differential equations is given by

$$\begin{cases} \frac{dS}{dt} & = \ \Lambda - \lambda S - dS, \\\\ \frac{dI}{dt} & = \ \lambda S - (d + \delta + \phi_1)I, \\\\ \frac{dR}{dt} & = \ \phi_1 I - dR, \\\\ \frac{dD}{dt} & = \ (d + \delta)I - \varepsilon D, \\\\ \frac{dP}{dt} & = \ \omega + \xi I + \theta D - \kappa P, \end{cases} \tag{14}$$

where $\lambda = \beta_1 I + \beta_2 D + \psi P$, and the appropriate initial conditions are given by

$$S(0) = S_0, \ I(0) = I_0, \ R(0) = R_0, \ D(0) = D_0, \ P(0) = P_0. \tag{15}$$

The birth rate of the susceptible individuals is recruited by the rate Λ, while the death rate is given by d. The susceptible individuals become infectious with the effective contact rate β_1 and β_2 with the deceased human individuals. The susceptible are able to attract the disease from the contaminated environment at a rate given by ψ. The death rate of the infected individuals due to Ebola virus is given by a rate δ, while the recovery from infection is ϕ_1. The deceased people can be directly buried during funerals at rate ε. At a rate of ω the environment is contaminated by the Ebola virus. At rates of ξ and θ the infected and deceased individuals, respectively, shed the virus in the environment. The virus decay of the Ebola virus from the population is given by parameter κ.

The sum of the first three equations of the Ebola disease model Equation (14) is given by

$$\frac{dN}{dt} = \ \Lambda - dN - \delta I, \tag{16}$$

where $N = S + I + R$ denotes the total alive human population. It should be noted that $\varepsilon \leq (d + \delta)$, which is an appropriate condition for the compartment D for which the model becomes relevant, otherwise the deceased human individuals will disappear and the model would be irrelevant. Further, the model given by Equation (14) is well posed and biologically feasible in the region given by

$$\Phi = \left\{ M \in \mathbb{R}_+^5 : N(t) \leq \frac{\Lambda}{d}, \ D \leq \frac{(d+\delta)\Lambda}{\varepsilon d}, P(t) = \frac{\varepsilon(d\omega + \Lambda\xi) + \theta\Lambda(\delta + d)}{d\kappa\varepsilon} \right\}, \tag{17}$$

where $M = (S(t), I(t), R(t), D(t), P(t))$.

4. Ebola Model in the Caputo Sense

The purpose of this section is to apply the proposed three operators on the Ebola disease model Equation (14). Initially, we will apply the Caputo derivative on the Ebola disease model, then, the Caputo–Fabrizio derivative, and finally the Atangana–Baleanu derivative. For each operator we will provide the solution procedure and the discussion on the graphical results in details. So, we start with the Caputo sense.

4.1. Ebola Model in the Caputo Sense

We can express the model given by Equation (14) in the Caputo derivative as follows:

$$\begin{cases} {}^{C}_{0}D_t^{\alpha}S &= \Lambda - \lambda S - dS, \\ {}^{C}_{0}D_t^{\alpha}I &= \lambda S - (d + \delta + \phi_1)I, \\ {}^{C}_{0}D_t^{\alpha}R &= \phi_1 I - dR, \\ {}^{C}_{0}D_t^{\alpha}D &= (d + \delta)I - \varepsilon D, \\ {}^{C}_{0}D_t^{\alpha}P &= \omega + \xi I + \theta D - \kappa P, \end{cases} \tag{18}$$

where $\lambda = \beta_1 I + \beta_2 D + \psi P$, and with the initial conditions, $S(0) = S_0$, $I(0) = I_0$, $R(0) = R_0$, $D(0) = D_0$, and $P(0) = P_0$.

4.2. Equilibrium Points

For the Ebola disease model Equation (18) in the Caputo sense, there is no disease-free equilibrium when $\omega > 0$ and we have the other equilibrium say, $E^* = (S^*, I^*, R^*, D^*, P^*)$, we have

$$\begin{cases} S^* = \frac{\Lambda - (d + \delta + \phi_1)I^*}{d}, \\[2mm] R^* = \frac{\phi_1 I^*}{d}, \\[2mm] D^* = \frac{(d+\delta)I^*}{\varepsilon}, \\[2mm] P^* = \frac{\theta I^*(d+\delta) + I^*\xi\varepsilon + \omega\varepsilon}{\kappa\varepsilon} \end{cases}$$

Using these values in the second equation of the model Equation (18), we have

$$C_2 I^{*2} - C_1 I^* - C_0 = 0, \tag{19}$$

where

$$C_2 = (d + \delta + \phi_1)\left(\beta_1 \kappa \varepsilon + \beta_2 \kappa (d + \delta) + \psi(\theta(d + \delta) + \xi \varepsilon)\right),$$

$$C_1 = -(-\kappa \Lambda \left(\beta_1 \varepsilon + \beta_2(d + \delta)\right) - \psi(\theta \Lambda(d + \delta) - \omega \varepsilon(d + \delta) + \Lambda \xi \varepsilon) + d\kappa \varepsilon(d + \delta))$$

$$-\varepsilon \phi_1(d\kappa + \omega \psi),$$

$$C_0 = \Lambda \omega \psi \varepsilon.$$

We have from the coefficient C_1,

$$C_1 = d\varepsilon \kappa(d + \delta + \phi_1)\left(\mathcal{R}_0 - 1 - \frac{\psi \omega}{d\kappa}\right),$$

where

$$\mathcal{R}_0 = \frac{\Lambda \beta_1}{d(d + \delta + \phi_1)} + \frac{\Lambda \beta_2(d + \delta)}{d\varepsilon(d + \delta + \phi_1)} + \frac{\psi \Lambda(\varepsilon \xi + \theta \delta + \theta d)}{d\varepsilon \kappa(d + \delta + \phi_1)}.$$

Considering the case when $\omega = 0$, we have

$$\begin{cases} S^* = \dfrac{\Lambda}{d\mathcal{R}_0}, \\[2mm] I^* = \dfrac{\Lambda(\mathcal{R}_0 - 1)}{(d + \delta + \phi_1)\mathcal{R}_0}, \\[2mm] R^* = \dfrac{\Lambda \phi_1(\mathcal{R}_0 - 1)}{d(d + \delta + \phi_1)\mathcal{R}_0}, \\[2mm] D^* = \dfrac{(\Lambda(d + \delta))(\mathcal{R}_0 - 1)}{\varepsilon(d + \delta + \phi_1)\mathcal{R}_0}, \\[2mm] P^* = \dfrac{\Lambda(\varepsilon \xi + \theta(d + \delta))(\mathcal{R}_0 - 1)}{\varepsilon \kappa(d + \delta + \phi_1)\mathcal{R}_0}, \end{cases}$$

and we have a disease-free equilibrium,

$$E_0 = \left(\frac{\Lambda}{d}, 0, 0, 0\right),$$

known as Ebola virus-free equilibrium.

4.3. Numerical Procedure for the Ebola Disease Model in the Caputo Sense

In the present subsection, we present the numerical scheme for the solution of the fractional Ebola disease model in the Caputo sense Equation (18). The present scheme that we use for the solution of the fractional Caputo nonlinear ordinary differential equation has been presented in [26,27]. The following procedure is presented

$$\begin{matrix}C\\0\end{matrix} D_t^{\alpha} z(t) = f(t, z(t)). \tag{20}$$

Using the fundamental theorem on Equation (20), we obtain

$$z(t) - z(0) = \frac{1}{\Gamma(\alpha)} \int_0^t f(\chi, z(\chi))(t - \chi)^{\alpha - 1} d\chi, \tag{21}$$

thus, at $t = t_{n+1}$, $n = 0, 1, ...$, the following is obtained

$$z(t_{n+1}) - z(0) = \frac{1}{\Gamma(\alpha)} \int_0^{t_{n+1}} (t_{n+1} - t)^{\alpha-1} f(t, z(t)) dt, \tag{22}$$

and

$$z(t_n) - z(0) = \frac{1}{\Gamma(\alpha)} \int_0^{t_n} (t_n - t)^{\alpha-1} f(t, z(t)) dt. \tag{23}$$

From Equations (23) and (22), we have

$$\begin{aligned}
z(t_{n+1}) &= z(t_n) + \underbrace{\frac{1}{\Gamma(\alpha)} \int_0^{t_{n+1}} (t_{n+1} - t)^{\alpha-1} f(t, z(t)) dt}_{A_{\alpha,1}} \\
&\quad - \underbrace{\frac{1}{\Gamma(\alpha)} \int_0^{t_n} (t_n - t)^{\alpha-1} f(t, z(t)) dt}_{A_{\alpha,2}}.
\end{aligned} \tag{24}$$

where

$$A_{\alpha,1} = \frac{1}{\Gamma(\alpha)} \int_0^{t_{n+1}} (t_{n+1} - t)^{\alpha-1} f(t, z(t)) dt, \tag{25}$$

and

$$A_{\alpha,2} = \frac{1}{\Gamma(\alpha)} \int_0^{t_n} (t_n - t)^{\alpha-1} f(t, z(t)) dt. \tag{26}$$

Using the Lagrange approximation for the function $f(t, z(t))$, we have

$$\begin{aligned}
P(t) &\simeq \frac{t - t_{n-1}}{t_n - t_{n-1}} f(t_n, z_n) + \frac{t - t_n}{t_{n-1} - t_n} f(t_{n-1}, z_{n-1}) \\
&= \frac{f(t_n, z_n)}{h} (t - t_{n-1}) - \frac{f(t_{n-1}, z_{n-1})}{h} (t - t_n).
\end{aligned} \tag{27}$$

The use of the above expression leads to

$$\begin{aligned}
A_{\alpha,1} &= \frac{f(t_n, z_n)}{h\Gamma(\alpha)} \int_0^{t_{n+1}} (t_{n+1} - t)^{\alpha-1} (t - t_{n-1}) dt \\
&\quad - \frac{f(t_{n-1}, z_{n-1})}{h\Gamma(\alpha)} \int_0^{t_{n+1}} (t_{n+1} - t)^{\alpha-1} (t - t_n) dt.
\end{aligned} \tag{28}$$

We have, after further simplification

$$\begin{aligned}
A_{\alpha,1} &= \frac{f(t_n, z_n)}{h\Gamma(\alpha)} \left[\frac{2h}{\alpha} t_{n+1}^\alpha - \frac{t_{n+1}^{\alpha+1}}{\alpha + 1} \right] \\
&\quad - \frac{f(t_{n-1}, z_{n-1})}{h\Gamma(\alpha)} \left[\frac{h}{\alpha} t_{n+1}^\alpha - \frac{1}{\alpha + 1} t_{n+1}^{\alpha+1} \right].
\end{aligned} \tag{29}$$

Similarly,

$$\begin{aligned}
A_{\alpha,2} &= \frac{1}{\Gamma(\alpha)} \int_0^{t_n} (t_n - t)^{\alpha-1} \left[\frac{f(t_n, z_n)}{h} (t - t_{n-1}) \right. \\
&\quad \left. - \frac{f(t_{n-1}, z_{n-1})}{h} (t - t_n) \right] dt.
\end{aligned} \tag{30}$$

Further simplifying, we get

$$
\begin{aligned}
\mathcal{A}_{\alpha,2} \;=\;& \frac{f(t_n, z_n)}{h\Gamma(\alpha)}\left[\frac{h}{\alpha}t_n^{\alpha} - \frac{t_n^{\alpha+1}}{\alpha+1}\right] \\
&+\frac{f(t_{n-1}, z_{n-1})}{h\Gamma(\alpha)}\left[\frac{1}{\alpha+1}t_n^{\alpha+1}\right].
\end{aligned}
\tag{31}
$$

We have the final approximate solution for the fractional nonlinear ordinary differential equation by substituting the Equations (30) and (31) into (24), given by

$$
\begin{aligned}
z(t_{n+1}) \;=\;& z(t_n) + \frac{f(t_n, z_n)}{h\Gamma(\alpha)}\left[\frac{2ht_{n+1}^{\alpha}}{\alpha} - \frac{t_{n+1}^{\alpha+1}}{\alpha+1} + \frac{h}{\alpha}t_n^{\alpha} - \frac{t_{n+1}^{\alpha+1}}{\alpha+1}\right] \\
&+\frac{f(t_{n-1}, z_{n-1})}{h\Gamma(\alpha)}\left[-\frac{h}{\alpha}t_{n+1}^{\alpha} + \frac{t_{n+1}^{\alpha+1}}{\alpha+1} + \frac{t_n^{\alpha+1}}{\alpha+1}\right].
\end{aligned}
\tag{32}
$$

The above scheme is used further for the solution of the Ebola disease model in the Caputo sense Equation (18) by considering the parameter values, $d = 0.05$, $\delta = 0.05$, $\phi = 0.06$, $\Lambda = 10$, $\epsilon = 0.008$, $\zeta = 0.004$, $\kappa = 0.03$, $\psi = 0.01$, $\beta_1 = 0.006$, $\beta_2 = 0.012$, $\varpi = 1$, and $\theta = 0.004$, and with various values of the fractional order parameter α. We have the graphical results for the numerical solution of the Ebola disease model Equation (18) in Figures 1–7. One can observe in Figures 1–7 by deceasing the value of α, the individuals infected with Ebola decreases while the population of infected individuals increases. We use this further to check the graphical results for the case when $\alpha = 0.3, 0.1$, then, one can see that infection is almost on the steady state, see Figures 6 and 7.

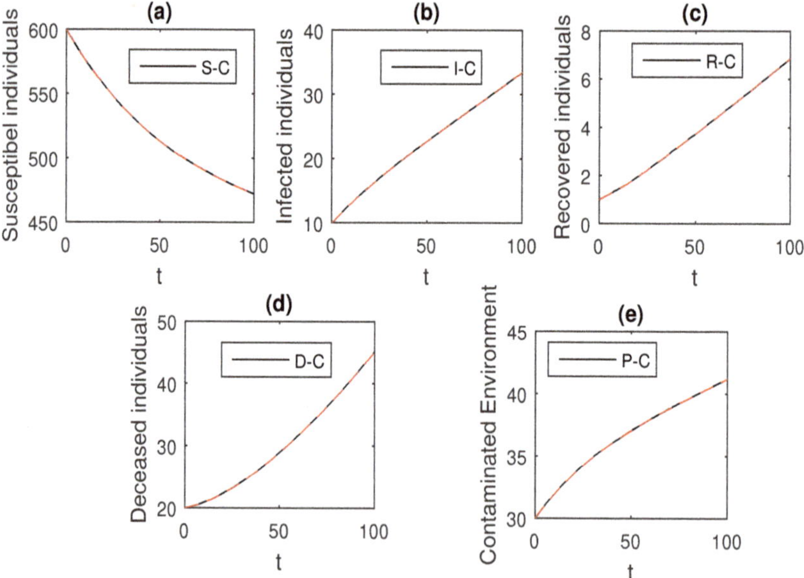

Figure 1. The graphical results show the dynamics of the Caputo derivative model (18), when $\alpha = 1$, where (**a**) Susceptible individuals, (**b**) Infected individuals, (**c**) recovered individuals, (**d**) deceased individuals, (**e**) Environment pathogens.

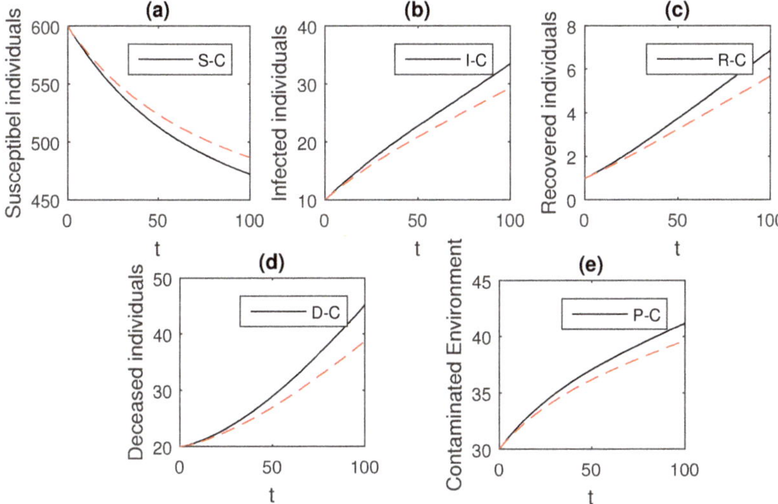

Figure 2. The graphical results show the dynamics of the Caputo derivative model (18), when $\alpha = 0.95$, where (**a**) Susceptible individuals, (**b**) Infected individuals, (**c**) recovered individuals, (**d**) deceased individuals, (**e**) Environment pathogens.

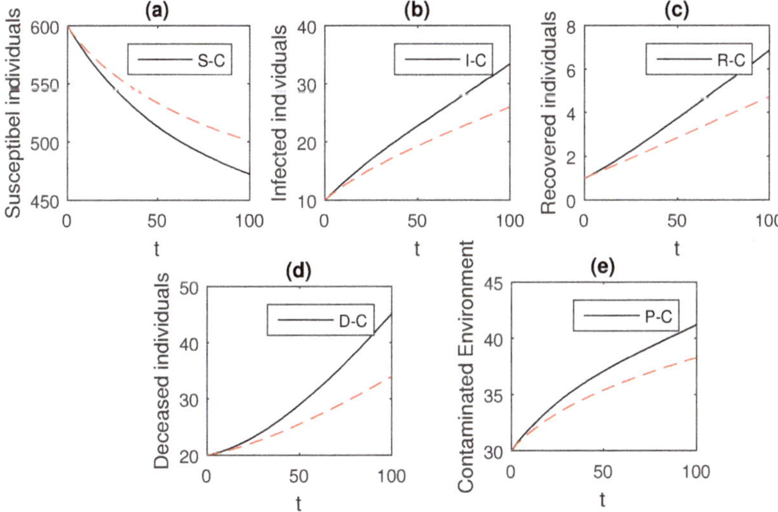

Figure 3. The graphical results show the dynamics of the Caputo derivative model (18), when $\alpha = 0.9$, where (**a**) Susceptible individuals, (**b**) Infected individuals, (**c**) recovered individuals, (**d**) deceased individuals, (**e**) Environment pathogens.

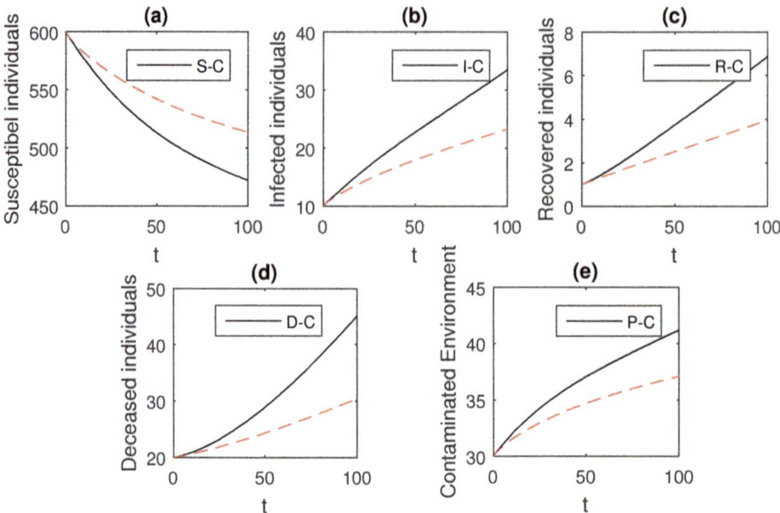

Figure 4. The graphical results show the dynamics of the Caputo derivative model (18), when $\alpha = 0.85$, where (**a**) Susceptible individuals, (**b**) Infected individuals, (**c**) recovered individuals, (**d**) deceased individuals, (**e**) Environment pathogens.

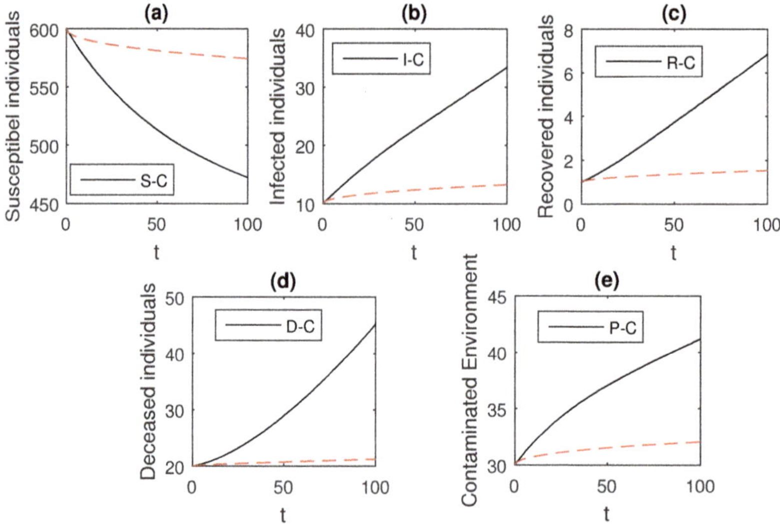

Figure 5. The graphical results show the dynamics of the Caputo derivative model (18), when $\alpha = 0.5$, where (**a**) Susceptible individuals, (**b**) Infected individuals, (**c**) recovered individuals, (**d**) deceased individuals, (**e**) Environment pathogens.

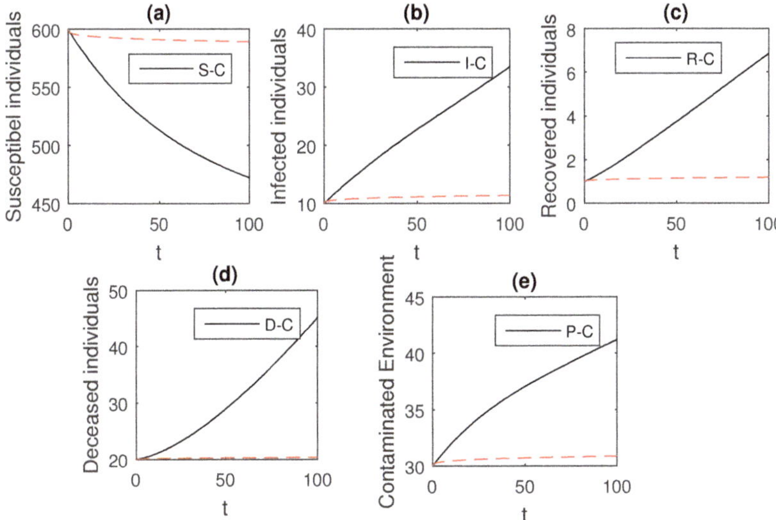

Figure 6. The graphical results show the dynamics of the Caputo derivative model (18), when $\alpha = 0.3$, where (**a**) Susceptible individuals, (**b**) Infected individuals, (**c**) recovered individuals, (**d**) deceased individuals, (**e**) Environment pathogens.

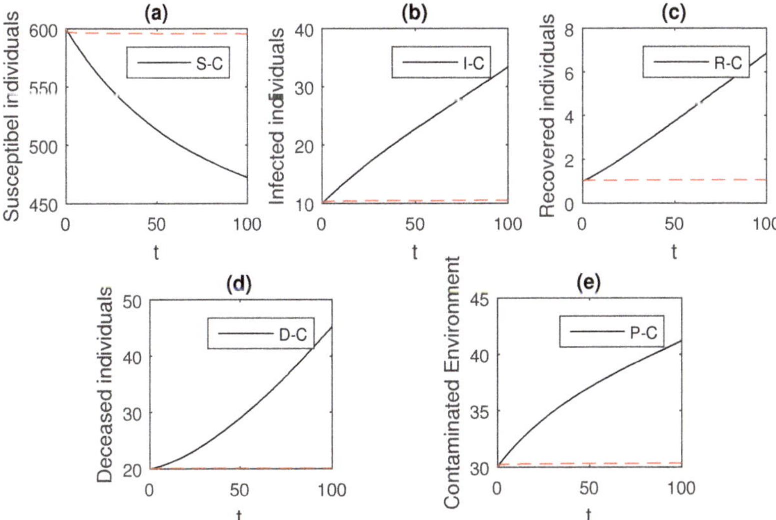

Figure 7. The graphical results show the dynamics of the Caputo derivative model (18), when $\alpha = 0.1$, where (**a**) Susceptible individuals, (**b**) Infected individuals, (**c**) recovered individuals, (**d**) deceased individuals, (**e**) Environment pathogens.

4.4. Ebola Model in the Caputo–Fabrizio Sense

We can express the model given by Equation (14) in Caputo–Fabrizio derivative as follows:

$$
\begin{cases}
{}^{CF}_{0}D^{\alpha}_{t}S &= \Lambda - \lambda S - dS, \\
{}^{CF}_{0}D^{\alpha}_{t}I &= \lambda S - (d + \delta + \phi_1)I, \\
{}^{CF}_{0}D^{\alpha}_{t}R &= \phi_1 I - dR, \\
{}^{CF}_{0}D^{\alpha}_{t}D &= (d + \delta)I - \varepsilon D, \\
{}^{CF}_{0}D^{\alpha}_{t}P &= \omega + \xi I + \theta D - \kappa P,
\end{cases}
\tag{33}
$$

where $\lambda = \beta_1 I + \beta_2 D + \psi P$, and with the initial conditions, $S(0) = S_0$, $I(0) = I_0$, $R(0) = R_0$, $D(0) = D_0$, and $P(0) = P_0$.

4.5. Numerical Solution for Caputo–Fabrizio Model

Here we present the numerical solution for the Caputo–Fabrizio model Equation (33) by using the scheme presented [27]. The following steps are taken as specified in one for the solution of Equation (33).

$$
S(t) - S(0) = \frac{(1-\alpha)}{B(\alpha)}\mathcal{F}_1(t,S) + \frac{\alpha}{B(\alpha)}\int_0^t \mathcal{F}_1(\zeta,S)d\zeta.
\tag{34}
$$

For $t = t_{n+1}$, $n = 0, 1, 2, \ldots$, we obtain

$$
S(t_{n+1}) - S_0 = \frac{1-\alpha}{B(\alpha)}\mathcal{F}_1(t_n,S_n) + \frac{\alpha}{B(\alpha)}\int_0^{t_{n+1}} \mathcal{F}_1(t,S)dt.
\tag{35}
$$

The successive terms difference is given as follows:

$$
S_{n+1} - S_n = \frac{1-\alpha}{B(\alpha)}\{\mathcal{F}_1(t_n,S_n) - \mathcal{F}_1(t_{n-1},S_{n-1})\} + \frac{\alpha}{B(\alpha)}\int_{t_n}^{t_{n+1}} \mathcal{F}_1(t,S)dt.
\tag{36}
$$

Over the close interval $[t_k, t_{(k+1)}]$, the function $\mathcal{F}_1(t,S)$ can be approximated by the interpolation polynomial

$$
P_k(t) \cong \frac{f(t_k,y_k)}{h}(t - t_{k-1}) - \frac{f(t_{k-1},y_{k-1})}{h}(t - t_k),
\tag{37}
$$

where $h = t_n - t_{n-1}$. Calculating the integral in Equation (36) using above polynomial approximation we get

$$
\begin{aligned}
\int_{t_n}^{t_{n+1}} \mathcal{F}_1(t,S)dt &= \int_{t_n}^{t_{n+1}} \frac{\mathcal{F}_1(t_n,S_n)}{h}(t - t_{n-1}) - \frac{\mathcal{F}_1(t_{n-1},S_{n-1})}{h}(t - t_n)dt \\
&= \frac{3h}{2}\mathcal{F}_1(t_n,S_n) - \frac{h}{2}\mathcal{F}_1(t_{n-1},S_{n-1}).
\end{aligned}
\tag{38}
$$

Substituting Equation (38) in (36) and after simplification we obtain

$$
S_{n+1} = S_n + \left(\frac{1-\alpha}{B(\alpha)} + \frac{3h}{2B(\alpha)}\right)\mathcal{F}_1(t_n,S_n) - \left(\frac{1-\alpha}{B(\alpha)} + \frac{\alpha h}{2B(\alpha)}\right)\mathcal{F}_1(t_{n-1},S_{n-1}).
\tag{39}
$$

In a similar way, for the rest of equations of system Equation (33) we obtain the recursive formula as below

$$
I_{n+1} = I_0 + \left(\frac{1-\alpha}{B(\alpha)} + \frac{3h}{2B(\alpha)}\right)\mathcal{F}_2(t_n,I_n) - \left(\frac{1-\alpha}{B(\alpha)} + \frac{\alpha h}{2B(\alpha)}\right)\mathcal{F}_2(t_{n-1},I_{n-1}),
$$

$$R_{n+1} = R_0 + \left(\frac{1-\alpha}{B(\alpha)} + \frac{3h}{2B(\alpha)}\right)\mathcal{F}_3(t_n, R_n) - \left(\frac{1-\alpha}{B(\alpha)} + \frac{\alpha h}{2B(\alpha)}\right)\mathcal{F}_3(t_{n-1}, R_{n-1}),$$

$$D_{n+1} = D_0 + \left(\frac{1-\alpha}{B(\alpha)} + \frac{3h}{2B(\alpha)}\right)\mathcal{F}_4(t_n, D_n) - \left(\frac{1-\alpha}{B(\alpha)} + \frac{\alpha h}{2B(\alpha)}\right)\mathcal{F}_4(t_{n-1}, D_{n-1}),$$

$$P_{n+1} = P_0 + \left(\frac{1-\alpha}{B(\alpha)} + \frac{3h}{2B(\alpha)}\right)\mathcal{F}_5(t_n, P_n) - \left(\frac{1-\alpha}{B(\alpha)} + \frac{\alpha h}{2B(\alpha)}\right)\mathcal{F}_5(t_{n-1}, P_{n-1}). \tag{40}$$

The numerical scheme presented above by considering the parameter values, $d = 0.05$, $\delta = 0.05$, $\phi = 0.06$, $\Lambda = 10$, $\epsilon = 0.008$, $\xi = 0.004$, $\kappa = 0.03$, $\psi = 0.01$, $\beta_1 = 0.006$, $\beta_2 = 0.012$, $\omega = 1$, and $\theta = 0.004$, we have the graphical results for the Ebola disease model in Caputo–Fabrizio model Equation (33). Various graphical results considering the fractional order parameter α are presented, see Figures 8–14. In these Figures, we obtain various graphical results for α, and we observe that by decreasing the value of α the infected compartments are decreasing well. Especially, when choosing $\alpha = 0.3,\ 0.1$, we can see that the number of infected individuals decreases rapidly.

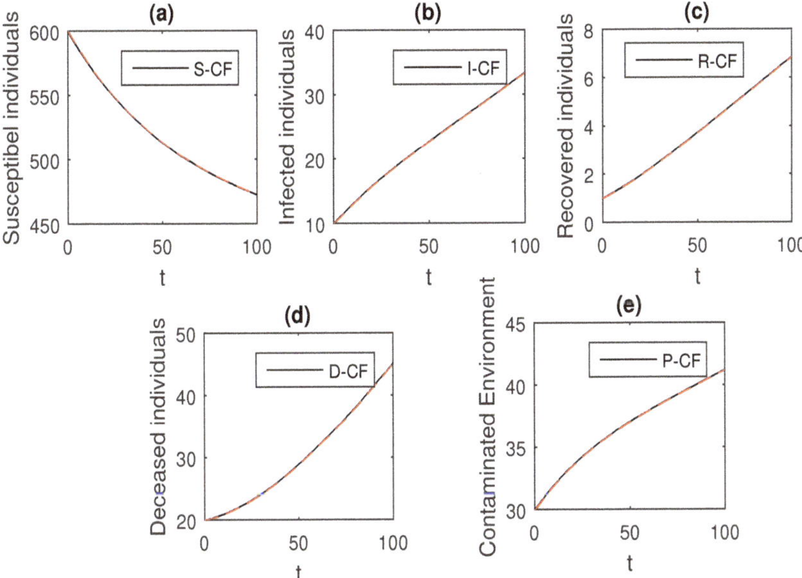

Figure 8. The graphical results show the dynamics of the Caputo–Fabrizio model (33), when $\alpha = 1$, where (**a**) Susceptible individuals, (**b**) Infected individuals, (**c**) recovered individuals, (**d**) deceased individuals, (**e**) Environment pathogens.

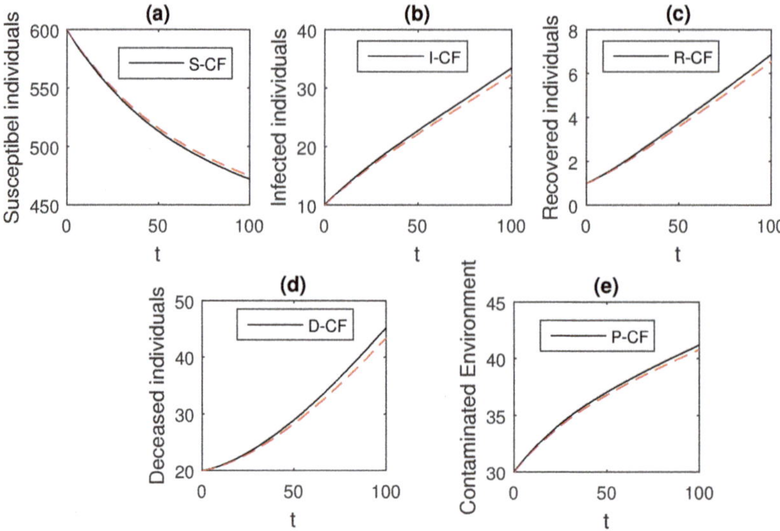

Figure 9. The graphical results show the dynamics of the Caputo–Fabrizio model (33), when $\alpha = 0.95$, where (**a**) Susceptible individuals, (**b**) Infected individuals, (**c**) recovered individuals, (**d**) deceased individuals, (**e**) Environment pathogens.

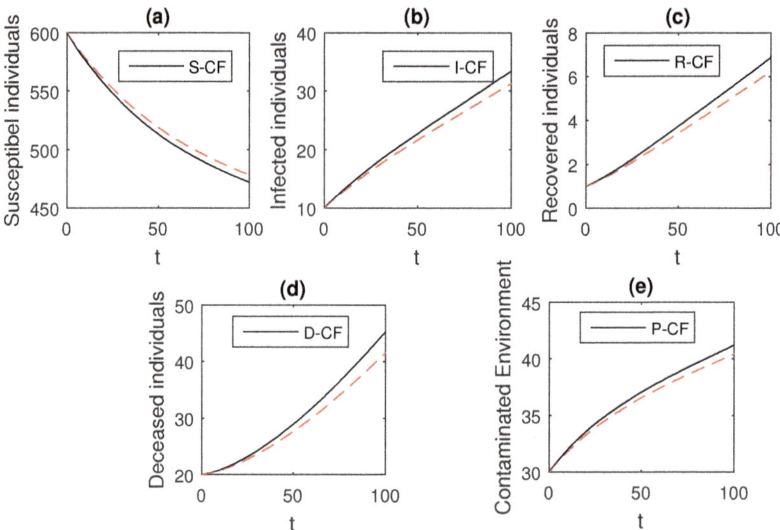

Figure 10. The graphical results show the dynamics of the Caputo–Fabrizio model (33), when $\alpha = 0.9$, where (**a**) Susceptible individuals, (**b**) Infected individuals, (**c**) recovered individuals, (**d**) deceased individuals, (**e**) Environment pathogens.

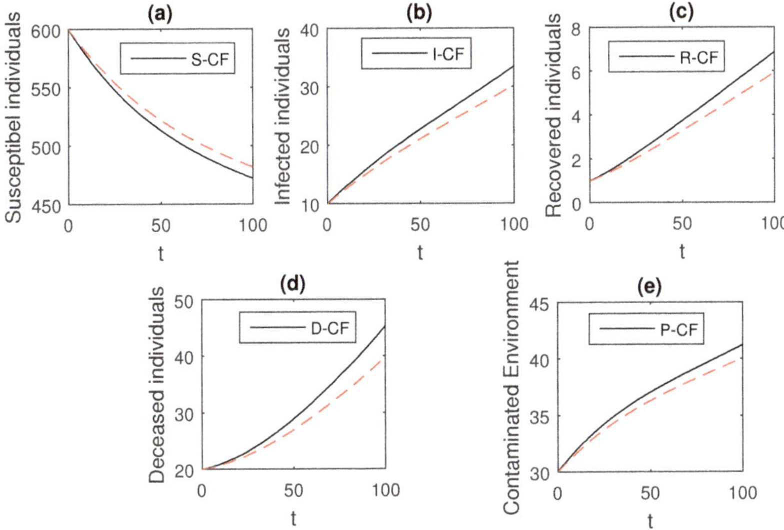

Figure 11. The graphical results show the dynamics of the Caputo–Fabrizio model (33), when $\alpha = 0.85$, where (**a**) Susceptible individuals, (**b**) Infected individuals, (**c**) recovered individuals, (**d**) deceased individuals, (**e**) Environment pathogens.

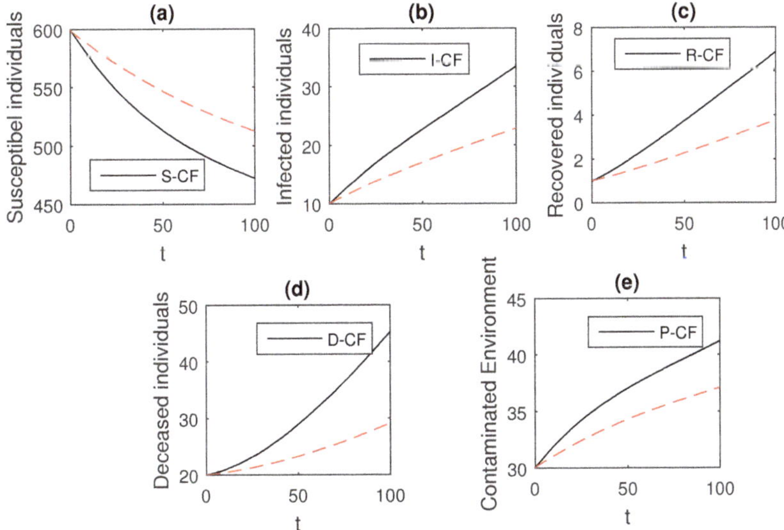

Figure 12. The graphical results show the dynamics of the Caputo–Fabrizio model (33), when $\alpha = 0.5$, where (**a**) Susceptible individuals, (**b**) Infected individuals, (**c**) recovered individuals, (**d**) deceased individuals, (**e**) Environment pathogens.

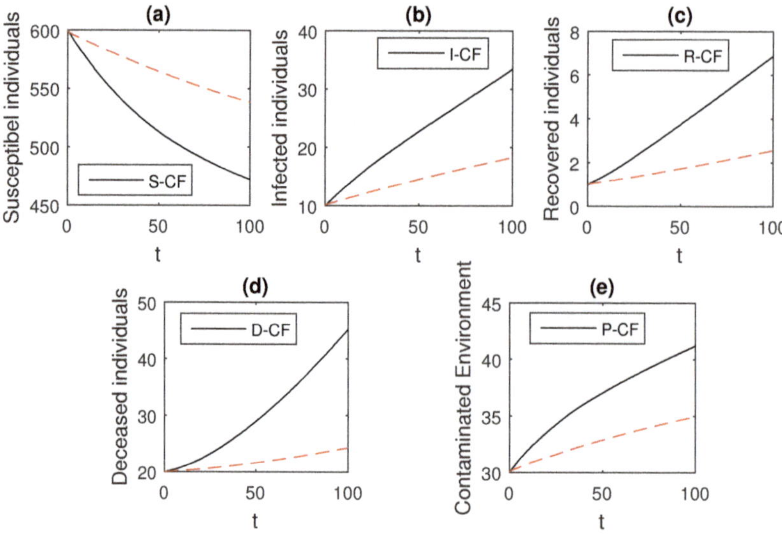

Figure 13. The graphical results show the dynamics of the Caputo–Fabrizio model (33), when $\alpha = 0.3$, where (**a**) Susceptible individuals, (**b**) Infected individuals, (**c**) recovered individuals, (**d**) deceased individuals, (**e**) Environment pathogens.

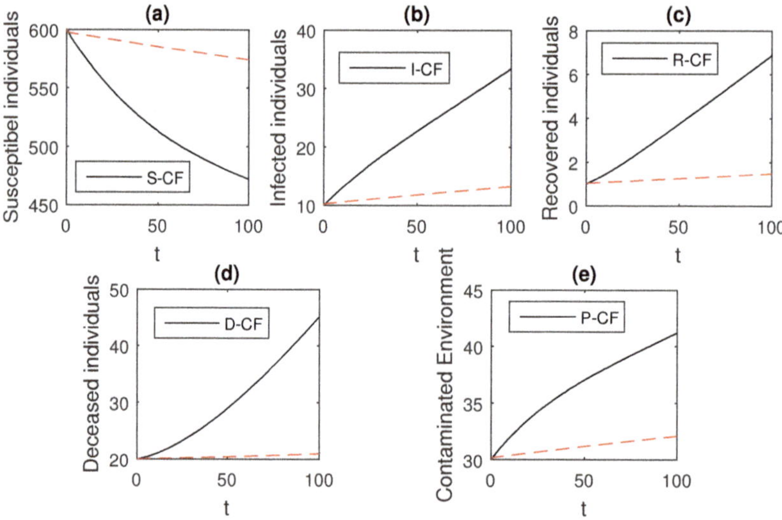

Figure 14. The graphical results show the dynamics of the Caputo–Fabrizio model (33), when $\alpha = 0.1$, where (**a**) Susceptible individuals, (**b**) Infected individuals, (**c**) recovered individuals, (**d**) deceased individuals, (**e**) Environment pathogens.

4.6. Ebola Model in the Atangana–Baleanu Sense

We can express the model given by Equation (14) in Atangana–Baleanu derivative as follows:

$$\begin{cases} {}^{ABC}_{0}D^{\alpha}_{t}S &= \Lambda - \lambda S - dS, \\ {}^{ABC}_{0}D^{\alpha}_{t}I &= \lambda S - (d + \delta + \phi_1)I, \\ {}^{ABC}_{0}D^{\alpha}_{t}R &= \phi_1 I - dR, \\ {}^{ABC}_{0}D^{\alpha}_{t}D &= (d + \delta)I - \varepsilon D, \\ {}^{ABC}_{0}D^{\alpha}_{t}P &= \varpi + \xi I + \theta D - \kappa P, \end{cases} \tag{41}$$

where $\lambda = \beta_1 I + \beta_2 D + \psi P$.

4.7. Existence of Solutions for the Atangana–Baleanu Model

It is obvious that the given model Equation (14) shows the dynamics of Ebola disease, which is described by a nonlinear system of differential equations, so it is not possible to obtain their exact solution but the existence of an approximate solution can be very effective if we show that the solution for the Ebola disease model Equation (41) under some conditions exists. To do this, we follow the results of fixed point theory for the given Ebola disease model Equation (41). We write the Ebola disease model given by Equation (41) for simplification purposes as follows:

$$\begin{cases} {}^{ABC}_{0}D^{\alpha}_{t}x(t) = \mathcal{F}(t, x(t)), \\ x(0) = x_0, \quad 0 < t < T < \infty, \end{cases} \tag{42}$$

where $x(t) = (S, I, R, D, P)$ represent the vector with state variables S, I, R, D, P and is a continuous vector function and can be defined as follows:

$$\mathcal{F} = \begin{pmatrix} \mathcal{F}_1 \\ \mathcal{F}_2 \\ \mathcal{F}_3 \\ \mathcal{F}_4 \\ \mathcal{F}_5 \end{pmatrix} = \begin{pmatrix} \Lambda - \lambda S - dS \\ \lambda S - (d + \delta + \phi_1)I \\ \phi_1 I - dR \\ (d + \delta)I - \varepsilon D \\ \varpi + \xi I + \theta D - \kappa P \end{pmatrix}. \tag{43}$$

The function $\mathcal{F}$ can be shown easily to satisfy the Lipschitz condition and can be represented as:

$$\|\mathcal{F}(t, x_1(t)) - \mathcal{F}(t, x_2(t))\| \leq \mathcal{K}\|x_1(t) - x_2(t)\|. \tag{44}$$

Now we have the results for the existence and uniqueness for the Ebola disease model in the Atangana–Baleanu derivative sense. We state and prove the following theorem:

Theorem 3. *The Ebola disease model in the Atangana–Baleanu form Equation (41) can have a unique solution under some conditions if the following holds*

$$\frac{(1 - \alpha)}{ABC(\alpha)}\mathcal{K} + \alpha\frac{T^{\alpha}_{max}}{ABC(\alpha)\Gamma(\alpha)}\mathcal{K} < 1. \tag{45}$$

Proof. The use of the Atangana–Baleanu fractional integration on model Equation (42) both sides, the following is obtained,

$$x(t) = x_0 + \frac{1 - \alpha}{ABC(\alpha)}\mathcal{F}(t, x(t)) + \frac{\alpha}{ABC(\alpha)\Gamma(\alpha)}\int_{0}^{t}(t - \eta)^{\alpha - 1}\mathcal{F}(\eta, x(\eta))d\eta. \tag{46}$$

Suppose $J = (0, T)$ and the operator $Y : \mathcal{C}(J, R^5) \to \mathcal{C}(J, R^5)$ defined by

$$Y[x(t)] = x_0 + \frac{1-\alpha}{ABC(\alpha)}\mathcal{F}(t, x(t)) + \frac{\alpha}{ABC(\alpha)\Gamma(\alpha)}\int_0^t (t-\eta)^{\alpha-1}\mathcal{F}(\eta, x(\eta))d\eta. \tag{47}$$

Then we can write Equation (46) as follows:

$$x(t) = Y[x(t)]. \tag{48}$$

We have, after applying the supremum norm on J,

$$\|x(t)\|_J = sup_{t \in J}\|x(t)\|, \quad x(t) \in \mathcal{C}. \tag{49}$$

Obviously, $\mathcal{C}(J, R^5)$ and the norm $\|.\|_J$ is a Banach space. Using the operator Equation (48), the following is presented

$$\|Y[x_1(t)] - Y[x_2(t)]\|_J \leq \left\| \frac{(1-\alpha)}{ABC(\alpha)}(\mathcal{F}(t, x_1(t)) - \mathcal{F}(t, x_2(t)) + \frac{\alpha}{ABC(\alpha)\Gamma(\alpha)} \times \right.$$
$$\left. \int_0^t (t-\eta)^{\alpha-1}(\mathcal{F}(\eta, x_1(\eta)) - \mathcal{F}(\eta, x_2(\eta)))d\eta \right\|_J. \tag{50}$$

Using the triangular inequality and Lipschitz condition presented in Equation (44) with some simplifications, we have

$$\|Y[x_1(t)] - Y[x_1(t)]\|_J \leq \left(\frac{(1-\alpha)\mathcal{K}}{ABC(\alpha)} + \frac{\alpha}{ABC(\alpha)\Gamma(\alpha)}\mathcal{K}T^\alpha \right)\|x_1(t) - x_2(t)\|_J. \tag{51}$$

Finally, we have

$$\|Y[x_1(t)] - Y[x_1(t)]\|_J \leq L\|x_1(t) - x_2(t)\|_J, \tag{52}$$

where

$$L = \frac{(1-\alpha)\mathcal{M}}{ABC(\alpha)} + \frac{\alpha}{ABC(\alpha)\Gamma(\alpha)}\mathcal{M}T^\alpha.$$

If the condition given by Equation (45) holds then the operator Y will be a contraction. Thus, the Banach fixed point theorem ensures that a unique solution for the Ebola disease model in the Atangana–Baleanu form Equation (41) exists, Equation (42). $\square$

4.8. Numerical Results for the Atangana–Baleanu Model and Simulation Results

In the present subsection we aim to obtain the numerical results for the Ebola disease model in the Atangana–Baleanu form given by Equation (41). First, we provide a numerical scheme in details and then show the graphical results for various values of the fractional order parameter α. The scheme given in [28] will be used to obtain the approximate solution of the Ebola disease model in the Atangana–Baleanu form Equation (41).

We write the model Equation (42) after using the fundamental theorem of fractional calculus:

$$w(t) - w(0) = \frac{(1-\alpha)}{ABC(\alpha)}\mathcal{F}(t, w(t)) + \frac{\alpha}{ABC(\alpha) \times \Gamma(\alpha)}\int_0^t \mathcal{F}(\phi, x(\phi))(t-\phi)^{\alpha-1}d\phi. \tag{53}$$

At $t = t_{n+1}, n = 0, 1, 2, ...$, we have

$$\begin{aligned} w(t_{n+1}) - w(0) &= \frac{1-\alpha}{ABC(\alpha)}\mathcal{F}(t_n, w(t_n)) + \\ &\quad \frac{\alpha}{ABC(\alpha) \times \Gamma(\alpha)} \int_0^{t_{n+1}} \mathcal{F}(\phi, w(\phi))(t_{n+1} - \phi)^{\alpha-1}d\phi, \\ &= \frac{1-\alpha}{ABC(\alpha)}\mathcal{F}(t_n, w(t_n)) + \\ &\quad \frac{\alpha}{ABC(\alpha) \times \Gamma(\alpha)} \sum_{j=0}^{n} \int_{t_j}^{t_{j+1}} \mathcal{F}(\phi, w(\phi))(t_{n+1} - \phi)^{\alpha-1}d\phi. \end{aligned} \tag{54}$$

The function $\mathcal{F}(\phi, w(\phi))$ can be approximated over $[t_j, t_{j+1}]$, using the interpolation polynomial

$$\mathcal{F}(\phi, w(\phi)) \cong \frac{\mathcal{F}(t_j, w(t_j))}{h}(t - t_{j-1}) - \frac{\mathcal{F}(t_{j-1}, w(t_{j-1}))}{h}(t - t_j). \tag{55}$$

Substituting in Equation (54) we get

$$\begin{aligned} w(t_{n+1}) &= w(0) + \frac{1-\alpha}{ABC(\alpha)}\mathcal{F}(t_n, w(t_n)) + \\ &\quad \frac{\alpha}{ABC(\alpha) \times \Gamma(\alpha)} \sum_{j=0}^{n} \left(\frac{\mathcal{F}(t_j, w(t_j))}{h} \int_{t_j}^{t_{j+1}} (t - t_{j-1})(t_{n+1} - t)^{\alpha-1}dt \right. \\ &\quad \left. - \frac{\mathcal{F}(t_{j-1}, w(t_{j-1}))}{h} \int_{t_j}^{t_{j+1}} (t - t_j)(t_{n+1} - t)^{\alpha-1}dt \right). \end{aligned} \tag{56}$$

After some calculation, we obtain the following:

$$\begin{aligned} w(t_{n+1}) &= w(t_0) + \frac{1}{ABC(\alpha)}\alpha\,\mathcal{F}(t_n, w(t_n)) + \\ &\quad \frac{\alpha}{ABC(\alpha)} \sum_{j=0}^{n} \\ &\quad \left(\frac{h^\alpha \mathcal{F}(t_j, w(t_j))}{\Gamma(\alpha+2)}((n+1-j)^\alpha(n-j+2+\alpha) - (n-j)^\alpha(n-j+2+2\alpha)) \right. \\ &\quad \left. - \frac{h^\alpha \mathcal{F}(t_{j-1}, w(t_{j-1}))}{\Gamma(\alpha+2)}((n+1-j)^{\alpha+1} - (n-j)^\alpha(n-j+1+\alpha)) \right). \end{aligned} \tag{57}$$

For the Ebola disease model we have the following results:

$$\begin{aligned} S(t_{n+1}) &= S(t_0) + \frac{1-\alpha}{ABC(\alpha)}\mathcal{F}_1(t_n, w(t_n)) + \\ &\quad \frac{\alpha}{ABC(\alpha)} \sum_{j=0}^{n} \\ &\quad \left(\frac{h^\alpha \mathcal{F}_1(t_j, w(t_j))}{\Gamma(\alpha+2)}((n+1-j)^\alpha(n-j+2+\alpha) - (n-j)^\alpha(n-j+2+2\alpha)) \right. \\ &\quad \left. - \frac{h^\alpha \mathcal{F}_1(t_{j-1}, w(t_{j-1}))}{\Gamma(\alpha+2)}((n+1-j)^{\alpha+1} - (n-j)^\alpha(n-j+1+\alpha)) \right), \end{aligned}$$

$$
\begin{aligned}
I(t_{n+1}) \;=\; & I(t_0) + \frac{1-\alpha}{ABC(\alpha)}\mathcal{F}_2(t_n, w(t_n)) + \\
& \frac{\alpha}{ABC(\alpha)}\sum_{j=0}^{n} \\
& \left(\frac{h^\alpha \mathcal{F}_2(t_j, w(t_j))}{\Gamma(\alpha+2)}\big((n+1-j)^\alpha(n-j+2+\alpha) - (n-j)^\alpha(n-j+2+2\alpha)\big)\right. \\
& \left.-\frac{h^\alpha \mathcal{F}_2(t_{j-1}, w(t_{j-1}))}{\Gamma(\alpha+2)}\big((n+1-j)^{\alpha+1} - (n-j)^\alpha(n-j+1+\alpha)\big)\right), \\
R(t_{n+1}) \;=\; & R(t_0) + \frac{1-\alpha}{ABC(\alpha)}\mathcal{F}_3(t_n, w(t_n)) + \\
& \frac{\alpha}{ABC(\alpha)}\sum_{j=0}^{n} \\
& \left(\frac{h^\alpha \mathcal{F}_3(t_j, w(t_j))}{\Gamma(\alpha+2)}\big((n+1-j)^\alpha(n-j+2+\alpha) - (n-j)^\alpha(n-j+2+2\alpha)\big)\right. \\
& \left.-\frac{h^\alpha \mathcal{F}_3(t_{j-1}, w(t_{j-1}))}{\Gamma(\alpha+2)}\big((n+1-j)^{\alpha+1} - (n-j)^\alpha(n-j+1+\alpha)\big)\right), \\
D(t_{n+1}) \;=\; & D(t_0) + \frac{1-\alpha}{ABC(\alpha)}\mathcal{F}_4(t_n, w(t_n)) + \\
& \frac{\alpha}{ABC(\alpha)}\sum_{j=0}^{n} \\
& \left(\frac{h^\alpha \mathcal{F}_4(t_j, w(t_j))}{\Gamma(\alpha+2)}\big((n+1-j)^\alpha(n-j+2+\alpha) - (n-j)^\alpha(n-j+2+2\alpha)\big)\right. \\
& \left.-\frac{h^\alpha \mathcal{F}_4(t_{j-1}, w(t_{j-1}))}{\Gamma(\alpha+2)}\big((n+1-j)^{\alpha+1} - (n-j)^\alpha(n-j+1+\alpha)\big)\right), \\
P(t_{n+1}) \;=\; & P(t_0) + \frac{1-\alpha}{ABC(\alpha)}\mathcal{F}_5(t_n, w(t_n)) + \\
& \frac{\alpha}{ABC(\alpha)}\sum_{j=0}^{n} \\
& \left(\frac{h^\alpha \mathcal{F}_5(t_j, w(t_j))}{\Gamma(\alpha+2)}\big((n+1-j)^\alpha(n-j+2+\alpha) - (n-j)^\alpha(n-j+2+2\alpha)\big)\right. \\
& \left.-\frac{h^\alpha \mathcal{F}_5(t_{j-1}, w(t_{j-1}))}{\Gamma(\alpha+2)}\big((n+1-j)^{\alpha+1} - (n-j)^\alpha(n-j+1+\alpha)\big)\right).
\end{aligned}
\tag{58}
$$

We using the above scheme for the numerical solution of the Ebola disease model Equation (41) and obtain the graphical results shown in Figures 15–21 by considering different values of the fractional order parameter α. In these Figures 15–21, by decreasing the values of the fractional order parameter α, the population of infected compartments decreases more efficiently for the cases of $\alpha = 0.3, 0.1$. One can see that the numerical results in the form of graphs obtained through the Atangana–Baleanu operator in comparison to the Caputo and Caputo–Fabrizio operator decrease the infection faster. So, from the above graphical results, it is suggested that the Atangana–Baleanu operator is more useful for infection elimination by decreasing the value of α.

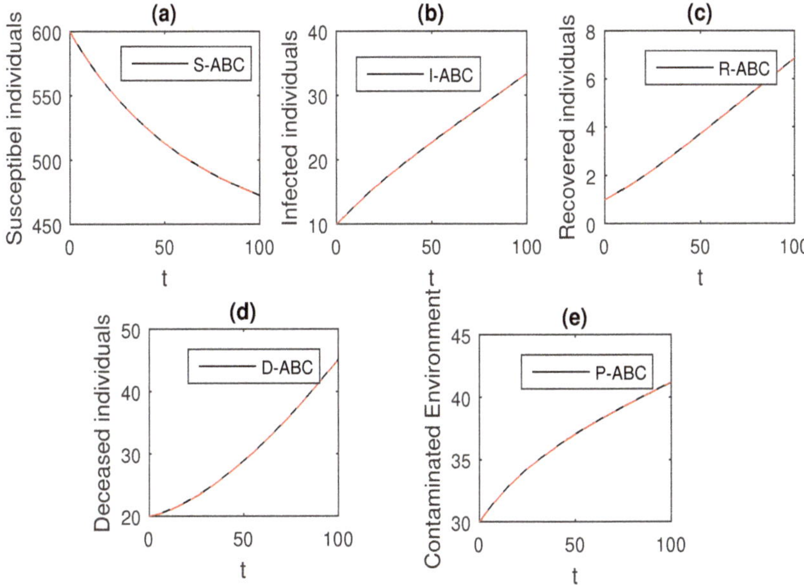

Figure 15. The graphical results show the dynamics of the Atangana–Baleanu model (41), when $\alpha = 1$, where (**a**) Susceptible individuals, (**b**) Infected individuals, (**c**) recovered individuals, (**d**) deceased individuals, (**e**) Environment pathogens.

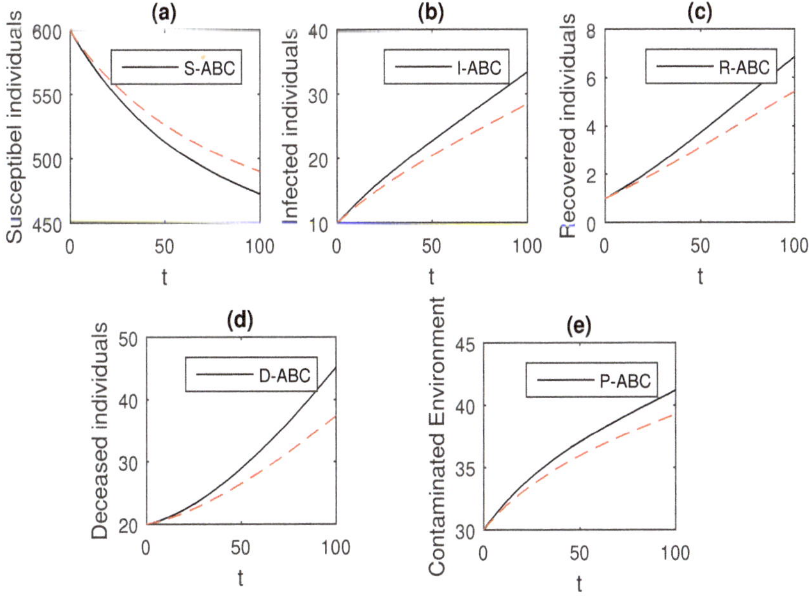

Figure 16. The graphical results show the dynamics of the Atangana–Baleanu model (41), when $\alpha = 0.95$, where (**a**) Susceptible individuals, (**b**) Infected individuals, (**c**) recovered individuals, (**d**) deceased individuals, (**e**) Environment pathogens.

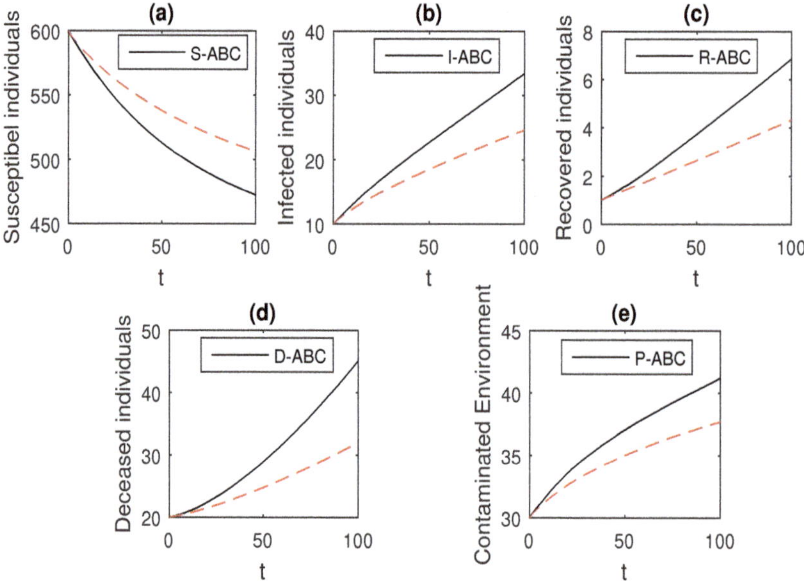

Figure 17. The graphical results show the dynamics of the Atangana–Baleanu model (41), when $\alpha = 0.9$, where (**a**) Susceptible individuals, (**b**) Infected individuals, (**c**) recovered individuals, (**d**) deceased individuals, (**e**) Environment pathogens.

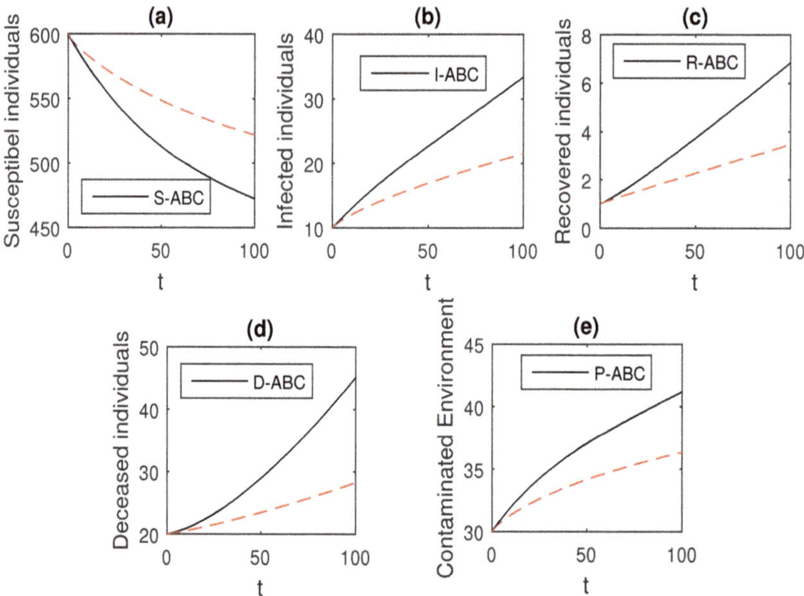

Figure 18. The graphical results show the dynamics of the Atangana–Baleanu model (41), when $\alpha = 0.85$, where (**a**) Susceptible individuals, (**b**) Infected individuals, (**c**) recovered individuals, (**d**) deceased individuals, (**e**) Environment pathogens.

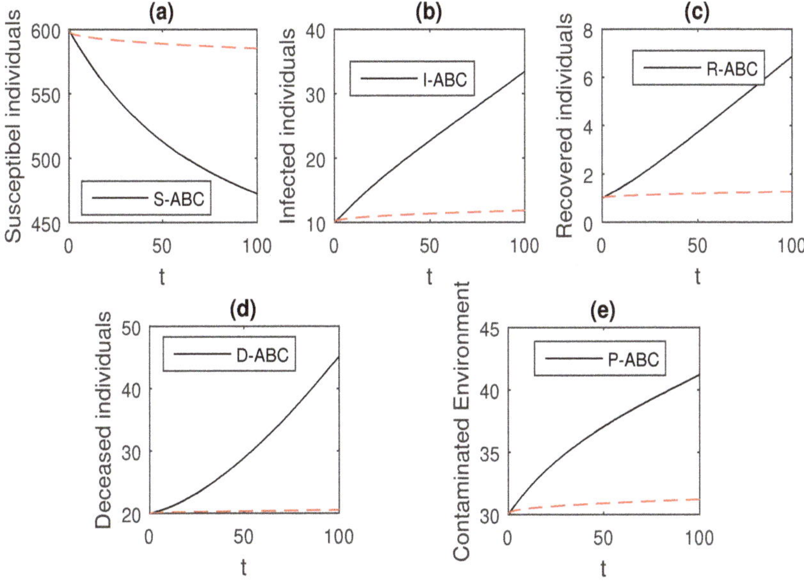

Figure 19. The graphical results show the dynamics of the Atangana–Baleanu model (41), when $\alpha = 0.5$, where (**a**) Susceptible individuals, (**b**) Infected individuals, (**c**) recovered individuals, (**d**) deceased individuals, (**e**) Environment pathogens.

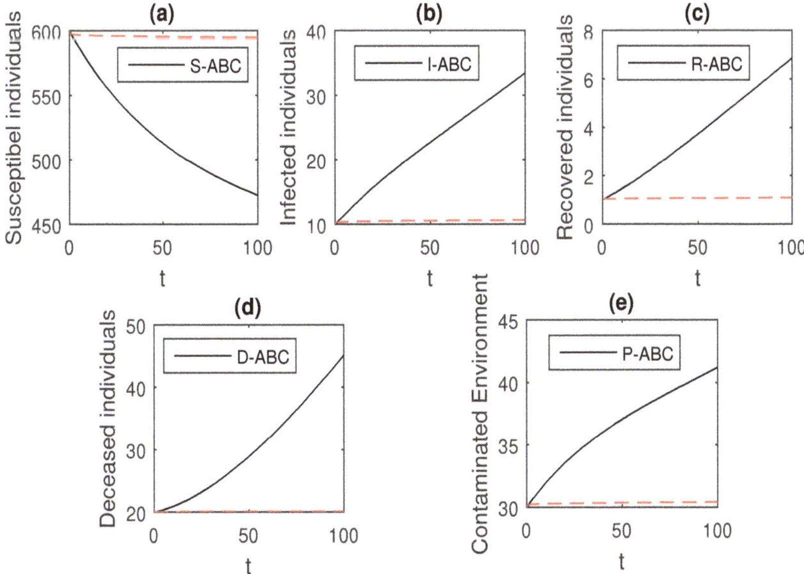

Figure 20. The graphical results show the dynamics of the Atangana–Baleanu model (41), when $\alpha = 0.3$, where (**a**) Susceptible individuals, (**b**) Infected individuals, (**c**) recovered individuals, (**d**) deceased individuals, (**e**) Environment pathogens.

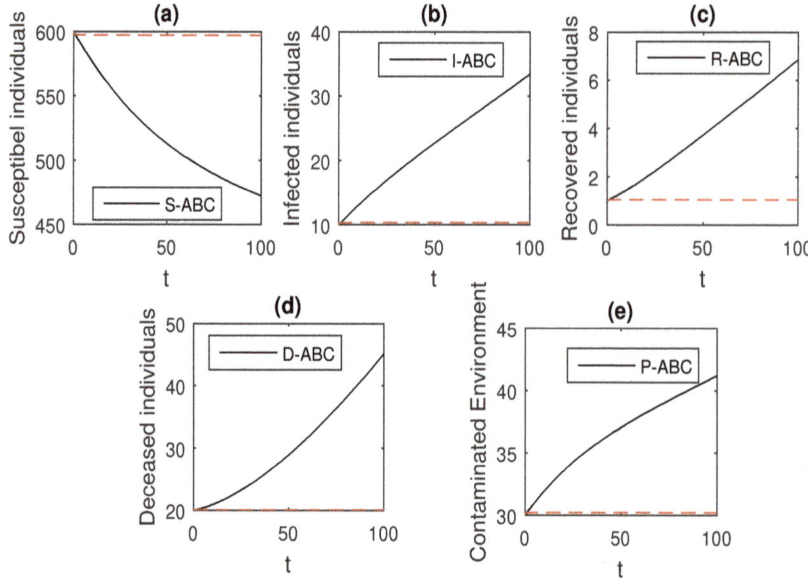

Figure 21. The graphical results show the dynamics of the Atangana–Baleanu model (41), when $\alpha = 0.1$, where (**a**) Susceptible individuals, (**b**) Infected individuals, (**c**) recovered individuals, (**d**) deceased individuals, (**e**) Environment pathogens.

4.9. Graphical Comparison of the Operators

Here, we provide comparison plots for the Caputo, Caputo–Fabrizio, and the Atangana–Baleanu operators. We considered various values of the fractional order parameter $\alpha = 1, 0.9, 0.7, 0.5, 0.3, 0.1$ and presented the graphical results for comparison (see Figures 22–27). By decreasing the value of α it can be seen that the number of infected individuals decreases well, compared to the Caputo–Fabrizio and Caputo derivatives. Especially, for the cases when $\alpha = 0.3, 0.1$, the Atangana–Baleanu derivative provides useful results for the infection elimination in comparison to the Caputo and the Caputo–Fabrizio operators.

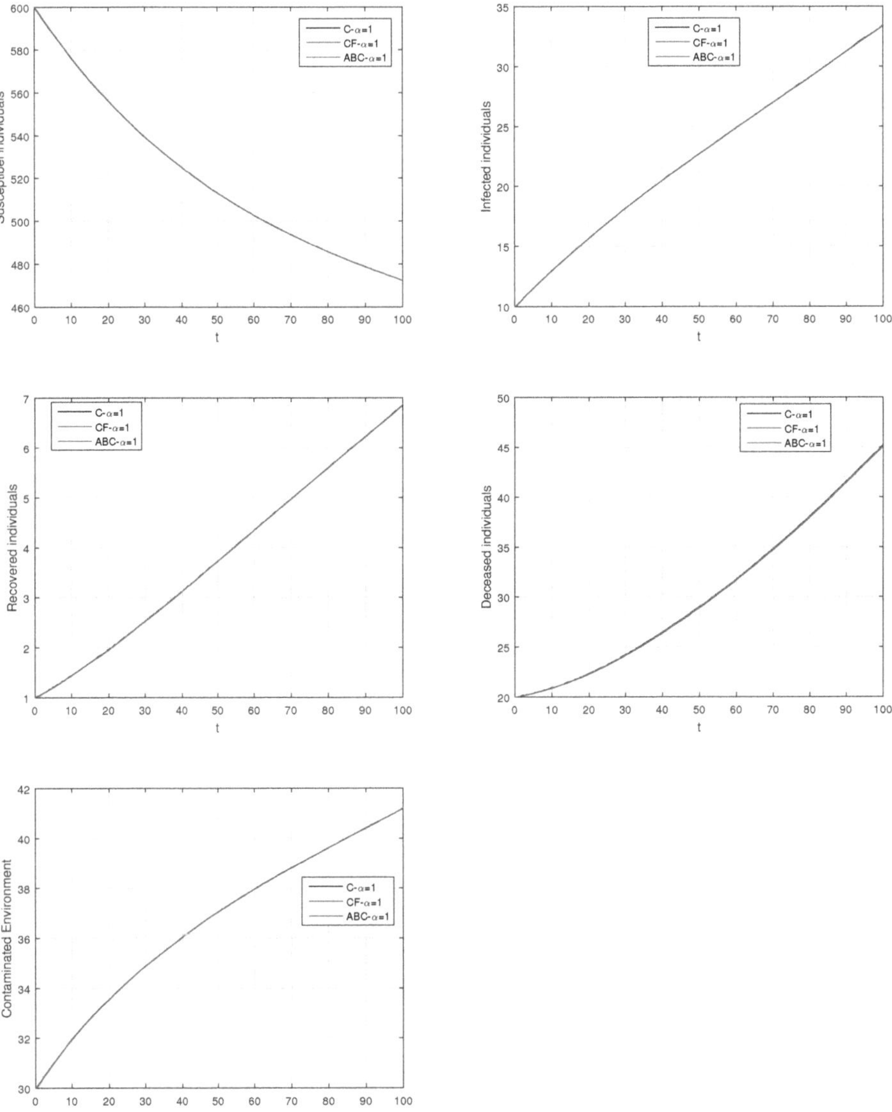

Figure 22. Comparison graphs for the Caputo, Caputo–Fabrizio, and the Atangana–Baleanu derivatives when $\alpha = 1$, where (**a**) Susceptible individuals, (**b**) Infected individuals, (**c**) recovered individuals, (**d**) deceased individuals, (**e**) Environment pathogens.

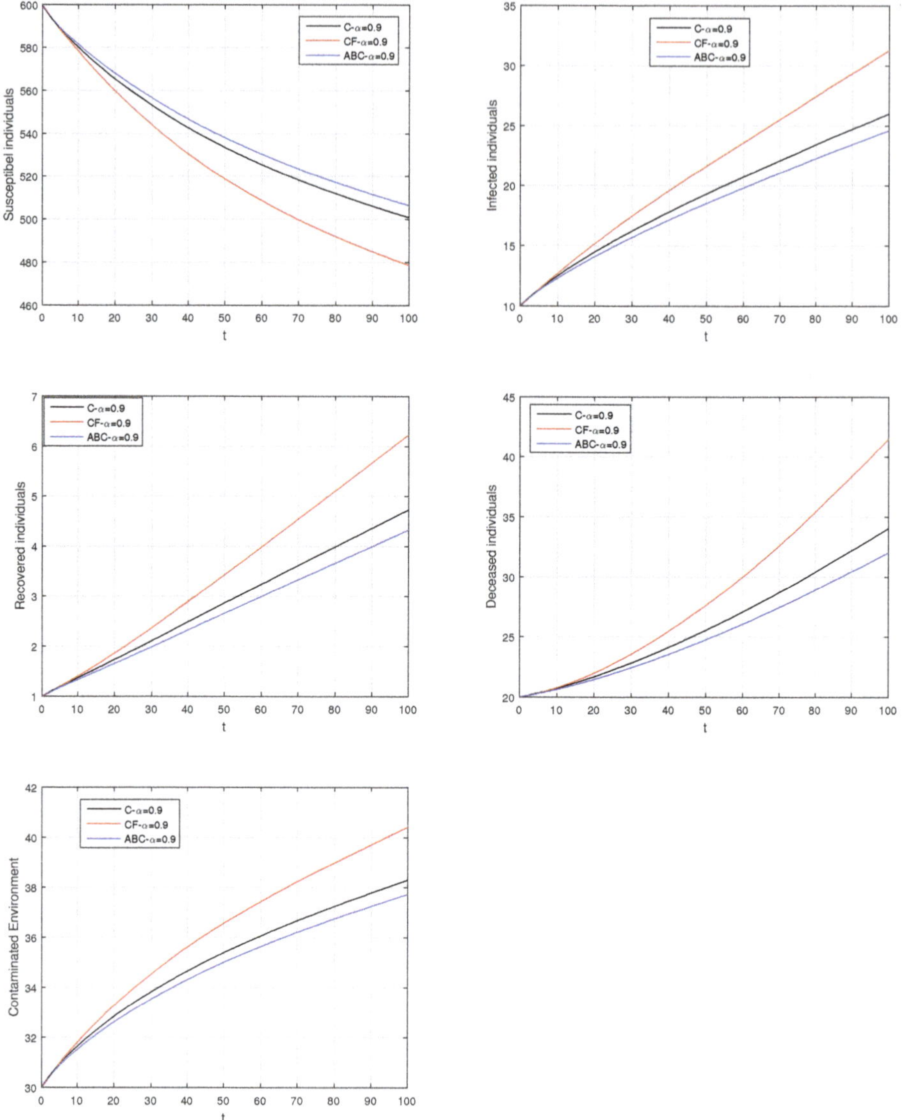

Figure 23. Comparison graphs for the Caputo, Caputo–Fabrizio, and the Atangana–Baleanu derivatives when $\alpha = 0.9$, where (**a**) Susceptible individuals, (**b**) Infected individuals, (**c**) recovered individuals, (**d**) deceased individuals, (**e**) Environment pathogens.

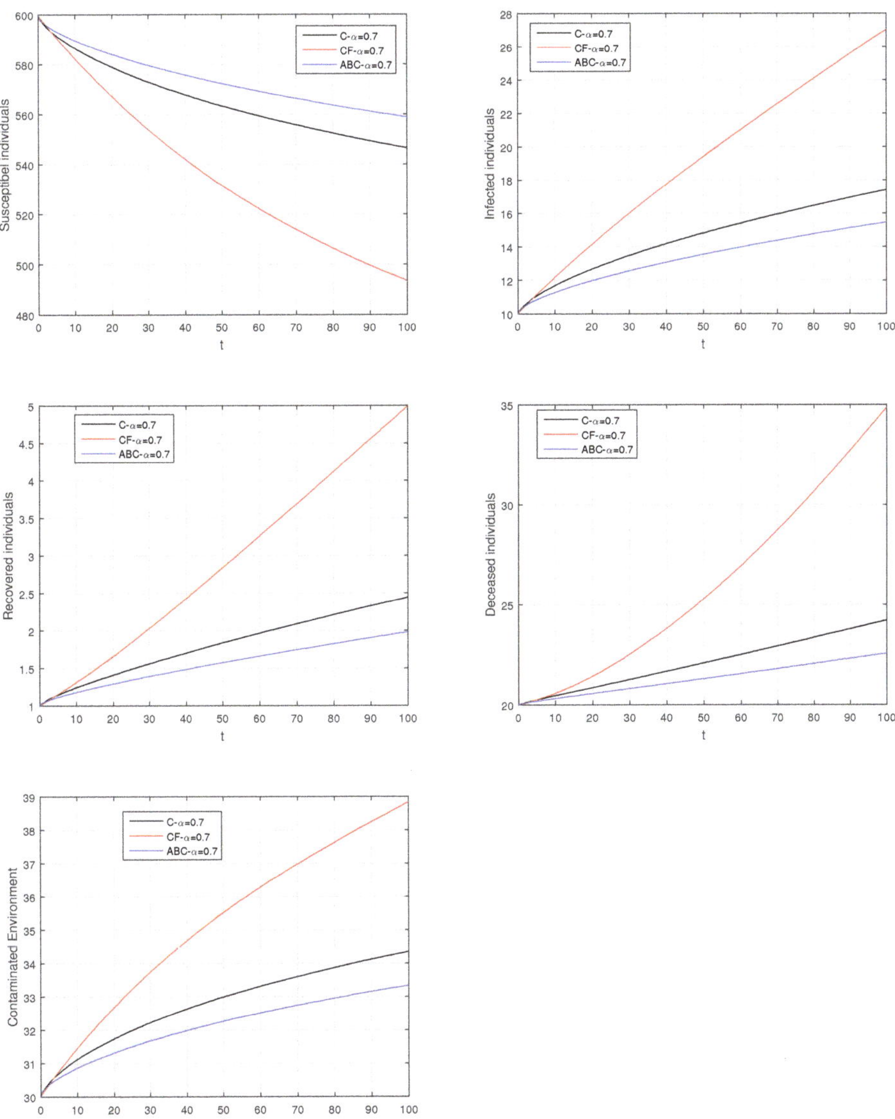

Figure 24. Comparison graphs for the Caputo, Caputo–Fabrizio, and the Atangana–Baleanu derivatives when $\alpha = 0.7$, where (**a**) Susceptible individuals, (**b**) Infected individuals, (**c**) recovered individuals, (**d**) deceased individuals, (**e**) Environment pathogens.

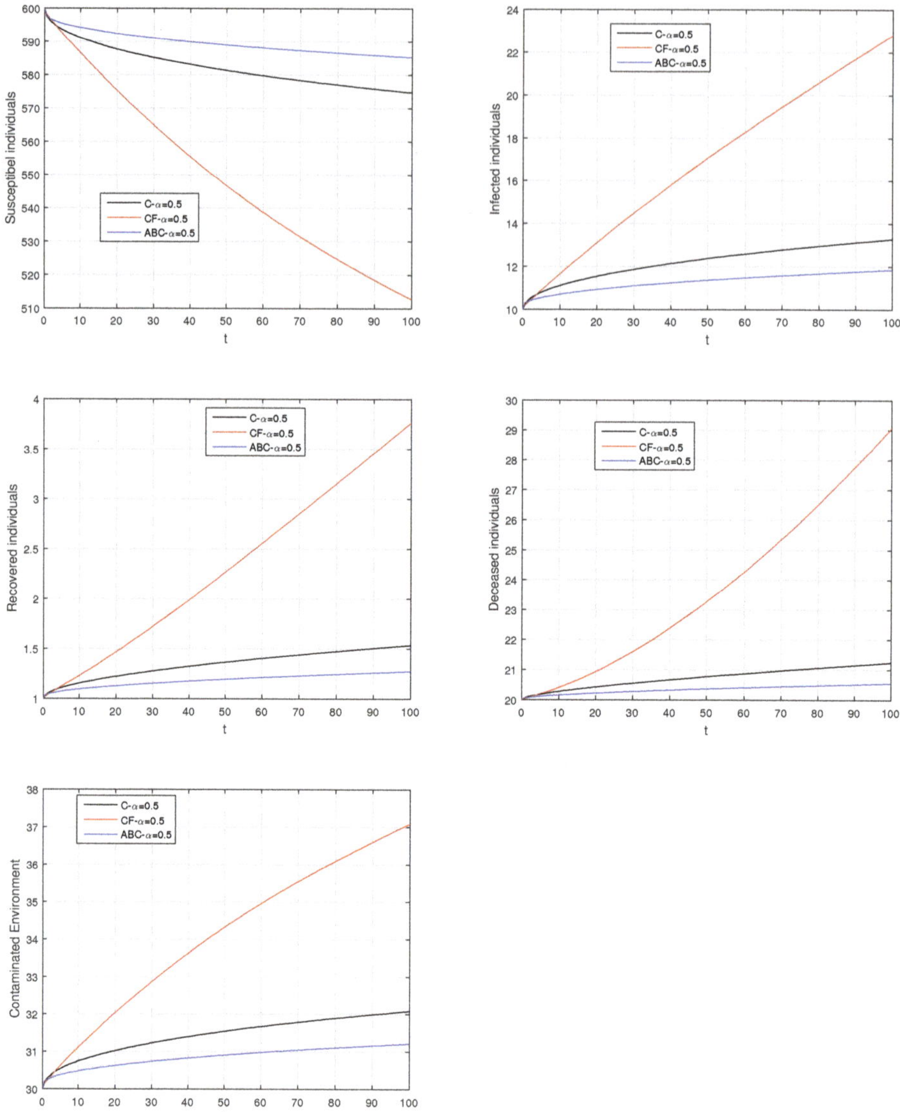

Figure 25. Comparison graphs for the Caputo, Caputo–Fabrizio, and the Atangana–Baleanu derivatives when $\alpha = 0.5$, where (**a**) Susceptible individuals, (**b**) Infected individuals, (**c**) recovered individuals, (**d**) deceased individuals, (**e**) Environment pathogens.

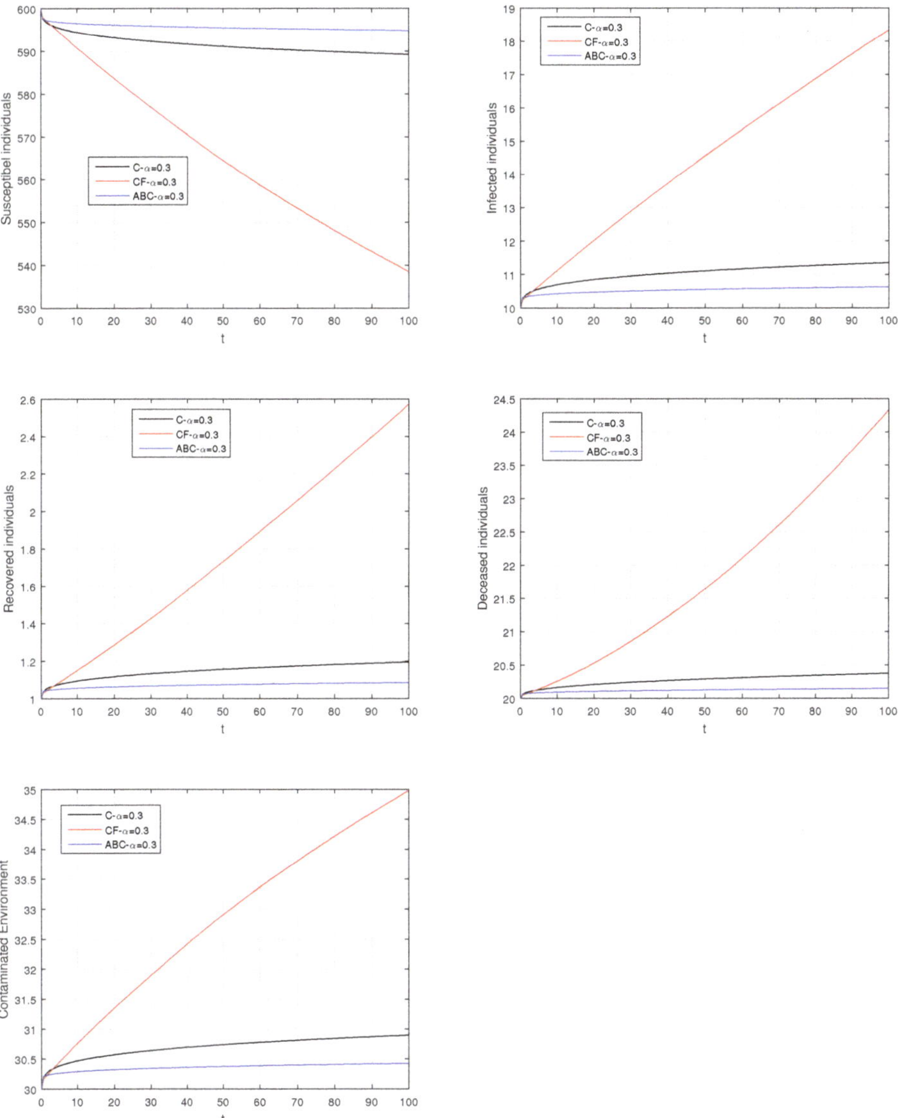

Figure 26. Comparison graphs for the Caputo, Caputo–Fabrizio, and the Atangana–Baleanu derivatives when $\alpha = 0.3$, where (**a**) Susceptible individuals, (**b**) Infected individuals, (**c**) recovered individuals, (**d**) deceased individuals, (**e**) Environment pathogens.

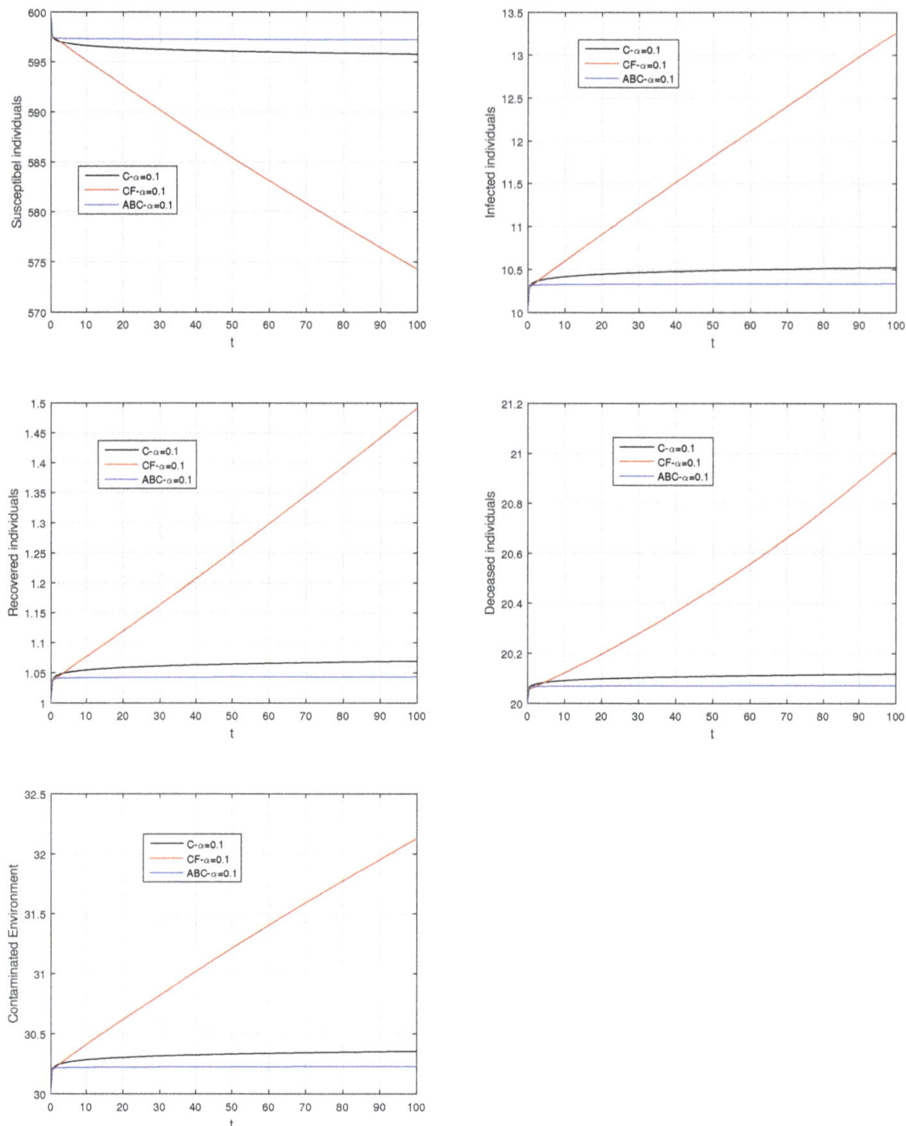

Figure 27. Comparison graphs for the Caputo, Caputo–Fabrizio, and the Atangana–Baleanu derivatives when $\alpha = 0.1$, where (**a**) Susceptible individuals, (**b**) Infected individuals, (**c**) recovered individuals, (**d**) deceased individuals, (**e**) Environment pathogens.

5. Conclusions

We presented the dynamics of an Ebola disease model in the framework of fractional calculus. We applied three fractional operators, which are the Caputo, Caputo–Fabrizio, and the Atangana–Baleanu models. Initially, we proposed an epidemic model available literature for Ebola disease and then applied the proposed operators. The Ebola disease model with the Caputo derivative is presented and an effective numerical scheme for the numerical solution was provided. We used many values for the fractional order parameters and obtained the graphical results. The same model is used further and applied to the Caputo–Fabrizio derivative and we then presented a numerical solution for their solution. The solution was obtained and presented in graphical shape with the use of various fractional order parameter values. Then the newly established derivative known as the Atangana–Baleanu derivative was successfully applied to the Ebola disease model. The Ebola disease model in the Atangana–Baleanu sense is used and the uniqueness and existence were presented. Then, we presented a numerical scheme for the solution and presented various graphical results for α. Comparisons of the proposed three operators for various values of the fractional order parameter $\alpha = 1, 0.9, 0.7, 0.5, 0.3, 0.1$ are presented and discussed. The comparison results show that the Atangana–Baleanu derivative is more helpful for disease elimination by decreasing the value of α, since the population of infected individuals decreased well. The use of three different fractional operators on the Ebola disease model suggests that the fractional order parameter greatly affects disease elimination for the non-integer case when decreasing α. Therefore, we suggest that the application of the various fractional derivatives on the present disease model shows the greater effectiveness of the arbitrary order derivative than that of the integer order model for the case of fractional order parameters.

Author Contributions: Conceptualization, K.M.A. and A.A.; methodology, A.A.; software, K.M.A.; validation, K.M.A. and A.A.; formal analysis, A.A.; investigation, A.A.; resources, K.M.A.; data curation, A.A.; writing–original draft preparation, A.A. and K.M.A.; writing–review and editing, K.M.A.; visualization, A.A.; supervision, A.A.; project administration, A.A.; funding acquisition, A.A.

Funding: No source of funding.

Acknowledgments: The authors are thankful to the anonymous reviewers and handling editor for the careful reading and suggestions.

Conflicts of Interest: No conflict of interests exists regarding the publishing of this work.

References

1. Ebola (Ebola Virus Disease). The Centers for Disease Control and Prevention. Available online: http://www.cdc.gov/ebola/resources/virus-ecology.html (accessed on 1 August 2014).
2. Bibby, K.; Casson, L.W.; Stachler, E.; Haas, C.N. Ebola virus persistence in the environment: State of the knowledge and research needs. *Environ. Sci. Technol. Lett.* **2015**, *2*, 2–6. [CrossRef]
3. Piercy, T.J.; Smither, S.J.; Steward, J.A.; Eastaugh, L.; Lever, M.S. The survival of floviruses in liquids, on solid substrates and in a dynamic aerosol. *J. Appl. Microbiol.* **2010**, *109*, 1531–1539. [PubMed]
4. Leroy, E.M.; Rouquet, P.; Formenty, P.; Souquière, S.; Kilbourne, A.; Froment, J.M.; Bermejo, M.; Smit, S.; Karesh, W.; Swanepoel, R.; et al. Multiple Ebola virus transmission events and rapid decline of central African wildlife. *Science* **2004**, *303*, 387–390. [CrossRef] [PubMed]
5. Leroy, E.M.; Kumulungui, B.; Pourrut, X.; Rouquet, P.; Hassanin, A.; Yaba, P.; Délicat, A.; Paweska, J.T.; Gonzalez, J.-P.; Swanepoel, R. Fruit bats as reservoirs of Ebola virus. *Nature* **2005**, *438*, 575–576. [CrossRef] [PubMed]
6. Althaus, C. Estimating the reproduction number of Ebola (EBOV) during outbreak in West Africa. *PLoS Curr.* **2014**. [CrossRef] [PubMed]
7. Chowell, G.; Hengartner, N.W.; Castillo-Chavez, C.; Fenimore, P.W.; Hyman, J.M. The basic reproductive number of Ebola and the e?ects of public health measures: The cases of Congo and Uganda. *J. Theor. Biol.* **2004**, *229*, 119–126. [CrossRef] [PubMed]
8. Fasina, F.O.; Shittu, A.; Lazarus, D.; Tomori, O.; Simonsen, L.; Viboud, C.; Chowell, G. Transmission dynamics and control of Ebola virus disease outbreak in Nigeria, July to September 2014. *Euro Surveill.* **2014**, *19*, 20920. Available online: http://www.eurosurveillance.org/ViewArticle.aspx?ArticleId=20920 (accessed on 1 August 2018). [CrossRef]
9. Fisman, D.; Khoo, E.; Tuite, A. Early epidemic dynamics of the Western African 2014 Ebola outbreak: Estimates derived with a simple two Parameter model. *PLoS Curr.* **2014**. [CrossRef] [PubMed]
10. Ivorra, B.; Ngom, D.; Ramos, A.M. Be-CoDiS: A mathematical model to predict the risk of human diseases spread between countries-validation and application to the 2014–2015 ebola virus disease epidemic. *Bull. Math. Biol.* **2015**, *77*, 1668–1704. [CrossRef]

11. Wang, X.-S.; Zhong, L. Ebola outbreak in West Africa: Real-time estimation and multiplewave prediction. *Math. Biosci. Eng.* **2015**, *12*, 1055–1063. [CrossRef]

12. Lekone, P.E.; Finkenstädt, B.F. Statistical inference in a stochastic epidemic SEIR model with control intervention: Ebola as a case study. *Biometrics* **2006**, *62*, 1170–1177. [CrossRef] [PubMed]

13. Berge, T.; Lubuma, J.M.-S.; Moremedi, G.M.; Morris, N.; Kondera-Shava, R. A simple mathematical model for Ebola in Africa. *J. Biol. Dyn.* **2017**, *11*, 42–74. [CrossRef]

14. Zhang, Z.; Liu, C.; Zhan, X.; Lu, X.; Zhang, C.; Zhang, Y. Dynamics of Information Diffusion and Its Applications on Complex Networks. *Phys. Rep.* **2016**, *651*, 1–34. [CrossRef]

15. Zhan, X.; Liu, C.; Zhou, G.; Zhang, Z.; Sun, G.; Zhu, J.J.H.; Jin, Z. Coupling dynamics of epidemic spreading and information diffusion on complex networks. *Appl. Math. Comput.* **2018**, *332*, 437–448. [CrossRef]

16. Liu, C.; Zhan, X.; Zhang, Z.; Sun, G.; Hui, P.M. How events determine spreading patterns: Information transmission via internal and external influences on social networks. *New J. Phys.* **2015**, *17*, 113045. [CrossRef]

17. Khan, M.A.; Ullah, S.; Okosun, K.O.; Shah, K. A fractional order pine wilt disease model with Caputo–Fabrizio derivative. *Adv. Differ. Equ.* **2018**, 410. [CrossRef]

18. Khan, M.A.; Ullah, S.; Chaos, M.F. A new fractional model for tuberculosis with relapse via Atangana–Baleanu derivative. *Chaos Solitons Fractals* **2018**, *116*, 227–238. [CrossRef]

19. Ullah, S.; Khan, M.A.; Farooq, M. A fractional model for the dynamics of TB virus *Chaos Solitons Fractals* **2018**, *116*, 63–71.

20. Ullah, S.; Khan, M.A.; Farooq, M. Modeling and analysis of the fractional HBV model with Atangana–Baleanu derivative. *Eur. Phys. J. Plus* **2018**, *133*, 313. [CrossRef]

21. Ullah, S.; Khan, M.A.; Farooq, M. A new fractional model for the dynamics of the hepatitis B virus using the Caputo–Fabrizio derivative. *Eur. Phys. J. Plus* **2018**, *133*, 237. [CrossRef]

22. Diethelm, K. A fractional calculus based model for the simulation of an outbreak of dengue fever. *Nonlinear Dyn.* **2013**, *71*, 613–619. [CrossRef]

23. Caputo, M.; Fabrizio, M. On the notion of fractional derivative and applications to the hysteresis phenomena. *Meccanica* **2017**, *52*, 3043–3052. [CrossRef]

24. Losada, J.; Nieto, J.J. Properties of the new fractional derivative without singular Kernel. *Progr. Fract. Differ. Appl.* **2015**, *1*, 87–92.

25. Atangana, A.; Baleanu, D. New fractional derivatives with nonlocal and non-singular kernel: Theory and application to heat transfer model. *arXiv* **2016**, arXiv:1602.03408.

26. Khan, M.A. Neglecting nonlocality leads to unrealistic numerical scheme for fractional differential equation: Fake and manipulated results. *Chaos* **2019**, *29*, 013144. [CrossRef]

27. Atangana, A.; Owolabi, K.M. New numerical approach for fractional differential equations. *Math. Model. Nat. Phenom.* **2018**, *13*. [CrossRef]

28. Toufik, M.; Atangana, A. New numerical approximation of fractional derivative with non-local and non-singular kernel: Application to chaotic models. *Eur. Phys. J. Plus* **2017**, *132*, 444. [CrossRef]

Article

On the Use of Entropy Issues to Evaluate and Control the Transients in Some Epidemic Models

Manuel De la Sen [1,*], **Raul Nistal** [1], **Asier Ibeas** [2] **and Aitor J. Garrido** [1,3]

[1] Institute of Research and Development of Processes IIDP, University of the Basque Country, Campus of Leioa, PO Box 48940 Leioa (Bizkaia), Spain; raul.nistal@gmail.com (R.N.); aitor.garrido@ehu.eus (A.J.G.)

[2] Department of Telecommunications and Systems Engineering, Universitat Autònoma de Barcelona, 08193 Barcelona, Spain; Asier.Ibeas@uab.cat

[3] Faculty of Engineering of Bilbao, University of the Basque Country, Rafael Moreno No. 3, 48013 Bilbao, Spain

* Correspondence: manuel.delasen@ehu.eus

Received: 13 March 2020; Accepted: 4 May 2020; Published: 9 May 2020

Abstract: This paper studies the representation of a general epidemic model by means of a first-order differential equation with a time-varying log-normal type coefficient. Then the generalization of the first-order differential system to epidemic models with more subpopulations is focused on by introducing the inter-subpopulations dynamics couplings and the control interventions information through the mentioned time-varying coefficient which drives the basic differential equation model. It is considered a relevant tool the control intervention of the infection along its transient to fight more efficiently against a potential initial exploding transmission. The study is based on the fact that the disease-free and endemic equilibrium points and their stability properties depend on the concrete parameterization while they admit a certain design monitoring by the choice of the control and treatment gains and the use of feedback information in the corresponding control interventions. Therefore, special attention is paid to the evolution transients of the infection curve, rather than to the equilibrium points, in terms of the time instants of its first relative maximum towards its previous inflection time instant. Such relevant time instants are evaluated via the calculation of an "ad hoc" Shannon's entropy. Analytical and numerical examples are included in the study in order to evaluate the study and its conclusions.

Keywords: Shannon entropy; epidemic model; transient behavior; vaccination and treatment intervention controls

---

1. Introduction

Some classical works by Boltzmann, Gibbs and Maxwell have defined entropy under a statistical framework. A useful entropy concept is the Shannon entropy since it is a basic tool to quantify the amount of uncertainty in many kinds of physical or biological processes [1–6]. It may be interpreted as a quantification of information loss [1–3,7–9]. On the other hand, entropy-based tools have been also proposed to evaluate the propagation of epidemics and related public control interventions (see, for instance, [10–17] and some of the references therein). There are also models whose basic framework relies on the use of entropy tools, as for instance [13–16]. It can be also pointed out that the control designs might be incorporated to some epidemic propagation and other biological problems, see, for instance, [18–27], and, in particular, for the synthesis of decentralized control in patchy (or network node-based) interlaced environments [24,27]. A typical situation is that of several towns each with its own health center, whose susceptible and infectious populations, apart from their coupled self-dynamics among their integrating subpopulations, might also mutually interact with the subpopulations of the neighboring nodes through in-coming and out-coming fluxes.

It can be pointed out that the knowledge or estimation of the transient behavior of the infection is very relevant for the hospital management of the disease since it is necessary to manage the availability of beds and other sanitary utensils and sanitary means, in general. The work by Wang et al. in [11] pays mainly attention to the description of the transient behavior of the evolution of epidemics rather than to the equilibrium states. The main purpose in that paper was to formulate the time interval occurring between the time instant of the maximum of the infection, which gives a relative maximum of the infection evolution through time (and which zeroes the first time-derivative of the infection function), and the time instant giving its previous inflection time instant. It turns out that the knowledge of the first part of the transient evolution is very relevant to fight against the initial exploding of the illness since any eventual control intervention is typically much more efficient as far as it is taken as quickly as possible. The model proposed in [11] is a time-varying differential equation of first-order describing the infectious population which is the unique explicit one in the model. It is also pointed out in that paper that the time-varying coefficient might potentially contain the supplementary environment information to make such an equation well-posed to practically describe a concrete disease evolution. An interesting point of that work is that the infection evolution is identified with a log-normal distribution whose parameterization is selected in such a way that the entropy production rate is maximized. The above proposed theoretical first-order model has been proved to be very efficient to describe the data of SARS 2003. Alternative interpretations of the entropy in terms of maximum entropy or maximum entropy rate are given, for instance, in [12–14] and some references therein.

This paper studies how to link the extension of the first-order differential system proposed in [11] for the study of infection propagations to epidemic models with more integrated coupled subpopulations (such as susceptible, immune, vaccinated etc.) by introducing the coupling and control information through the time-varying coefficient which drives the basic differential equation model. It is considered relevant the control of the infection along its transient to fight more efficiently against a potential initial exploding transmission. Note that the disease-free and endemic equilibrium points and their stability properties depend on the concrete parameterization while they admit a certain design monitoring by the choice of the control and treatment gains and the use of feedback information in the corresponding controls. See, for instance [19,27]. Therefore, special attention is paid to the transients of the infection curve evolution in terms of the time instants of its first relative maximum towards its previous inflection time instant since there is a certain gap in the background literature concerning the study of such transients. The ratio of such time instants is later on considered subject to some worst-case uncertainty relations via the calculation and analysis of an "ad hoc" Shannon's entropy. Note that entropy issues have been considered in the study of biological, evolution and epidemic models by incorporating techniques of information theory. See, for instance [11–13,28–32]. It is well-known that the entropy production theorems might be classified according to a generalized sequence of stable thermodynamic states. Also, the thermodynamic equilibrium, which is characterized by the absence of gradients of state or kinematic variables, is in a state of maximum entropy and zero entropy production [33,34]. Furthermore, linear non-equilibrium processes are associated with entropy production so that the entropy concept may be also invoked in transient processes [35]. On the other hand, it may be pointed out that uncertainties can appear in the characterization of the infection evolution through time, even in deterministic models, due to parameterization uncertainties, fluxes of populations or existing uncertainties in the initial conditions. Other mathematical techniques of interest which combine analytical and numerical issues have been also been applied to the analysis and discussion of epidemic models with eventual support of mathematical techniques on homotopy analysis and distribution functions as, forinstance, the log-normal distribution [36,37]. For instance, in [38], the SIR and SIS epidemic models are solved through the homotopy analysis method. A one-parameter family of series solutions is obtained which gives a method to ensure convergent series solutions for those kinds of models. On the other hand, in [39], the analytic solutions of an SIR epidemic model are investigated in parametric form. It is also found that the generalization of a SIR model including births and mortality with vital dynamics might be reduced to an Abel-type which greatly simplify the analysis.

The paper is organized as follows: Section 2 gives an extension of the basic model of [11] to be then compared in subsequent sections with some existing models with several subpopulations. Such a model only considers the infection evolution through time and it is based on the action of two auxiliary non-negative functions which define appropriately the time-varying coefficient which defines the first-order differential equation of the infection evolution. The model includes, as particular case, that of the abovementioned reference where both such auxiliary functions are identical to the time argument. Particular choices of those functions make it possible to consider alternative effects linked to the basic model like, for instance, the influence on the infectious subpopulation of other coupled subpopulations in more general models like, for instance, the susceptible, exposed, recovered or vaccinated ones. It is also possible to include the control effects through such a varying coefficient, if any, like for instance, the vaccination and treatment controls. Some basic formal results are stated and proved mainly concerning with the first relative maxima and inflection time instants of the infection curve through time. The above two time instants are relevant to take appropriate control interventions to fight against an initially exploding infectious disease.

Section 3 links the basic model of Section 2 with some known epidemic models which integrate more subpopulations than just the infectious one, like for instance, the susceptible and recovered subpopulations, The time-varying coefficient driving the infection evolution is defined explicitly for each of the discussed epidemic models. Basically, it is taken in mind that some relevant information of higher-order differential epidemic models concerning the transient trajectory solution can be captured by a parameter-dependent and time-varying coefficient which drives a first-order differential equation to the light of the basic model of Section 2. So, the time-varying coefficient describing the infection evolution depends in those cases of the remaining subpopulations integrated in the model. The maximum and inflection time instants are characterized for some given examples involving epidemic models of several subpopulations. In particular, the last one of the discussed theoretical examples includes the effects of vaccination and treatment intervention controls generated by linear feedback of the susceptible and infectious subpopulations, respectively. Later on, Section 4 investigates the entropy associated with the infection accordingly to the generalizations of Section 2 concerning the specific structure of the time-varying coefficient describing the infection dynamics and its links with the theoretical examples discussed in Section 3. The error of the entropy related to the reference one associated with the log-normal distribution is estimated. In practice, that property can be interpreted in terms of public medical and social interventions which control the disease propagation when introducing the controls of the last example discussed in Section 3. The second part of Section 4 is devoted to linking the entropy and inflection and maximum infection time instants and their reached values of the discussed multi-population structures to their counterparts of the maximum dissipation rate being associated to the formulation of a simpler model based on the log-normal distribution and one-dimensional infection dynamics. Some numerical tests are performed for comparisons of the entropies and its width of the basic model with two of the discussed examples in the previous sections which involve the presence of more than one integrated subpopulations. Finally, conclusions end the paper.

Notation

$$R_+ = \{r \in R : r > 0\}; \ R_{0+} = \{r \in R : r \geq 0\} = R_+ \cup \{0\}$$

$$Z_+ = \{r \in Z : r > 0\}; \ Z_{0+} = \{r \in Z : r \geq 0\} = Z_+ \cup \{0\}$$

$$\bar{n} = \{0, 1, \cdots, n\}$$

2. The Basic Model Description and Some Related Technical Results

Since disease propagation can be interpreted as a thermodynamic system, it can be assumed that the rate of increase or decrease is proportional to the infection at the previous day following the approach of modelling the rate of chemical reactions, [11]. Thus, assume that the infection evolution obeys the following time-varying differential equation:

$$\dot{I}(t) = \alpha(t)I(t); \; I(0) = I_0 > 0 \tag{1}$$

where $\alpha : \mathbf{R}_{0+} \to \mathbf{R}_{0+}$ is continuous and time differentiable on $(0, +\infty)$. The particular structure of the varying coefficient $\alpha(t)$ depends on the balances between the spreading mechanism and the exerted controls during the public intervention. Such a coefficient contains the available information related to the incorporation of all the control mechanisms and the coupling dynamics between the infectious populations and the remaining interacting ones such as the susceptible, immune or vaccinated ones. By taking time-derivatives with respect to time in (1), one gets:

$$\begin{aligned}\ddot{I}(t) &= \dot{\alpha}(t)I(t) + \alpha(t)\dot{I}(t) \\ &= \big(\dot{\alpha}(t)/\alpha(t) + \alpha(t)\big)\dot{I}(t) \\ &= \big(\dot{\alpha}(t) + \alpha^2(t)\big)I(t); \; \dot{I}(0) = \dot{I}_0 = \alpha(0)I_0\end{aligned} \tag{2}$$

It is proposed in [11] to consider two relevant time instants in the disease evolution, namely:

(1)   The inflection time instant of $I(t)$ which is the date in the infection evolution at which the controlling actions take effect on the evolution. Typically, this time instant is the undulation point date in the evolution of $I(t)$, that is the zero of $\ddot{I}(t)$, provided that the first non-zero derivative of $I^{(n)}(t) = \frac{d^n I(t)}{dt^n}$, $n \geq 3$ occurs for some even n since this last condition ensures that the undulation time instant is the inflection time instant.

(2)   The critical time instant at which the spread rate turns from initial growing to decrease which can be empirically attributed to the global influence of the control interventions. This time instant is a relative maximum of $I(t)$ and it satisfies the constraints $\dot{I}(t) = 0$ and $\ddot{I}(t) < 0$ under the reasonable assumption that $\dot{I}_0 > 0$.

It turns out that, along the whole disease evolution, several successive inflection points and relative maxima can happen. The subsequent result which is concerned with the non-negativity, boundedness and asymptotic vanishing property of the infection as time tends to infinity and its two first- time derivatives is immediate from the above expressions (1) and (2):

Theorem 1. *The following properties hold:*

(i)   *The infection population and its two first-time derivatives obey the following time evolution equations:*

$$I(t) = e^{\int_0^t \alpha(\tau)d\tau}I_0; \dot{I}(t) = \alpha(t)e^{\int_0^t \alpha(\tau)d\tau}I_0; \ddot{I}(t) = \big(\dot{\alpha}(t) + \alpha^2(t)\big)e^{\int_0^t \alpha(\tau)d\tau}I_0; \forall t \in \mathbf{R}_{0+} \tag{3}$$

(ii) $I(t) > 0; \forall t \in \mathbf{R}_{0+}$ *if and only if* $I_0 \geq 0$; *and* $I(t) = 0; \forall t \in \mathbf{R}_{0+}$ *if and only if* $I_0 = 0$.

(iii)   *If* $+\infty > I_0 \geq 0$ *then* $I(t) \leq KI_0 < +\infty$; $\forall t \in \mathbf{R}_{0+}$ *for some* $K \in \mathbf{R}_+$ *if and only if* $\alpha : \mathbf{R}_{0+} \to \mathbf{R}_{0+}$ *is such that* $\int_0^t \alpha(\tau)d\tau \leq K < +\infty$; $\forall t \in \mathbf{R}_{0+}$.

(iv) $I(t) \to 0$ *as* $t \to +\infty$ *for any given finite* I_0 *if and only if* $\lim\limits_{t \to +\infty} \int_0^t \alpha(\tau)d\tau = -\infty$.

(v)   *If* $+\infty > I_0 \geq 0$ *and* $\int_0^t \alpha(\tau)d\tau \leq K < +\infty$; $\forall t \in \mathbf{R}_{0+}$ *for some* $K \in \mathbf{R}_{0+}$ *then* $\left|\dot{I}(t)\right| < +\infty$; $\forall t \in \mathbf{R}_{0+}$ *if and only if, for some* $K_1 \in \mathbf{R}_+$, $\left|\alpha(t)\right| \leq K_1 < +\infty$; $\forall t \in \mathbf{R}_{0+}$. *If* $+\infty > I_0 \geq 0$ *then* $\left|\dot{I}(t)\right| < +\infty$; $\forall t \in \mathbf{R}_{0+}$ *if and only if* $\left|\alpha(t)\right|e^{\int_0^t \alpha(\tau)d\tau} \leq K_2 < +\infty$; $\forall t \in \mathbf{R}_{0+}$, *for some* $K_2 \in \mathbf{R}_+$ *provided that* $\alpha(0) < +\infty$.

(vi) $\dot{I}(t) \to 0$ as $t \to +\infty$ *for any given finite* I_0 *if and only if* $\lim_{t \to +\infty}\left(\alpha(t)e^{\int_0^t \alpha(\tau)d\tau}\right) = 0$. *If* $\alpha : R_{0+} \to R_{0+}$

   is bounded and $I(t) \to 0$ *as* $t \to +\infty$ *then* $\dot{I}(t) \to 0$ *as* $t \to +\infty$.

(vii) *If* $+\infty > I_0 \geq 0$ *then* $\left|\ddot{I}(t)\right| < +\infty$; $\forall t \in R_+$ *if and only if* $\left|\dot{\alpha}(t) + \alpha^2(t)\right|e^{\int_0^t \alpha(\tau)d\tau} \leq K_3 < +\infty$;

   $\forall t \in R_{0+}$, *for some* $K_3 \in R_+$. $\ddot{I}(t) \to 0$ *as* $t \to +\infty$ *for any given finite* I_0 *if and only if*
   $\lim_{t \to +\infty}\left(\left(\dot{\alpha}(t) + \alpha^2(t)\right)e^{\int_0^t \alpha(\tau)d\tau}\right) = 0$. *If* $\left(\alpha + \dot{\alpha}\right) : R_{0+} \to R_{0+}$ *is bounded and* $I(t) \to 0$ *as* $t \to +\infty$

   then $\ddot{I}(t) \to 0$ *as* $t \to +\infty$.

Note that $\alpha(t)$ (respectively, $\alpha(t) + \dot{\alpha}(t)$) is infinity at $t = 0$ while it is bounded for $t > 0$, as it happens for instance with the α—function proposed in [11], then $\dot{I}(t)$ (respectively, $\ddot{I}(t)$) is still bounded under the conditions of Theorem 1 (v) (respectively, Theorem 1 (vii)) on R_+.

Note also that the vanishing infection condition of Theorem 1 typically occurs under convergence of the solution to the disease-free equilibrium point if the disease reproduction number is less than one [19,22–24,27,29,30,36]. However, it can happen that the infection oscillates around some stable equilibrium or that it converges to a nonzero positive constant defining the corresponding component of the endemic equilibrium steady-state as it is discussed in the next result.

Corollary 1. *The following properties hold:*

(i) *Assume that there exists some* $C \in R_+$ *such that* $\int_0^t \alpha(\tau)d\tau \to C$ *as* $t \to +\infty$ *and that* $\alpha(t), \dot{\alpha}(t) \to 0$
   as $t \to +\infty$. *Then,* $I(t) \to e^C I_0$, $\dot{I}(t) \to 0$ *and* $\ddot{I}(t) \to 0$ *as* $t \to +\infty$.

(ii) *Assume that* $\int_0^t \alpha(\tau)d\tau \to C$ *as* $t \to +\infty$ *and that* $\alpha : R_{0+} \to R_{0+}$ *is uniformly continuous. Then,*
   $\alpha(t) \to 0$, $I(t) \to e^C I_0$ *and* $\dot{I}(t) \to 0$ *as* $t \to +\infty$. *Assume, in addition, that* $\dot{\alpha} : R_{0+} \to R_{0+}$ *is*
   uniformly continuous. Then $\ddot{I}(t) \to 0$ *as* $t \to +\infty$.

Proof of Property (i). Follows directly from (1)–(3). On the other hand, since $\alpha : R_{0+} \to R_{0+}$ is uniformly continuous and the limit $\lim_{t \to +\infty}\int_0^t \alpha(\tau)d\tau = C$ exists and it is finite then $\alpha(t) \to 0$ as $t \to +\infty$ (Barbalat's Lemma) and $I(t) \to e^C I_0$ as $t \to +\infty$ from (3), $\dot{I} : R_+ \to R_{0+}$ is bounded, since being continuous, it cannot diverge in finite time, and $\dot{I}(t) \to 0$ as $t \to +\infty$ from (1). If, furthermore, $\dot{\alpha} : R_{0+} \to R_{0+}$ is uniformly continuous and, since $\lim_{t \to +\infty}\int_0^t \dot{\alpha}(\tau)d\tau = \lim_{t \to +\infty}\alpha(t) - \alpha(0) = -\alpha(0)$ then $\dot{\alpha}(t) \to 0$ as $t \to +\infty$ (again from Barbalat's Lemma). Since $\alpha(t), \dot{\alpha}(t) \to 0$ as $t \to +\infty$ then $\ddot{I}(t) \to 0$ as $t \to +\infty$ from (2). $\square$

Let us introduce the following definitions and lemma of usefulness for the proof of the subsequent theorem [36]:

Definition 1. *Let* $f : R \to R$ *be everywhere continuous and twice differentiable at* $t_0 \in R$. *Then,* t_0 *is an undulation point (or pre-inflection point) of* f *if* $\ddot{f}(t_0) = 0$.

   Inflection points of the continuous and twice-differentiable $f : R \to R$ *are the undulation points of the function where the curvature changes its sign, that is, points of change of local convexity to local concavity or vice-versa. They are also the isolated extrema of* $\ddot{f} : R \to R$. *A well-known technical definition and a related result on inflection points follow:*

Definition 2. *Let* $f : R \to R$ *be everywhere continuous and twice differentiable at* $t_0 \in R$ *which is an isolated extremum of* $\ddot{f}$ *(that is, a local maximum or minimum, and also an undulation point of* $\ddot{f}$ *as a result).*

Lemma 1. *The following properties hold:*

(i)   Let $f : R \to R$ be everywhere continuous and twice differentiable at $t_0 \in R$. Then, t_0 is an inflection point of f if $\ddot{f}(t + \varepsilon)\ddot{f}(t - \varepsilon) < 0$ for some sufficiently small $\varepsilon \in R_+$.

(ii)   Let $f : R \to R$ be everywhere continuous and an odd number $k(\geq 3)$ -times differentiable, within a neighborhood of $t_0 \in R$ which is an undulation point of f satisfying $f^{(j)}(t_0) = 0$ for $j = 2, 3, \ldots k - 1$ and $f^{(k)}(t_0) \neq 0$. Then, t_0 is an inflection point of f.

The subsequent result has a very technical proof leading to the basic result that the zeros at finite time instants of $\dot{I}(t)$ and $\ddot{I}(t)$ alternate if $I(t)$ is sufficiently smooth and $\alpha(t)$ is sufficiently smooth. In order to simplify the result proof, it is assumed, with no loss in generality, that the disease dynamics (1)–(2) has no equilibrium points such that the zeros under study are isolated.

Theorem 2. *Assume that the function* $\alpha : R_{0+} \to R_{0+}$ *defined by* $\alpha(t) = -\frac{c \ln(g(t)/E)}{h(t)}$, *where* c, $E \in R_+$ *and* $g, h : R_{0+} \to R_{0+}$ *are everywhere continuous and time-differentiable such that* $g(0) = 0$ *with* $\lim\limits_{t \to 0} \frac{\ln(g(t)/E)}{h(t)} \leq -\varepsilon$ *for some* $\varepsilon \in R_{0+}$, *and furthermore,* $\alpha : R_{0+} \to R_{0+}$ *fulfills the constraints:*

$$\alpha(D_i) = 0; \; \dot{\alpha}(L_i) = -a^2(L_i) \tag{4}$$

$$\frac{h(L_i)\dot{g}(L_i) - \ln(g(L_i)/E)\dot{h}(L_i)g(L_i)}{g(L_i)\ln^2(g(L_i)/E)} = K > 0; \; \forall L_i \in L_S \cap \left[0, \bar{L}\right] \tag{5}$$

for any given positive real number $\bar{L}$, *with* $D_i \in D_S$ *and* $L_i \in L_S$, *where* $D_S = \{D \in R_+ : \alpha(D) = g(D) = E\} \subset R_{0+}$ *and* $L_S = \{L \in R_+ : \dot{\alpha}(L) + a^2(L) = 0\} \subset R_{0+}$ *are assumed to be nonempty and of zero Lebesgue measure. Then, the following properties hold:*

(i) $g(L_i) = E \Leftrightarrow h(L_i)\dot{g}(L_i) > 0$, *equivalently,* $D_S \cap L_S = \varnothing$.

(ii)   *(a)* $\mathrm{card} L_S = \mathrm{card} D_S + \vartheta$ *with* $\vartheta = \{0, 1\}$, *and (a) if* $\mathrm{card}(D_S) = \mathrm{card}(L_S) \leq \aleph_0$ *(with* $\aleph_0$ *denoting the infinite cardinality of denumerable sets) then* $L_i < D_i < L_{i+1}$; $\forall i \in Z_{0+}$ *for any pairs* D_i, $D_{i+1} \in D_S$ *and* L_i, $L_{i+1} \in L_S$ *fulfilling* $(D_i, D_{i+1}) \cap D_S = \varnothing$ *and* $(L_i, L_{i+1}) \cap L_S = \varnothing$, *(b) if* $1 \leq \mathrm{card}(L_S) = \mathrm{card}(D_S) - 1 = \ell < \infty$ *then* $L_i < D_i < L_{i+1}$ *for* $i \in \overline{\ell - 1}$. $\alpha : R_{0+} \to R_{0+}$ *is subject to the constraint* $c = K$, $\dot{I}(D_i) = \ddot{I}(L_i) = 0$; $\forall L_i \in L_S \cap [0, L]$ *and* $D_i > L_i$.

(iii) $\alpha : R_{0+} \to R_{0+}$ *is subject to* $\dot{I}(D_i) = \ddot{I}(L_i) = 0$; $\forall D_i \in D_S \cap [0, L]$, $\forall L_i \in L_S \cap [0, L]$ *and* $D_1 > L_1$ *for any* $I_0 > 0$.

Proof. First, note that $\dot{I}(D) = \ddot{I}(L) = 0$; $\forall D \in D_S$, $\forall L \in L_S$ since $\alpha(D) = 0$ even if $I(D) \neq 0$. On the other hand, L_S is the set of undulation points of $I : R_{0+} \to R_{0+}$ and it is clear that D_S is contained in the set of relative maximum and minimum points of $I(t)$. The properties (i)–(iii) are now proved:

Proof of Property (i). It is now proved that D_S is the set of extreme points of $I(t)$ which is disjoint to its set of undulation points L_S. Assume, on the contrary, that there is some $D \notin D_S$ such that $\dot{I}(D) = 0$. Then, $I(D) = 0$ since $\alpha(D) \neq 0$, and then the disease-free equilibrium point is reached in finite time contradicting the fact that $\alpha(t) = -\frac{c \ln(g(t)/E)}{h(t)}$ is only zero at finite time for a discrete set of time instants satisfying $g(t) = E$ so that $\dot{I}(D) = 0$ if and only if $D \in D_S$. Then, $I(D) = \dot{I}(D) = \ddot{I}(D) = 0$ is a disease-free equilibrium point which is reached in finite time which contradicts the given hypothesis. So, it is easy to see that L_S and D_S are discrete sets of non-negative real time instants which can be strictly ordered. Note also from (1)–(2) that:

$$\alpha(D_i) = -\frac{c \ln(g(D_i)/E)}{h(D_i)} = 0; \; \forall D_i \in D_S \tag{6}$$

$$\dot{\alpha}(L_i) = -a^2(L_i); \; \forall L_i \in L_S \tag{7}$$

If $D_i \leq D < +\infty$ then $g(D_i) = E$ since $c \neq 0$. Also, $\alpha(t) = -\frac{cln(g(t)/E)}{h(t)}$ and, if $L_i \in L_S$ and since $h(L_i) > 0$, one has:

$$\alpha^2(L_i) = -\dot{\alpha}(L_i) = \frac{d}{dt}\left[\frac{cln(g(t)/E)}{h(t)}\right]_{t=L_i}$$

$$= c\frac{h(L_i)\dot{g}(L_i)/g(L_i)-ln(g(L_i)/E)\dot{h}(L_i)}{h^2(L_i)} = c^2\frac{ln^2(g(L_i)/E)}{h^2(L_i)} \tag{8}$$

$$c = \frac{h(L_i)\dot{g}(L_i) - ln(g(L_i)/E)\dot{h}(L_i)g(L_i)}{g(L_i)ln^2(g(L_i)/E)} > 0 \tag{9}$$

Now, if there is some $L_i \in L_S \cap D_S$, equivalently, $L_S \cap D_S \neq \varnothing$, then $g(L_i) = E \Leftrightarrow h(L_i)\dot{g}(L_i) \neq 0$ from (9) since $c \neq 0$ and $ln(g(L_i)/E) = 0$ and, furthermore, one gets from (8) that $\dot{\alpha}(L_i) \neq 0$ since $g(L_i) = E$. But one also has that $\dot{\alpha}(L_i) = \alpha(L_i) = 0$, since $\dot{\alpha}(L_i) = -\alpha^2(L_i)$; $\forall L_i \in L_S$ from the first identity of (8). Then, $0 \neq \dot{\alpha}(L_i) = 0$ is a contradiction so that $L_i \notin L_S \cap D_S$. Equivalently, $D_S \cap L_S = \varnothing$. Property (i) has been proved. $\square$

Proof of Property (ii). Since $\dot{\alpha}(L_i) = -\alpha^2(L_i)$ then $\ddot{\alpha}(L_i) = -2\alpha(L_i)\dot{\alpha}(L_i)$ so that:

$$\ddot{I}(L_i) = \big(\dot{\alpha}(L_i) + \alpha^2(L_i)\big)I(L_i) = 0$$

$$\dddot{I}(L_i) = \big(\ddot{\alpha}(L_i) + 2\alpha(L_i)\dot{\alpha}(L_i)\big)I(L_i) + \big(\dot{\alpha}(L_i) + \alpha^2(L_i)\big)\dot{I}(L_i)$$

$$= \big(\ddot{\alpha}(L_i) - 2\alpha^3(L_i)\big)I(L_i)$$

Since the zeros of $\alpha(t)$ and those of its first time- derivative do not coincide since $D_S \cap L_S = \varnothing$ (from Property (i)), it turns out that the two sets of respective zeros alternate if there are not two zeros of $\alpha(t)$ within any open time interval of two consecutive zeros of $\dot{\alpha}(t)$ or vice-versa. One proceeds by contradiction arguments by assuming two cases which are both rebutted.

Case 1: Assume that there are two consecutive zeros of $\dot{I}(t)$ between two consecutive zeros of $\ddot{I}(t)$, then satisfying the constraint $0 \leq L_i < D_i < D_{i+1} < L_{i+1}$ for some two consecutive time instants D_i, D_{i+1} in D_S and two consecutive time instants L_i, L_{i+1} in L_S so that $\alpha(D_i) = \alpha(D_{i+1}) = \ddot{I}(L_i) = \ddot{I}(L_{i+1}) = 0$. Assume that $I(t) = 0$ for some $t \in (D_i, D_{i+1})$ then $\dot{I}(t) = \alpha(t)I(t) = 0$ so that $t \in D_S$ and then D_i, D_{i+1} are not consecutive time instants in D_S and this case has to be excluded from further reasoning. Now, assume that $I(t) \neq 0$ for all $t \in (D_i, D_{i+1})$ and $\dot{\alpha}(t) \neq 0$, otherwise, if $\dot{\alpha}(t) = 0$ then $t \in D_s$ and D_i, D_{i+1} are not consecutive time instants in D_S. Thus, $\alpha(t) = \alpha(D_i) + \int_{D_i}^t \dot{\alpha}(\tau)d\tau = \int_{D_i}^t \dot{\alpha}(\tau)d\tau$ for all $t \in (D_i, D_{i+1})$. Since $\dot{\alpha}(t) \neq 0$ for all $t \in (D_i, D_{i+1})$, it has no sign change in (D_i, D_{i+1}) so that $\lim\limits_{t \to D_{i+1}^-} \alpha(t) \neq 0$ and since $\alpha : R_{0+} \to R_{0+}$ is continuous then $\alpha(D_{i+1}) \neq 0$ which contradicts that $D_{i+1} \in D_S$. It has been proved that Case 1 is impossible $0 \leq L_i < D_i < D_{i+1} < L_{i+1}$ cannot happen.

Case 2: Assume now that there are two consecutive zeros of $\ddot{I}(t)$ between two consecutive zeros of $\dot{I}(t)$, that is $0 \leq D_i < L_i < L_{i+1} < D_{i+1}$ for some consecutive time instants D_i, D_{i+1} in D_S and some two consecutive time instants L_i, L_{i+1} in L_S. Then, $\alpha(t) \neq 0$ for all $t \in (L_i, L_{i+1})$ since, otherwise, there exists some $t \in (L_i, L_{i+1})$ such that $t \in D_S$, and then the previously claimed constraint $0 \leq D_i < L_i < L_{i+1} < D_{i+1}$ does not hold, and also $\dot{\alpha}(t) \neq -\alpha^2(t) < 0$ for all $t \in (L_i, L_{i+1})$ since, otherwise, there exists some $t \in (L_i, L_{i+1})$ such that $t \in L_S$ and then L_i and L_{i+1} are not two consecutive time instants in L_S as claimed. Also, note that.

$\dot{\alpha}(L_i) + \alpha^2(L_i) = \dot{\alpha}(L_i) + \alpha^2(L_i) = 0$ with $\alpha(L_i) \neq 0$ and $\alpha(L_{i+1}) \neq 0$ since $L_i, L_{i+1} \notin D_S$. But then, by continuity arguments on $\dot{\alpha}(t) + \alpha^2(t)$, there is a change of sign point $t \in (L_i, L_{i+1})$ which zeroes this function which contradicts $\dot{\alpha}(t) \neq -\alpha^2(t) < 0$ for all $t \in (L_i, L_{i+1})$. Then, Case 2 is impossible so that $0 \leq D_i < L_i < L_{i+1} < D_{i+1}$ cannot happen and Property (ii) has been proved. $\square$

Proof of Property (iii). Assume that, contrarily to the statement, $D_1 \leq L_1$. If $L_1 = D_1$ then $\dot{I}(L_1) = \ddot{I}(L_1)$ and the equilibrium point is reached in finite time what is impossible, since $I_0 > 0$, for a non-trivial solution of a continuous-time first-order differential equation with continuous-time parameterization. Then, $L_1 = D_1$ is impossible. Now, assume that $L_1 > D_1$ and $0 = \dot{I}(L_1) = \dot{I}(D_1) + \int_{L_1}^{D_1} \ddot{I}(\tau)d\tau = \int_{L_1}^{D_1} \ddot{I}(\tau)d\tau$ with $\ddot{I}(L_1) = 0$ and then it exists some $L_2 \in (L_1, D_1)$ such that $\ddot{I}(L_2) = 0$ and $L_2 \in L_S$. As a result, there is $D_1 > L_2 > L_1$ and then there are two consecutive undulation time instants what contradicts Property (ii). As a result, $D_1 > L_1$ as claimed. $\square$

Remark 1. *In Theorem 2, note that the sets D_S and L_S have the following properties:*

They are nonempty so that there is at least one $D \in D_S$ such that $\alpha(D) = 0$ implying that $\dot{I}(D) = 0$ and at least one $L \in L_S$ such that $\dot{\alpha}(L) = -\alpha^2(L)$ implying that $\ddot{I}(L) = 0$. Otherwise, the infection could converge asymptotically to zero as time goes to infinity but it would not have finite zeros,

They are sets of zero Lebesgue measure so that they are denumerable discrete sets of strictly ordered isolated real points, for any real numbers,

They fulfill that $cardL_S = cardD_S + \vartheta$ with $\vartheta = \{0, 1\}$ so that they are of either identical finite or infinite cardinal or the cardinal of L_S is finite and exceeds that of D_S by one,

If $\vartheta = 0$ then $card(D_S) = card(L_S) \leq \aleph_0$, that is, if both sets have infinity cardinal or identical finite one then any ordered points of L_S and D_S alternate.

On the other hand, note that:

Equation (4) establishes that D_S is the set of zeros of $\alpha(t)$. At those zeros, the first-time derivative of the infection function is zeroed from (1) without such a function being necessarily zero while on the other hand, Equation (5) is a nonzero real constant for any finite undulation time instant $L_i \leq \bar{L}$ of $I : \mathbf{R}_{0+} \to \mathbf{R}_{0+}$ zeroing the second derivative of the infection function according to (2) which holds if $c = K$ from (5). The fact that (5) is constant follows easily under periodicity conditions of the same or integer multiple/submultiple periods of $g(t)$ and $h(t)$.

Since $\alpha : \mathbf{R}_{0+} \to \mathbf{R}_{0+}$ has no finite zero coincident with a zero of its first time-derivative, by hypothesis, then $g(L_i) = E \Leftrightarrow h(L_i)\dot{g}(L_i) \neq 0$ since $c \neq 0$ from inspection of (8)–(9). This is equivalent to $D_S \cap L_S = \varnothing$, that is, the finite zeros which make zero $\dot{I}(t)$ and which do not make zero $I(t)$ do not make zero either $\ddot{I}(t)$. However, $\ddot{I}(t) = 0$ if $I(t) = \dot{I}(t) = 0$ from (2), provided that $\alpha : \mathbf{R}_{0+} \to \mathbf{R}_{0+}$ is twice everywhere continuously differentiable in $[0, +\infty)$ but this can only happen as time tends to infinity for certain structures of $g(t)$ and $h(t)$. Note that the constraint (5) also implies that the auxiliary functions $g, h : \mathbf{R}_{0+} \to \mathbf{R}_{0+}$ used to define the function $\alpha : \mathbf{R}_{0+} \to \mathbf{R}_{0+}$ in (1) fulfill the constraint $h(L_i)\dot{g}(L_i) \neq ln(g(L_i)/E)\dot{h}(L_i)g(L_i); \forall L_i \in L_S$.

By examining Definitions 1 and 3 and Lemma 1, it turns out that the set L_S of undulation points of $I(t)$ includes but, maybe non-properly, the set of its inflection points. However, it suffices to give some further weak conditions on $\alpha : \mathbf{R}_{0+} \to \mathbf{R}_{0+}$, that is, on $g, h : \mathbf{R}_{0+} \to \mathbf{R}_{0+}$ to guarantee that every undulation point of $I(t)$ is also an inflection point. Some such conditions are discussed in the next corollary.

Corollary 2. *The following properties hold:*

(i)   *Assume that:*

$$\limsup_{\varepsilon \to 0^+} [\theta(L_i + \varepsilon)\theta(L_i - \varepsilon)] < 0; \ \forall L_i \in L_S$$

   where:

$$\theta(t) = h(t)\dot{g}(t) - g(t)\dot{h}(t)ln(g(t)/E); \ \forall t \in \mathbf{R}_{0+}$$

   Then, the set L_S of undulation points of $I(t)$ is the set of its inflection points.

(ii) *Assume that $f, g : \mathbf{R}_{0+} \to \mathbf{R}_{0+}$ are twice continuously differentiable at each undulation point $L_i \in L_S$. Then, the sets of undulation points and that of the inflection points of $I(t)$ coincide if*

$$\frac{h^3(L_i)\left(g(L_i)\,\ddot{g}(L_i) - \dot{g}^2(L_i)\right)}{h^4(L_i)g^2(L_i)} \neq \frac{2}{h^3(L_i)}\left(ln^3\frac{g(L_i)}{E} + \dot{h}(L_i)ln\frac{g(L_i)}{E}\right); \; \forall L_i \in L_S$$

Proof. Note that $\ddot{I}(t) = \left(\alpha^2(t) + \dot{\alpha}(t)\right)I(t)$, $\forall t \in \mathbf{R}_{0+}$ so that $\ddot{I}(L_i \pm \varepsilon) = \left(\alpha^2(L_i \pm \varepsilon) + \dot{\alpha}(L_i \pm \varepsilon)\right)I(L_i \pm \varepsilon)$. Since $L_i > 0$, $g(t)h(t) > 0$ if $t > 0$ and $\lim\limits_{\varepsilon \to 0} I(L_i \pm \varepsilon) = I(L_i)$, since $I(t)$ is continuous, one gets that $\lim\limits_{\varepsilon \to 0^+}\sup\left[\ddot{I}(L_i + \varepsilon)\ddot{I}(L_i - \varepsilon)\right] < 0$ if and only if $\lim\limits_{\varepsilon \to 0^+}\sup[\theta(L_i + \varepsilon)\theta(L_i - \varepsilon)] < 0$. Property (i) has been proved.

On the other hand, if $f, g : \mathbf{R}_{0+} \to \mathbf{R}_{0+}$ are twice continuously differentiable at each undulation point $L_i \in L_S$ of $I(t)$, then $\ddot{f}, \ddot{g}$ exist in L_S. Then, defining $\hat{\alpha}(t) = -c^{-1}\alpha(t) = \frac{ln(g(t)/E)}{h(t)}$; $\forall t \in \mathbf{R}_{0+}$ yields:

$$\dot{\hat{\alpha}}(t) = \frac{h(t)\dot{g}(t) - g(t)\dot{h}(t)ln(g(t)/E)}{h^2(t)g(t)}; \; \forall t \in L_S$$

$$\ddot{\hat{\alpha}}(t) = \frac{h^3(t)\left(g(t)\ddot{g}(t) - \dot{g}^2(t)\right)/g^2(t) + \dot{h}(t)\dot{g}(t)/g(t) - 2h(t)\dot{h}(t)ln(g(t)/E)}{h^4(t)}; \; \forall t \in L_S$$

$$\ddot{I}(t) = \left(\alpha^2(t) + \dot{\alpha}(t)\right)I(t) \Rightarrow \ddot{I}(t) = 0 \text{ with } \alpha^2(t) = -\dot{\alpha}(t) \text{ and } I(t) > 0; \; \forall t \in L_S$$

$$\dddot{I}(t) = \left(\alpha^2(t) + \dot{\alpha}(t)\right)\dot{I}(t) + \left(2\alpha(t)\dot{\alpha}(t) + \ddot{\alpha}(t)\right)I(t) \Rightarrow \dddot{I}(t) = \left(2\alpha(t)\dot{\alpha}(t) + \ddot{\alpha}(t)\right)I(t) = \left(\ddot{\alpha}(t) - 2\alpha^3(t)\right)I(t); \; \forall t \in L_S$$

Since $I(t) > 0$; $\forall t \in \mathbf{R}_{0+}$ then $\dddot{I}(t) \neq 0$; $\forall t \in L_S$ if and only if $\ddot{\alpha}(t) \neq 2\alpha^3(t)$; $\forall t \in L_S$, equivalently, if and only if $\ddot{\hat{\alpha}}(t) \neq 2\hat{\alpha}^3(t)$, $\forall t \in L_S$ which is fully equivalent to the condition of Property (ii). The proof is complete. $\square$

Remark 2. *Note that Theorem 2 applies, in particular, to the case when there are equilibrium points with the initial conditions being distinct from such points. It can be also extended by including the above case by redefining finite discrete sets of the zeros of $\dot{I}(t)$ and $\ddot{I}(t)$ $D_S \to D_S \cap [0, L]$, $L_S \to L_S \cap [0, L]$ for any given $L \in [0, \infty)$ in the sense that the eventual zeros at finite time of $\dot{I}(t)$ and $\ddot{I}(t)$ alternate although an equilibrium points has not still been reached provided that it exists.*

Inspired in Theorem 2, some conditions are discussed in the next result which imply that the first undulation point of the infection evolution function (i.e., the first zero of its second-time derivative) precedes the first zero of its first time-derivative. It is not required that the infection has necessarily a disease-free equilibrium point or that it might be oscillatory leading to successive zeros of its time-derivative along time.

Theorem 3. *Assume that the function $\alpha(t) = -\frac{cln(g(t)/E)}{h(t)}$, where c, $E \in \mathbf{R}_+$ and $g, h : \mathbf{R}_{0+} \to \mathbf{R}_{0+}$ are everywhere continuous and time-differentiable and satisfy the constraints:*

(1) $g(t) < E$; $\forall t \in [0, D)$, $g(D) = E$

(2) $\dot{g}(0) < \left(\dfrac{cln^2(g(0)/E)}{h(0)} - \dfrac{|ln(g(0)/E)|}{h(0)\dot{h}(0)}\right)g(0)$

(3) $g(t) > 0$ and $h(t) > 0$ if $t > 0$

(4) $\dfrac{ln(g(0)/E)}{h(0)} \neq 0$

Assume also that $I_0 > 0$. Then, $min\big(I(t), \dot{I}(t)\big) > 0$; $\forall t \in [0, D)$; $\dot{I}(D) = 0$ and there is some $L \in (0, D)$ such that $\ddot{I}(t) \neq 0$; $\forall t \in [0, L)$ and $\ddot{I}(L) = 0$.

Proof. Note from the definition of $\alpha(t)$, (1), (2) and the given constraints 1 and 2 that

$$\dot{\alpha}(0) = -\frac{c}{h(0)}\left(\frac{1}{h(0)\dot{h}(0)}\left|\ln\frac{g(0)}{E}\right| + \frac{\dot{g}(0)}{g(0)}\right)$$

$\alpha(0) > 0$, since $0 \le g(0) < E$, $\alpha(D) = 0$, since $g(D) = E$), $\alpha^2(0) + \dot{\alpha}(0) > 0$, from the condition 2 since $\alpha(0) > 0$ and since $\alpha : R_{0+} \to R_{0+}$ is continuous and time-differentiable since $g, h : R_{0+} \to R_{0+}$ are everywhere continuous and time-differentiable. Note also that, from the given assumptions and constraints, $min\big(I_0, \dot{I}_0, \ddot{I}_0\big) > 0$ since $I_0 > 0$ by hypothesis, $\dot{I}_0 = \alpha(0) I_0 > 0$ and $\ddot{I}_0 = \big(\alpha^2(0) + \dot{\alpha}(0)\big)I_0 > 0$. Furthermore, $\dot{I}(D) = \alpha(D)I(D) = 0$. From the constraint 3 and the continuity of $g, h : R_{0+} \to R_{0+}$, one has that $\alpha, \dot{\alpha}, \ddot{\alpha} : R_{0+} \to R_{0+}$ are continuous and bounded on $(0, +\infty)$, $\dot{I}(t) > 0$; $\forall t \in [0, D)$ and $\ddot{I}(t) > 0$; $\forall t \in [0, L_0)$ and some $L_0 \in R_+$. Furthermore since $c > 0$ and $\frac{\ln(g(0)/E)}{h(0)} \neq 0$, from the constraint 4, $g(t) < E$; $\forall t \in [0, D)$, from the constraint 1, and $g(t) > 0$ and $h(t) > 0$ if $t > 0$, from the constraint 3. Then $\alpha(t) > 0$; $\forall t \in [0, D)$. Since $g, h : R_{0+} \to R_{0+}$ are continuous and positive on any bounded interval $[0, T)$ then $\alpha(t)$ is positive and finite on $[0, D)$. It is now proved that $t = D$ is the first zero of $\dot{I}(t)$. Assume that this is not the case so that there is some $D_1 < D$ such that $\dot{I}(D_1) = 0$, with $\alpha(D_1) \neq 0$, and $\dot{I}(t) > 0$; $\forall t \in [0, D_1)$. Then $I(D_1) = \dot{I}(D_1) = \dot{I}(D_1) = 0$ from (2) and the infection extinguishes in a finite time $D_1 < D$. This leads to a contradiction since $I(D_1) = I_0 + \int_0^{D_1} \dot{I}(\tau)d\tau > 0$ since $I_0 > 0$ and $\dot{I}(t) > 0$; $\forall t \in [0, D_1)$. Therefore, if $D_1 < D$ such that $\dot{I}(D_1) = 0$ then $I(D_1) > 0$. But then $\alpha(D_1) = \dot{I}(D_1)/I(D_1) = 0$ from (1) which contradicts that $\alpha(t) \neq 0$; $\forall t \in [0, D)$. As a result, $t = D$ is the first zero of $\dot{I}(t)$ and there is no $D_1 < D$ such that $I(D_1) = 0$. Since $I, \dot{I} : R_{0+} \cap [0, D] \to R_{0+}$ are continuous with $\dot{I}(t) > 0$; $\forall t \in [0, D)$ and $\dot{I}(D) = 0$ and $\ddot{I}(t)$; $\forall t \in [0, L_0)$ and some $L_0 \in R_+$ then there is some $L \in (0, D)$ such that $\ddot{I}(L) = 0$. Assume that this is not the case. Then, $0 = \dot{I}(D) = \dot{I}_0 + \int_0^D \ddot{I}(\tau)d\tau > 0$. Hence, a contradiction arises. Thus, there is some $L \in (0, D)$ such that $\ddot{I}(L) = 0$. $\square$

Remark 3. *Note that, under all the conditions of Theorem 3, $\alpha(t) > 0$; $\forall t \in [0, D)$ and $\alpha(D) = 0$. Furthermore, the first zero of $\dot{I}(t) = 0$ occurs at $t = D$, there is no $t < D$ such that $I(t) = \dot{I}(t) = 0$ and there is some $L < D$ such that $\ddot{I}(L) = 0$.*

The following example describes the basic model proposed in [11] under a first-order differential equation for the infection evolution without any entropy considerations at this stage:

Example 1. *The function $\alpha(t) = -c\ln(t/D)/t$, for some $D > 0$, proposed in [11] satisfies all the conditions of Theorem 3 with $h(t) = g(t) = t$ and $E = D$. It satisfies, in addition, that $\alpha(0) = +\infty$. This function satisfies also the given further conditions of Theorem 2 $g(0) = h(0) = 0$ with $\lim\limits_{t \to 0} \frac{\ln(g(t)/E)}{h(t)} \le -\varepsilon$.*

Note that the condition $\alpha(0) > 0$ of Theorem 3 avoids that $\dot{I}_0 = 0$ if $I_0 \neq 0$ so that $t = 0$ is a zero of $\dot{I}(t)$.

It can be argued that the proposed basic model (1) is a very simple time-varying differential equation of first-order which describes the infective population time-evolution. Note that the use of appropriate particular structures in the definition of the time-varying coefficient $\alpha(t)$ can take care of the eventual incorporation of the necessary supplementary environment information to make such an equation well-posed to practically describe a concrete disease evolution through time. The incorporation which can be incorporated is the eventual couplings of the infectious subpopulation with another ones (such as the susceptible, recovered or vaccinated subpopulations and their associated dynamics) or the information about the feedback information controls in more elaborated models. The next section develops some work in this direction.

3. Further Examples of Linking the Basic Model to Some Existing Epidemic Models Incorporating Other Subpopulations

The infection description via (1) assumes implicitly that it has a first-order dynamics. It has been argued that $\alpha(t)$ in (1) contains the information about the controls and other coupled subpopulations influencing the disease evolution through time. It can be of interest to discuss its application to infection descriptions described by differential equations of orders higher than one which is a very common situation in disease transmission mathematical models.

It is now seen how a well-known epidemic model can be also discussed under the point of view of Theorem 3. In the subsequent example, the above characterization, based on the first zero of infection evolution time-derivative and on the undulation point of the infection evolution, is used for a model with three subpopulations via an appropriate choice of $g(t)$ and $h(t)$ in the definition of $\alpha(t)$.

Example 2. *Consider the following SIR model without demography [30]:*

$$\dot{S}(t) = -\beta S(t)I(t); \; \dot{I}(t) = (\beta S(t) - \gamma)I(t); \; \dot{R}(t) = \gamma I(t); \; \forall t \in \mathbf{R}_{0+} \tag{10}$$

where $S(t)$, $I(t)$ and $R(t)$ are, respectively, the susceptible, infectious and recovered (or immune) subpopulations, under nonzero initial conditions being subject to $\min(S(0), I(0), R(0)) \geq 0$, where β is the coefficient transmission rate and γ is the removal or recovery rate (its inverse γ^{-1} being the average infectious period). The mathematical study of this model and their variants is not easy as seen in [30,40]. First, note that the total population $N(t) = S(t) + R(t) + I(t) = S_0 + R_0 + I_0$; $\forall t \in \mathbf{R}_{0+}$ is constant for all time. The basic reproductive ratio (or reproduction number) is $R_* = \beta/\gamma$ and, if $S_0 \leq R_*^{-1}$, then $\dot{I}_0 \leq 0$ while if $S_0 > R_*^{-1}$, it becomes endemic for all time since $\dot{I}_0 > 0$. The solution of (10) becomes in closed form:

$$S(t) = e^{-\beta \int_0^t I(\tau)d\tau} S_0; \; I(t) = e^{\int_0^t (\beta S(\tau) - \gamma)d\tau} I_0; \; R(t) = S_0 + R_0 + I_0 - S(t) - I(t); \; \forall t \in \mathbf{R}_{0+} \tag{11}$$

Note that by combining the above equations that:

$$S(t) = e^{-\beta I_0 \int_0^t e^{\int_0^\tau (\beta S(\sigma) - \gamma)d\sigma} d\tau} S_0; \; I(t) = e^{\int_0^t (\beta e^{-\beta \int_0^\tau I(\sigma)d\sigma} S_0 - \gamma)d\tau} I_0 \tag{12}$$

Note from (11) that $S : \mathbf{R}_{0+} \to \mathbf{R}_{0+}$ is non-increasing so that there exists a susceptible equilibrium subpopulation $S_e = \lim_{t\to\infty} S(t) \leq S_0$ for any given non-negative initial conditions. Note also from (10) that $\dot{N}(t) = 0$ and then $N(t) = N_0$; $\forall t \in \mathbf{R}_{0+}$ Note that If $I_0 = 0$ then $I(t) = 0$, $S(t) = S_0$ and $R(t) = R_0 = N_0 - S_0$; $\forall t \in \mathbf{R}_{0+}$. We examine three cases for $I_0 > 0$:

Case (a) if $S_0 < R_*^{-1}$ then $S(t) \leq S_0$ and $\beta S(t) - \gamma < 0$; $\forall t \in \mathbf{R}_{0+}$, then $I(t) \to 0, S(t) \to S_e$ and $R(t) \to R_e = N_0 - S_e$ as $t \to \infty$. Since $S : \mathbf{R}_{0+} \to \mathbf{R}_{0+}$ is non-increasing, $S_e \leq S_0 < R_*^{-1}$. This implies that $\lim_{t\to\infty} \int_0^t (\beta S(\tau) - \gamma)d\tau = -\infty$ and $\dot{I}(t) = -\lambda(t)I(t) \leq -\lambda_a I(t)$, $I(t) \to 0$ at exponential rate as $t \to \infty$ for some $\lambda_a > 0$ from (10) and (11) since $I_0 - I(t) \geq \lambda_a \int_0^t I(\tau)d\tau$ so that $\int_0^\infty I(\tau)d\tau \leq I_0/\lambda_a < +\infty$. Then, $I : \mathbf{R}_{0+} \to \mathbf{R}_{0+}$ is integrable on $[0, \infty)$. Thus, $C = \beta \int_0^\infty I(t)dt < +\infty$ so that $S_e = e^{-\beta \int_0^\infty I(t)dt} S_0 = e^{-C} S_0 > 0$ (then there is a nonzero susceptible equilibrium level) and $R_e = N_0 - S_e < N_0$.

Case (b) if $S_0 = R_*^{-1}$ then $S(t) \to S_e \leq S_0 = \gamma/\beta$ as $t \to \infty$ since $S : \mathbf{R}_{0+} \to \mathbf{R}_{0+}$ is non-increasing and then it converges to S_e satisfying $0 \leq S_e \leq S_0$. By inspection of the second equation of (11), it also follows that $I(t) \to I_e$ and $R(t) \to R_e$ as $t \to \infty$ satisfying $I_e \geq 0$ and $R_e \geq 0$. Assume that $I_e > 0$ then $S_e = 0$ from the first equation of (11). But if $S_e = 0$ then $I_e = 0$ since then $I : \mathbf{R}_{0+} \to \mathbf{R}_{0+}$ is strictly decreasing on $[t_a, \infty)$ for some finite $t_a > 0$ from the second equation of (11). Hence, a contradiction to

$I_e > 0$ follows implying that $I_e = 0$ if $S_e = 0$. Now, assume that $\gamma/\beta > S_e > 0$. Then, from the second equation of (11), $I(t) \to I_e = 0$ as $t \to \infty$. But then $S_e > 0$, from the first equation of (12), since $\gamma/\beta > S_e$ if $I_0 > 0$ and then $R_e = N_0 - S_e$. From the second equation of (12) and, under a similar reasoning as that of Case a, $I : \mathbf{R}_{0+} \to \mathbf{R}_{0+}$ is integrable on $[0, \infty)$ and $S_e > 0$. In summary, if $S_0 = R_*^{-1} = \gamma/\beta$ and $I_0 > 0$ then $I(t) \to 0, S(t) \to 0$ and $R(t) \to N_0 = S_0 + R_0 + I_0$ as $t \to \infty$ in the same way as in Case a if $S_0 \leq R_*^{-1}$.

Case (c) if $S_0 > R_*^{-1}$ then $\dot{I}_0 > 0$ from (10) and $S : \mathbf{R}_{0+} \to \mathbf{R}_{0+}$ is increasing on some interval $[0, t_0]$. The fact that $I : \mathbf{R}_{0+} \to \mathbf{R}_{0+}$ is strictly increasing on some initial time interval is of interest from the point of view of hospital management of availability of beds and other sanitary specific means in the event that the disease might have a relevant number of seriously infected individuals. Since $S : \mathbf{R}_{0+} \to \mathbf{R}_{0+}$ is non-increasing then either $I(t) \to I_e = S_0 + I_0 + R_0 = N_0$, $S(t) \to S_e = 0$ and $R(t) \to R_e = 0$ as $t \to \infty$ or $S(t) \to S_e \in \left(0, R_*^{-1}\right]$ as $t \to \infty$ from (11) since $S : \mathbf{R}_{0+} \to \mathbf{R}_{0+}$ is non-increasing. The firs possibility $I(t) \to I_e = N_0$ is unfeasible since from the first equation of (11) $I(t) \to \infty$ as $t \to \infty$. Then, $S(t) \to S_e \in \left(0, R_*^{-1}\right]$ as $t \to \infty$. Now, first, assume that $S_e \in \left(\gamma/\beta, R_*^{-1}\right]$. Then, from the first equation of (12), $\dot{S}(t) \to 0$ as $t \to \infty$. Then, $S_e = 0$ which contradicts that $S_e > \gamma/\beta$, As a result, $0 \leq S_e \leq \gamma/\beta$. Now, assume that $S_e = 0$. Then, from (11), $I(t) \to I_e = 0$ and $I : \mathbf{R}_{0+} \to \mathbf{R}_{0+}$ being square-integrable, and following a similar argument as that of Cases a–b, one again concludes that $S_e > 0$ so that $S_e \in (0, \gamma/\beta]$ and $R_e = N_0 - S_e$, as a result. But, since $S_e \leq \gamma/\beta$ then $I_e = 0$ from (11) since $I : \mathbf{R}_{0+} \to \mathbf{R}_{0+}$ is strictly decreasing after some finite time instant t_0 and integrable on $[0, \infty)$ and a following again the reasoning of Cases a–b, one concludes that $S_e > 0$. As a result, if $S_0 > R_*^{-1}$ and $I_0 > 0$, then $I_e = 0$, $S_e > 0$ and $R_e = N_0 - S_e$. Thus, the relevant conclusions on the disease- free equilibrium point which is a disease- free one are similar for the three above cases.

On the other hand, since $S : \mathbf{R}_{0+} \to \mathbf{R}_{0+}$ it exists a finite $t = D > 0$ such that $S(D) = R_*^{-1} = \gamma/\beta$ and $\dot{I}(D) = \alpha(D)I(D) = (\beta S(D) - \gamma)I(D) = 0, I(D) = e^{\int_0^D (\beta S(\tau) - \gamma)d\tau} I_0 \neq 0$, if $I_0 \neq 0$ and, furthermore,

$$\begin{aligned}
\ddot{I}(D) &= \left(\beta \dot{S}(D) - \gamma\right)I(D) + (\beta S(D) - \gamma)\dot{I}(D) \\
&= \left(\beta \dot{S}(D) - \gamma\right)I(D) = -\beta^2 S(D)I^2(D) - \gamma I(D) \\
&= -\gamma\left(\beta e^{\int_0^D (\beta S(\tau) - \gamma)d\tau} I_0 + 1\right)e^{\int_0^D (\beta S(\tau) - \gamma)d\tau} I_0 < 0
\end{aligned} \tag{13}$$

and also:

$$\ddot{I}_0 = -\beta^2 S_0 I_0^2 + (\beta S_0 - \gamma)\dot{I}_0 = I_0\left[(\beta S_0 - \gamma)^2 - \beta^2 S_0 I_0\right] \tag{14}$$

and $\ddot{I}_0 > 0$ under the reasonable assumption that I_0 is sufficiently small (the initial numbers of infectious is usually very small in practice) satisfying $I_0 < \frac{(\beta S_0 - \gamma)^2}{\beta^2 S_0}$. As a result, there is some time instant $L \in (0, D)$ such that $\ddot{I}(L) = 0$ so that it is an undulation point of $I : \mathbf{R}_{0+} \to \mathbf{R}_{0+}$. As a result, we find that if the basic reproduction number exceeds unity then the infection curve corresponding to the endemic solution has a minimum at a larger time instant that the one defining its undulation point. That situation corresponds to the situation of small initial infection force with reproduction number greater than one. On the other hand, if $\ddot{I}_0 \leq 0$, then $\dot{I}_0 > 0$ does not hold.

Comparing the infectious subpopulation evolution to (1) and the structure of the function in Theorem 3 yields:

$$\alpha(t) = \beta S(t) - \gamma = -\frac{c\ln(g(t)/E)}{h(t)} \tag{15}$$

$$\dot{\alpha}(t) = \beta \dot{S}(t) = -c\frac{d}{dt}\left(\frac{\ln(g(t)/E)}{h(t)}\right) \tag{16}$$

$$= -\beta^2 S(t)I(t) = -\frac{c}{h(t)}\left(\frac{1}{h(t)\dot{h}(t)}\left|\ln\frac{g(t)}{E}\right| + \frac{\dot{g}(t)}{g(t)}\right); \tag{17}$$

$\forall t \in \mathbf{R}_{0+}$. If one defines $g(t) = t$; $\forall t \in \mathbf{R}_{0+}$ and $h(t) = \frac{c\ln(t/E)}{\gamma - \beta S(t)}$; $\forall t \in \mathbf{R}_{0+}$, then $h(t) = \frac{c|\ln(t/E)|}{\beta S(t) - \gamma}$; $\forall t \in \mathbf{R}_{0+}$. It is easy to verify that these functions satisfy the conditions of Theorem 3.

In the case when the reproduction number is less than unity and it is an upper-bound of the normalized susceptible population, each primary infection generates, in average, less than one secondary one so that the infection extinguishes asymptotically. According to this particular model, also the susceptible subpopulation extinguishes asymptotically. See Case a referred to (11). Thus, the disease-free equilibrium point is $\left(S^*_{df}, I^*_{df}, R^*_{df}\right)^T = (0, 0, N)^T$. In this case, $I(t), \dot{I}(t), \ddot{I}(t) \to 0$ as $t \to \infty$ but there are no finite time instants of minimum and undulation of the infectious curve to the light of Theorem 3.

However, we can have a practical visualization of the disease removal by defining a design quadruple $(k_1, k_2, k_3, \varepsilon) \in \mathbf{R}^4_+$ and the following cut associate time instants:

$$t_{Ii}(k_i, \varepsilon) = min\left(\tau \in \mathbf{R}_{0+} : \left|\frac{dI^{(i-1)}}{dt}\right| \le k_i \varepsilon : t \in [\tau, +\infty)\right); \ i = 1, 2, 3 \tag{18}$$

Note that $t_{I2}(k_2, \varepsilon)$ and $t_{I3}(k_3, \varepsilon)$ generalize the roles of the time instants D and L, that is, the finite minimum infection and undulation time instants, respectively, within prescribed margins when those time instants do not exist.

Example 3. *Consider Case a of Example 2 so that $S(t) \le S_0 < \gamma/\beta$ leading to $I(t) \to 0, S(t) \to S_e > 0$ and $R(t) \to R_e = N_0 - S_e$ as $t \to \infty$ and $I(t) > 0, \dot{I}(t) < 0$ and $\ddot{I}(t) < 0$ are strictly decreasing on $[0, +\infty)$. Take prescribed constants $\varepsilon \in (0,1)$ $k_i \ge 1$ for $i = 1,2,3$. The solution trajectory converges to the disease-free equilibrium point at exponential rate. Then, one gets by combining (10)–(12) and (18) that:*

$$\left|\int_0^{t_{I1}} (\gamma - \beta S(\tau))d\tau\right| = \left|\int_0^{t_{I1}} \left(\gamma - \beta e^{-\beta \int_0^\sigma I(\sigma)d\sigma} S_0\right)d\tau\right| \le \ln I_0 - \ln k_1 + |\ln \varepsilon|; \ \forall t \in \mathbf{R}_{0+} \tag{19}$$

$$\left(\gamma - \beta e^{-\beta \int_0^{t_{I2}} I(\tau)d\tau} S_0\right)e^{-\int_0^{t_{I2}} (\gamma - \beta e^{-\beta \int_0^\tau I(\sigma)d\sigma} S_0)d\tau} I_0 \le k_2 \varepsilon; \ \forall t \in \mathbf{R}_{0+} \tag{20}$$

$$\left[\beta^2 S(t)I(t) - (\beta S(t) - \gamma)^2\right]I(t) \le k_3 \varepsilon, \ \forall t \in \mathbf{R}_{0+} \tag{21}$$

implying that:

$$t_{I1} = min\left(t \in \mathbf{R}_{0+} : \gamma t - \beta S_0 \int_0^t e^{-\beta \int_0^\sigma I(\sigma)d\sigma} d\tau = \ln I_0 - \ln k_1 + |\ln \varepsilon|\right) \ge \frac{1}{\gamma}(\ln I_0 - \ln k_1 + |\ln \varepsilon|) \tag{22}$$

$$\begin{aligned} 2min(k_2 \varepsilon I_0, \beta S_0)e^{-\beta \int_0^{t_{I2}} I(\tau)d\tau} \\ \le \eta(t_{I2}) = k_2 \varepsilon e^{\int_0^{t_{I2}} (\gamma - \beta S(\tau))d\tau} I_0 + \beta e^{-\beta \int_0^{t_{I2}} I(\tau)d\tau} S_0 \\ \le (k_2 \varepsilon I_0 + \beta S_0)e^{\int_0^{t_{I2}} (\gamma - \beta S(\tau))d\tau} I_0 \end{aligned} \tag{23}$$

which leads to:

$$\begin{aligned} e^{\int_0^{t_{I2}} (\gamma - \beta S(\tau) + \beta I(\tau))d\tau} \ge \frac{2min(k_2 \varepsilon I_0, \beta S_0)}{(k_2 \varepsilon I_0 + \beta S_0)I_0} \\ \Rightarrow t_{I2} \ge max\left(t > 0 : \int_0^t (\gamma - \beta S(\tau) + \beta I(\tau))d\tau\right) = \ln\left[\frac{2min(k_2 \varepsilon I_0, \beta S_0)}{(k_2 \varepsilon I_0 + \beta S_0)I_0}\right] \end{aligned} \tag{24}$$

$$\begin{aligned} e^{\int_0^{t_{I2}} (\beta S(\tau) - \beta I(\tau) - \gamma)d\tau} \le \frac{(k_2 \varepsilon I_0 + \beta S_0)I_0}{2min(k_2 \varepsilon I_0, \beta S_0)} \\ \Rightarrow t_{I2} < min\left(t > 0 : \int_0^t (\beta S(\tau) - \beta I(\tau) - \gamma)d\tau\right) = \ln\left[\frac{(k_2 \varepsilon I_0 + \beta S_0)I_0}{2min(k_2 \varepsilon I_0, \beta S_0)}\right] \end{aligned} \tag{25}$$

and:

$$-k_3 \varepsilon \le \ddot{I}(t) = (\beta S(t) - \gamma)\dot{I}(t) + \beta \dot{S}(t)I(t) = \left[(\gamma - \beta S(t))^2 - \beta^2 S(t)I(t)\right]I(t) \le k_3 \varepsilon$$

what implies that $\left|\ddot{I}(t)\right| \le k_3\varepsilon$; $\forall t \in [t_{13}, \infty)$ *such that:*

$$t_{13} \ge max\left(t > 0 : \left[(\gamma - \beta S(t))^2 - \beta^2 S(t)I(t)\right]I(t)\right) \ge -k_3\varepsilon,$$

$$t_{13} \le min\left(t > 0 : \left[(\gamma - \beta S(t))^2 - \beta^2 S(t)I(t)\right]I(t)\right) \le k_3\varepsilon$$

Example 4. *Consider the following SIS model with vaccination and antiviral or antibiotic controls:*

$$\dot{S}(t) = \gamma I(t) - \beta S(t)I(t) - k_V S(t); \quad \dot{I}(t) = (\beta S(t) - \gamma - k_T)I(t); \quad \forall t \in \mathbf{R}_{0+} \tag{26}$$

subject to $S(0) = S_0$, $I(0) = I_0$ *with* $min(S_0, I_0) \ge 0$ *where the vaccination and treatment feedback controls on the susceptible and infectious are, respectively,* $V(t) = k_V S(t)$ *and* $T(t) = k_T I(t)$ *with* $min(k_V, k_T) \ge 0$. *If it is assumed that the total population* $N(t) = N_0 = S_0 + I_0$; $\forall t \in \mathbf{R}_{0+}$ *is constant through time then there is a complementary recovered (or immune) subpopulation present which obeys the differential equation* $\dot{R}(t) = k_V S(t) + k_T I(t)$ *with* $R(0) = R_0 = 0$. *The solution is:*

$$\begin{aligned} S(t) &= e^{-\int_0^t (\beta I(\tau)+k_V)d\tau} S_0 + \gamma \int_0^t e^{-\int_\tau^t (\beta I(\sigma)+k_V)d\sigma} I(\tau)d\tau \\ &= e^{-k_V t} S_0 - \int_0^t e^{-k_V(t-\tau)}(\beta S(\tau) - \gamma)I(\tau)d\tau \end{aligned} \tag{27}$$

$$I(t) = e^{\beta \int_0^t S(\tau)d\tau} e^{-(\gamma + k_T)t} I_0 \tag{28}$$

$$R(t) = \int_0^t (k_V S(\tau) + k_T I(\tau))d\tau \tag{29}$$

The following result links the above SIS model with a complementary recovered subpopulation to the generic one (1) under a minimum number of initial susceptible and sufficiently large number of initial infectious with initial growing rate.

Theorem 4. *Assume that* $S_0 > \frac{\gamma+k_T}{\beta}$, $I_0 < 1 + \frac{1}{\gamma}(k_T + k_V S_0)$ *and* $\dot{I}_0 > \frac{\beta|\dot{S}_0|I_0}{\beta S_0 - \gamma - k_T}$. *Then, the following properties hold:*

(i) $\dot{S}_0 < 0$ *and* $\ddot{I}_0 > 0$,

(ii) $S(t)$ *is strictly decreasing on* $[0, t_{Smin}]$ *with* $t_{Smin} = min(t \in \mathbf{R}_{0+} : S(t) = \gamma/\beta)$,

(iii) $I(t)$ *is strictly increasing on* $[0, t_{Imax}]$, *and* $I_{max} = I(t_{max}) = max(I(t) : t \in [0, t_{Imax}], t_{Imax} = min(t \in \mathbf{R}_{0+} : S(t) = (\gamma + k_T)/\beta))$ *with* $t_{Imax} \ge t_{Smin}$,

(iv) *There is* $t_{und} < t_{Imax}$ *which is an undulation and, furthermore, strict inflection time instant of* $I(t)$,

(v)   *Assume, in addition, that* I_0 *is large enough to satisfy* $I_0 > \frac{(\gamma+k_T)k_V}{(\gamma-\beta(\gamma+k_T))} e^{-\beta \int_0^{t_{Imax}} S(\tau)d\tau} e^{(\gamma + k_T)t_{Imax}}$. *Then, the epidemic model (26) can be written in the form (1) on* $[0, t_{Imax}]$ *with the following function* $\alpha : [0, t_{Imax}] \to \mathbf{R}_{0+}$:

$$\alpha(t) = \beta\left(e^{-\int_0^t (\beta I(\tau)+k_V)d\tau} S_0 + \gamma \int_0^t e^{-\int_\tau^t (\beta I(\sigma)+k_V)d\sigma} I(\tau)d\tau\right) - \gamma - k_T; \quad t \in [0, t_{Imax}] \tag{30}$$

which is of the form $\alpha(t) = -\frac{c\ln(g(t)/E)}{h(t)}$ *with* $g : [0, t_{Imax}] \to [0, E]$; $\forall t \in [0, t_{Imax}]$ *and any given* $E \in \mathbf{R}_+$ *and* $h(t) = \dfrac{c|\ln(g(t)/E)|}{\beta\left(e^{-\int_0^t (\beta I(\tau)+k_V)d\tau} S_0 + \gamma \int_0^t e^{-\int_\tau^t (\beta I(\sigma)+k_V)d\sigma} I(\tau)d\tau\right) - \gamma - k_T}$; $\forall t \in [0, t_{Imax}]$.

(vi) *The equilibrium points are* $S_1^* = I_1^* = 0$, $R_1^* = N_0$ *if* $k_V \ne 0$ *and* $k_T \ge 0$, *and* $S_2^* = \frac{\gamma+k_T}{\beta}$, $I_2^* = 0$ *and* $R_2^* = N_0 - \frac{\gamma+k_T}{\beta}$ *which is only reachable if* $k_V = 0$ *since, otherwise,* $I_2^* = -\frac{k_V}{k_T}\frac{\gamma+k_T}{\beta} < 0$.

Proof. Since $S_0 > \frac{\gamma+k_T}{\beta}$ and $I_0 < 1 + \frac{1}{\gamma}(k_T + k_V S_0)$ then $\dot{I}_0 > 0$ and $\dot{S}_0 < 0$. Also, $\ddot{I}_0 = \beta S_0 I_0 + (\beta S_0 - \gamma - k_T)\dot{I}_0 = (\beta S_0 - \gamma - k_T)\dot{I}_0 - \beta|\dot{S}_0|I_0 > 0$ if $\dot{I}_0 > \frac{\beta|\dot{S}_0|I_0}{\beta S_0 - \gamma - k_T}$. Property (i) has been proved. Furthermore, $S_0 > \frac{\gamma+k_T}{\beta} \geq \frac{\gamma}{\beta}$ implies from (27) that $S(t)$ is strictly decreasing on $[0, t\prime]$ where $t\prime = min(t \in \boldsymbol{R}_{0+} : S(t) = \gamma/\beta)$ what proves Property (ii) with $t_{Smin} = t\prime$. On the other hand and since $S : \boldsymbol{R}_{0+} \to \boldsymbol{R}_{0+}$ is continuous, there exists some $t'' \in [0, t\prime]$ such that $S(t'') = \frac{\gamma+k_T}{\beta}$ with $t'' = t\prime$ if and only if $k_T = 0$. From (26), $\dot{I}(t'') = 0$ and $\dot{I}(t) > 0$ for $t \in [0, t'')$ since $\dot{I}_0 > 0$. On the other hand, one has from (26) and (28) that:

$$\ddot{I}(t'') = (\beta S(t'') - \gamma - k_T)\dot{I}(t'') + \beta\dot{S}(t'')I(t'')$$

$$= \beta[\beta(\gamma - \beta S(t''))I(t'') - k_V S(t'')]I(t'')$$

$$= -[\beta^2 k_T I(t'') + k_V(\gamma + k_T)]I(t'')$$

$$= -[\beta^2 k_T e^{\beta \int_0^t S(\tau)d\tau} e^{-(\gamma+k_T)t}I_0 + k_V(\gamma + k_T)]e^{\beta \int_0^{t''} S(\tau)d\tau} e^{-(\gamma+k_T)t''}I_0 < 0$$

and $I(t)$ has a relative maximum I_{max} at $t = t'' = t_{Imax}$ which is also the absolute maximum on $[0, t_{max}]$. Property (iii) has been proved. Note also that since $\ddot{I}(t)$ is continuous and $\ddot{I}_0 > 0$, there exists some $t_{und} < t''$ such that t_{und} is an undulation point of $I(t)$. Note furthermore that

$$\ddot{I}(t_{und}) = (\beta S(t_{und}) - \gamma - k_T)\dot{I}(t_{und}) + \beta\dot{S}(t_{und})I(t_{und}) = 0$$

From Lemma 1(i), $\ddot{I}(t_{und} - \varepsilon)\ddot{I}(t_{und} + \varepsilon) < 0; \forall \varepsilon \in \boldsymbol{B}(0, r)$ and some $r \in \boldsymbol{R}_+$ implies that t_{und} is also an inflection time instant of $I(t)$. The equivalent logic contrapositive proposition establishes that:

$$\left[\forall r \in \boldsymbol{R}_+, \exists \varepsilon \in [0, r] : \ddot{I}(t_{und} - \varepsilon)\ddot{I}(t_{und} + \varepsilon) \geq 0\right] \Rightarrow [t_{und} \text{ is not an inflection time instant of } I(t)]$$

Then, if $\ddot{I}(t_{und} - \varepsilon)\ddot{I}(t_{und} + \varepsilon) < 0; \forall \varepsilon \in \boldsymbol{B}(0, r)$ and some $r \in \boldsymbol{R}_+$ then t_{und} is in fact an inflection time instant of $I(t)$. Assume that there is some arbitrarily small $\varepsilon \in \boldsymbol{R}_+$ such that $\ddot{I}(t_{und} - \varepsilon)\ddot{I}(t_{und} + \varepsilon) \geq 0$
Then:
$\dot{I}(t_{und} + \varepsilon) = \dot{I}(t_{und}) + \int_0^\varepsilon \ddot{I}(t_{und} + \tau)d\tau; \dot{I}(t_{und} - \varepsilon) = \ddot{I}(t_{und}) + \int_0^{-\varepsilon} \ddot{I}(t_{und} + \tau)d\tau$.
Since $\ddot{I}(t)$ is continuous on $[t_{und} - \varepsilon, t_{und} + \varepsilon]$ and one gets that

$$\dot{I}(t_{und} + \varepsilon) - \dot{I}(t_{und} - \varepsilon) = \int_0^\varepsilon \ddot{I}(t_{und} + \tau)d\tau - \int_0^{-\varepsilon} \ddot{I}(t_{und} + \tau)d\tau$$

It is known that $0 < \varepsilon_I \leq \dot{I}(t_{und}) < \dot{I}_0$ so that, for some arbitrarily small $\varepsilon \in \boldsymbol{R}_+$ such that $\ddot{I}(t_{und} - \varepsilon)\ddot{I}(t_{und} + \varepsilon) \geq 0$, there are $\varepsilon_1 \in [0, \varepsilon]$ and $\varepsilon_2 \in \boldsymbol{R}_+$ with $-\varepsilon_2 \in [-\varepsilon, 0]$ such that the following joint constraints hold:

(1) $\dot{I}(t_{und} + \tau) > 0; \forall \tau \in [-\varepsilon_2, \varepsilon_1] \subset [-\varepsilon, \varepsilon]$ with $\dot{I}(t)$ being strictly increasing on $[-\varepsilon_2, \varepsilon_1]$.
(2) $\int_0^{\varepsilon_1} \ddot{I}(t_{und} + \tau)d\tau = \int_0^{-\varepsilon_2} \ddot{I}(t_{und} + \tau)d\tau$

Then, one gets from Condition 2 that:

$$\dot{I}(t_{und} + \varepsilon_1) - \dot{I}(t_{und} - \varepsilon_2) = \int_0^{\varepsilon_1} \ddot{I}(t_{und} + \tau)d\tau - \int_0^{-\varepsilon_2} \ddot{I}(t_{und} + \tau)d\tau = 0$$

so that $\dot{I}(t)$ is not strictly increasing on $[-\varepsilon_2, \varepsilon_1]$, hence a contradiction. As a result, the undulation time instant t_{und} of $I(t)$ is also a strict inflection time instant of $I(t)$ since $\ddot{I}(t_{und}) \neq 0$ since Lemma 1 (ii) holds and the first zero of $\dot{I}(t)$ occurs at $t_{Imax} > t_{und}$. Property (iv) has been proved. To prove Property (v), note that Equation (30) follows from (26)–(27). Now, we equalize (30) to (1) to get admissible functions $g, h : R_{0+} \to R_{0+}$ leading to:

$$\alpha(t) = \beta\left(e^{-\int_0^t (\beta I(\tau)+k_V)d\tau} S_0 + \gamma \int_0^t e^{-\int_\tau^t (\beta I(\sigma)+k_V)d\sigma} I(\tau)d\tau\right) - \gamma - k_T = -\frac{c\ln(g(t)/E)}{h(t)} \tag{31}$$

and note that $\alpha(0) = \beta S_0 - \gamma - k_T > 0$. Note also that $\alpha(0) = \frac{+\infty}{h(0)}$ from the use of (31) in (30) implies that $h(0) = 0$ irrespective of $g(t)$ while $g(t)$ is chosen arbitrary and continuous time-differentiable subject to $g(0) = 0$ and $\alpha(t_{Imax}) = 0$, $g(t_{Imax}) = E$ (so that $\ln(g(t_{Imax})/E) = 0$) with $h(t) = \frac{c/E}{\beta\gamma I(t) - \beta(\beta I(t) + k_V)S(t)}$ for $t \in [0, t_{Imax}]$.

Now, note that $h(t_{Imax})$ is a primary $(0/0)$—type indetermination which is resolved through L´Hôpital rule leading to:

$$h(t_{Imax}) = \frac{c/g(t_{Imax})}{\beta \dot{S}(t_{Imax})} = \frac{c/E}{\beta\gamma I(t_{Imax}) - \beta(\beta I(t_{Imax}) + k_V)S(t_{Imax})}$$
$$= \frac{c/(\beta E)}{\gamma I(t_{Imax}) - (\beta I(t_{Imax}) + k_V)(\gamma + k_T)}$$

Since $I(t_{Imax}) = e^{\beta\int_0^{t_{Imax}} S(\tau)d\tau} e^{-(\gamma+k_T)t} I_0$ then for sufficiently large I_0 such that

$$I_0 > \frac{(\gamma + k_T)k_V}{(\gamma - \beta(\gamma + k_T))} e^{-\beta\int_0^{t_{Imax}} S(\tau)d\tau} e^{(\gamma+k_T)t_{Imax}}$$

then:

$$h(t) = \frac{c\left|\ln(g(t)/E)\right|}{\beta\left(e^{-\int_0^t (\beta I(\tau)+k_V)d\tau} S_0 + \gamma\int_0^t e^{-\int_\tau^t (\beta I(\sigma)+k_V)d\sigma} I(\tau)d\tau\right) - \gamma - k_T}$$
$$= \frac{c\ln(g(t)/E)}{\gamma + k_T - \beta\left(e^{-k_V t}S_0 - \int_0^t e^{-k_V(t-\tau)}(\beta S(\tau) - \gamma)I(\tau)d\tau\right)}$$

fulfilling, in particular:

$$h(t_{Imax}) = \frac{c/(\beta E)}{(\gamma - \beta(\gamma + k_T))It_{Imax} - (\gamma + k_T)k_V}$$
$$= \frac{c/(\beta E)}{(\gamma - \beta(\gamma + k_T))e^{\beta\int_0^{t_{Imax}} S(\tau)d\tau}e^{-(\gamma+k_T)t_{Imax}}I_0 - (\gamma + k_T)k_V} > 0$$

Property (v) has been proved. Property (vi) is obvious by zeroing (26). □

Example 4 is tested numerically in the sequel with the following data $\beta = 30$, $\gamma = 50$ years^{-1}, implying that the average infectious period is $T_\gamma = 365/50 = 7.3$ days, $k_V = 1$ and $k_T = 50$. The time scale of the figures is in a scale of years accordingly with the above numerical values. In Figure 1, the solution trajectories of all the subpopulation are shown with the constraints of Theorem 4 being fulfilled by the initial conditions, in particular $S_0 > \frac{\gamma+k_T}{\beta}$, $I_0 = 1 - S_0$ and $R_0 = 0$ so that N_0 is normalized to unity. It is seen that the infectious subpopulation trajectory has a maximum at a finite time and that the state trajectory solution converges asymptotically to an endemic equilibrium point. In Figure 2, the state trajectory solution is shown with $N_0 = 1$ when $S_0 = (\gamma + k_T)/\beta$ which violates the conditions of Theorem 4 with $\dot{I}_0 = 0$. In this case, there is no relative maximum of the infectious subpopulation at finite time. In both situations, it has been observed by extending the overall simulation time that the susceptible and the infectious subpopulations converge asymptotically to zero while the recovered subpopulation converges to unity as time tends to infinity. The controls are suppressed in Figure 3 with $N_0 = 1$. In this case, the recovered subpopulation may be deleted from the model since it is unnecessary while being identically zero. The infectious and susceptible subpopulations are in an endemic equilibrium point for all time so that the infection results to be permanent in the sense that it cannot be asymptotically removed. See Theorem 4(vi) for the case $k_V = 0$. Figure 4 exhibits a trajectory solution which agrees with Theorem 4 while there is no normalization of the initial conditions to unity. In this case, the maximum of the infectious subpopulation at a finite time becomes very apparent.

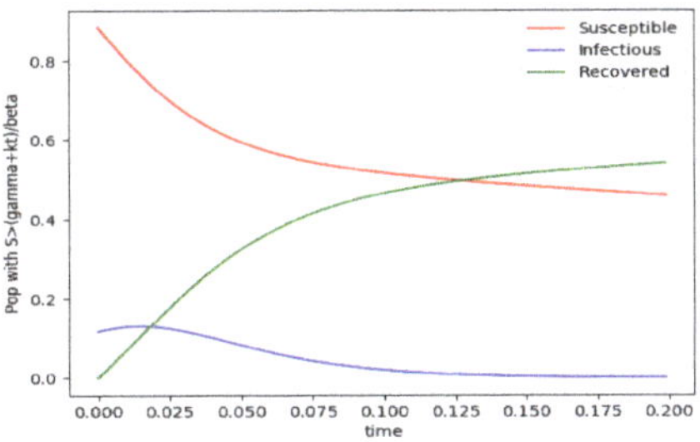

Figure 1. $N_0 = 1$ and the initial conditions constraints of Theorem 4 hold with $\dot{I}_0 > 0$.

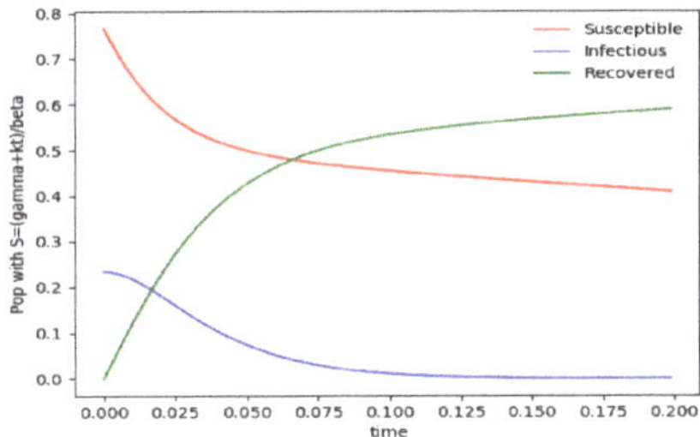

Figure 2. $N_0 = 1$ and the initial conditions constraints of Theorem 4 fail with $\dot{I}_0 = 0$

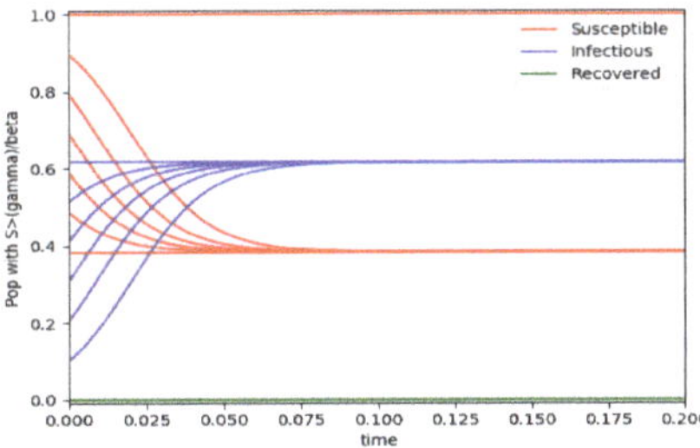

Figure 3. $N_0 = 1$ and the initial conditions constraints of Theorem 4 hold with no controls used.

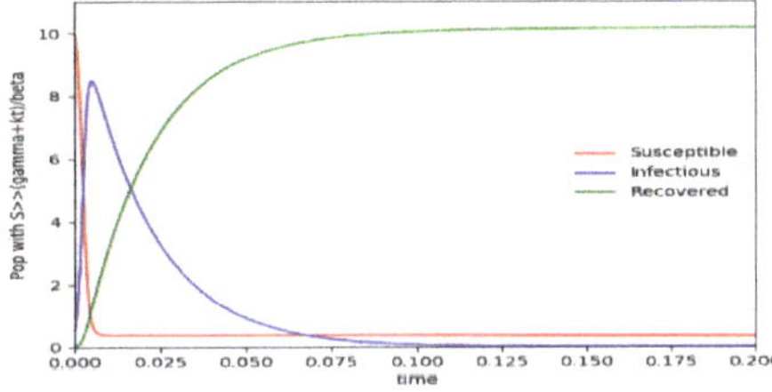

Figure 4. $S_0 > 1$, $I_0 > 1$ (unnormalized to unity total population) and the initial conditions constraints of Theorem 4 hold with $\dot{I}_0 > 0$.

4. Links with Entropy and Maximum Dissipation Mechanism Issues

4.1. Comparison of the Epidemic Model and Reference Model Information Entropies

Since (1) is a scalar equation, a valid solution for the particular model-dependent time-varying coefficient $a(t) = -cln(g(t)/E)/h(t)$ of Theorem 2 and Theorem 3 is, according to Theorem 1:

$$I(t) = e^{-c\int_0^t h^{-1}(\tau)\,\ln(g(\tau)/E)d\tau}I_0; \ t \in \mathbf{R}_{0+} \tag{32}$$

Under the particular constraints $E = D, c = (1 - ln(L/D))/ln^2(L/D)$ and $g(t) = h(t) = t$, it is got in [11] that $a(t) = \left[(ln(L/D) - 1)/ln^2(L/D)\right]t^{-1}ln(t/D)$ and (32), namely:

$$I_p(t) = e^{(ln(L/D)-1)/ln^2(L/D)\int_0^t \tau^{-1}ln(\tau/D)d\tau}I_0; \ t \in \mathbf{R}_+ \tag{33}$$

approaches the log-normal distribution:

$$I_r(t) = \frac{k}{\sqrt{2\pi}\,\sigma_r t}e^{-\frac{(ln\,t-\mu_r)^2}{2\sigma_r^2}}; \ t \in \mathbf{R}_+ \tag{34}$$

for reference values $D = D_r$ and $L = L_r$ of the maximum and inflection reference time instants where $\mu_r = lnD_r + \sigma_r^2$ and σ_r is given by the principle of extreme entropy production rate, typically $\sigma_r \approx 0.408$ gives the width of the distribution function for the maximum dissipation rate for the usual definition of the Shannon entropy. The main reason for the limitation of such a width is that the medical and social interventions are a dissipation mechanism which controls and limits the disease propagation. Comparing (33) and (34), one gets that $k = \sqrt{2\pi}\,\sigma_r^3 I_0$ after solving the indetermination $0/0$ at $t = 0$ via L´Hôpital rule leading to the "infection reference evolution" $I_r(t) = I_p(t)$, that is by equalizing (23) and (24), under the above set of particular constraints, where:

$$I_r(t) = t^{-1}\sigma_r^2 e^{-\frac{(lnt-lnD_r-\sigma_r^2)^2}{2\sigma_r^2}}I_0; \ t \in \mathbf{R}_+ \tag{35}$$

Now, equalize $I(t) = I_r(t) + \widetilde{I}(t); \forall t \in \mathbf{R}_+$ for some perturbation function $\widetilde{I}: \mathbf{R}_+ \to \mathbf{R}_{0+}$ resulting to be from (32) and (35) for $I_0 > 0$:

$$\widetilde{I}(t) = \left(e^{-c\int_0^t h^{-1}(\tau)\,\ln(g(\tau)/E)d\tau} - t^{-1}\sigma_r^2 e^{-\frac{(lnt-lnD_r-\sigma_r^2)^2}{2\sigma_r^2}}\right)I_0; \ t \in \mathbf{R}_+ \tag{36}$$

The Shannon entropy of the infection $S_I(\eta)$ results to be given by the following Riemann-Stieljes integral which quantifies the entropy error $\widetilde{S}_I(\eta)$ of that associated with any given model related to the entropy of the "infection reference evolution" given by the log-normal function $S_{I_r}(\eta) = S_{I_r}(\eta,\sigma_r)$ for the given reference width value $\sigma_r = \sqrt{1/2\eta}$:

$$\begin{aligned}
S_I(\eta) &= -\int_0^\infty t^{1-\eta}I(t)ln\left(t^{1-\eta}I(t)\right)dt^\eta \\
&= -\int_0^\infty t^{1-\eta}I(t)((1-\eta)lnt + lnI(t))dt^\eta \\
&= -\int_0^\infty t^{1-\eta}\left(I_r(t) + \widetilde{I}(t)\right)\left((1-\eta)lnt + ln\left(I_r(t)\left(1 + I_r^{-1}(t)\widetilde{I}(t)\right)\right)\right)dt^\eta \\
&= -\int_0^\infty t^{1-\eta}\left(I_r(t) + \widetilde{I}(t)\right)\left((1-\eta)lnt + lnI_r(t) + ln\left(1 + I_r^{-1}(t)\widetilde{I}(t)\right)\right)dt^\eta \\
&= S_{I_r}(\eta) - \int_0^\infty t^{1-\eta}I_r(t)ln\left(1 + I_r^{-1}(t)\widetilde{I}(t)\right)dt^\eta - \int_0^\infty t^{1-\eta}\widetilde{I}(t)lnI(t)dt^\eta - (1-\eta)\int_0^\infty t^{1-\eta}\widetilde{I}(t)lntdt^\eta \\
&= S_{I_r}(\eta) - \int_0^\infty t^{1-\eta}I_r(t)ln\left(1 + I_r^{-1}(t)\widetilde{I}(t)\right)dt^\eta - (1-\eta)\int_0^\infty t^{1-\eta}\widetilde{I}(t)lntdt^\eta \\
&\quad + \int_0^\infty t^{1-\eta}I_r(t)\left(1 - I_r^{-1}(t)I(t)\right)lnI(t)dt^\eta \\
&= S_{I_r}(\eta) + \widetilde{S}_I(\sigma); \ t \in \mathbf{R}_+
\end{aligned} \tag{37}$$

after using $I(t) = I_r(t)\left(1 + I_r^{-1}(t)\widetilde{I}(t)\right)$ and its equivalent expression $\widetilde{I}(t) = -I_r(t)\left(1 - I_r^{-1}(t)I(t)\right)$, where the reference entropy based on the identification of the log-normal function (34) with the solution of (1), that is, (33), yields for $\sigma_r = \sqrt{1/2\eta}$:

$$\begin{aligned} S_{I_r}(\eta) &= -\int_0^\infty t^{1-\eta} I_r(t) ln\left(t^{1-\eta} I_r(t)\right) dt^\eta \\ &= \eta\left(ln\left(\sqrt{\tfrac{\pi}{\eta}}\right) + \eta\left(ln D_r + \tfrac{1}{2\eta}\right) + \tfrac{1}{2}\right) \end{aligned} \tag{38}$$

after converting the Riemann-Stieljes integral (39) in a Riemann integral via differentiation of dt^η by using (35). Note that it is assumed that both current and reference entropies are evaluated for the same parameter η which is typically chosen as $\eta = 3$. At the same time, it is assumed that the maximum dissipation rate proportional to the maximum rate of entropy production is governed by the width of the distribution function σ. So the current model can potentially have a value $\sigma \neq \sigma_r$. See [11] for the normalized case obtained for $I_0 = 1$, and, also one gets the following entropy error:

$$\begin{aligned} \widetilde{S}_I(\eta) &= -\int_0^\infty t^{1-\eta}\left[ln\left((I(t)/I_r(t))^{I_r(t)}\right) + ln\left(I(t)^{I(t)-I_r(t)}\right)\right] dt^\eta - (1-\eta)\int_0^\infty t^{1-\eta}\widetilde{I}(t) ln t\, dt^\eta \\ &= -\int_0^\infty t^{1-\eta}\left[ln\left((I(t)/I_r(t))^{I_r(t)}\left(I(t)^{I(t)-I_r(t)}\right)\right)\right] dt^\eta + (1-\eta)\int_0^\infty t^{1-\eta}I_r(t)\left(1 - I_r^{-1}(t)I(t)\right) ln t\, dt^\eta \\ &= -\eta\int_0^\infty ln\left(\frac{I(t)^{I(t)}}{I_r(t)^{I_r(t)}}\right) dt + \eta(\eta-1)\int_0^\infty ln\left(t^{I_r(t)-I(t)}\right) dt \\ &= -\eta\int_0^\infty ln\left(\frac{I(t)^{I(t)}}{I_r(t)^{I_r(t)} t^{(\eta-1)(I_r(t)-I(t))}}\right) dt;\ t \in R_+ \end{aligned} \tag{39}$$

It turns out obvious that the integrand of (39) is identically zero if $\widetilde{I}(t) \equiv 0$, so that $I(t) \equiv I_r(t)$, leading to $\widetilde{S}_I(\eta) \equiv 0$. The expression (37), subject to (38)–(39), parameterizes the incremental entropy with the same parameter η which parameterizes the reference entropy $S_{I_r}(\eta_r)$. Now, define the error:

$$\delta(t) = \frac{I(t)^{I(t)}}{I_r(t)^{I_r(t)} t^{(\eta-1)(I_r(t)-I(t))}} - 1;\ t \in R_{0+} \tag{40}$$

so that $\widetilde{S}_I(\eta) \equiv 0$ if $\delta(t) \equiv 0$ and, expanding $ln\left(\frac{I(t)^{I(t)}}{I_r(t)^{I_r(t)}}\right)$ via the Newton- Mercator series for the logarithm, leads to:

$$ln\left(\frac{I(t)^{I(t)}}{I_r(t)^{I_r(t)} t^{(\eta-1)(I_r(t)-I(t))}}\right) = ln(1+\delta(t)) = \delta(t) + \sum_{n=2}^\infty \frac{(-1)^{n+1}}{n}\delta^n(t);\ t \in R_{0+} \tag{41}$$

and such a series converges to $ln(1+\delta(t))$ for all $t \in R_{0+}$ provided that $\delta(t) \in (-1, 1]$, equivalently, $\widetilde{I}(t) \in (-I_r(t), I_r(t)]$; $\forall t \in R_{0+}$; $\forall t \in R_{0+}$. Thus, the following description in linear and higher-order additive terms of the entropy error follows from (40)–(41) into (39):

$$\widetilde{S}_I(\eta) = \widetilde{S}_{IL}(\eta) + \widetilde{\widetilde{S}}_I(\eta);\ t \in R_{0+} \tag{42}$$

where:

$$\widetilde{S}_{IL}(\eta) = -\eta\int_0^\infty \left(\frac{I(t)^{I(t)}}{I_r(t)^{I_r(t)} t^{(\eta-1)(I_r(t)-I(t))}} - 1\right) dt;\ t \in R_{0+} \tag{43}$$

$$\widetilde{\widetilde{S}}_I(\eta) = -\eta\left(\sum_{n=2}^\infty \frac{(-1)^{n+1}}{n}\left(\frac{I(t)^{I(t)}}{I_r(t)^{I_r(t)} t^{(\eta-1)(I_r(t)-I(t))}} - 1\right)^n dt\right);\ t \in R_{0+} \tag{44}$$

The subsequent results hold related to the case when the error between the infectious functions of the model and the reference one associated to the log-normal function converges asymptotically to zero as time tends to infinity. The first result, stated separately by convenience concerned its proof, discusses the simplest case for $\eta = 1$.

Proposition 1. *Assume that $\eta = 1$ and $\displaystyle\lim_{t\to+\infty}\left|\int_0^t \ln\left(\frac{I(\tau)^{I(\tau)}}{I_r(\tau)^{I_r(\tau)}}\right)d\tau\right| < +\infty$.*

then, $\widetilde{S}_I(1) < +\infty$ for all $t \in \mathbf{R}_{0+}$ and $\displaystyle\lim_{t\to+\infty}(I(t) - I_r(t)) = 0$.

Proof. Note from (39) that $\widetilde{S}_I(1) = \int_0^\infty \ln\left(\frac{I(t)^{I(t)}}{I_r(t)^{I_r(t)}}\right)dt < +\infty$. Since the function $\frac{I(t)^{I(t)}}{I_r(t)^{I_r(t)}}$ is uniformly continuous on $\mathbf{R}_{0+}$ and $\displaystyle\lim_{t\to+\infty}\left|\int_0^t \ln\left(\frac{I(\tau)^{I(\tau)}}{I_r(\tau)^{I_r(\tau)}}\right)d\tau\right| < +\infty$ then $\ln\frac{I(t)^{I(t)}}{I_r(t)^{I_r(t)}} \to 0$ as $t \to +\infty$ from Barbalat´s lemma and then $\frac{I(t)^{I(t)}}{I_r(t)^{I_r(t)}} \to 1$ as $t \to +\infty$. It is clear that a limit solution which satisfies this constraint is $\displaystyle\lim_{t\to+\infty}(I(t) - I_r(t)) = 0$. It is now proved that no alternative limiting constraint on the pair $(I(t), I_r(t))$ as $t \to +\infty$ is compatible with $\displaystyle\lim_{t\to+\infty}\frac{I(t)^{I(t)}}{I_r(t)^{I_r(t)}} = 1$. Assume that $\displaystyle\liminf_{t\to+\infty}\left|I(t) - I_r(t)\right| > 0$ It can happen that:

(a) $\displaystyle\liminf_{t\to+\infty}(I(t) - I_r(t)) > 0$. Then, $\displaystyle\liminf_{t\to+\infty}\ln\frac{I(t)^{I(t)}}{I_r(t)^{I_r(t)}} = \liminf_{t\to+\infty}(I(t)\ln I(t) - I_r(t)\ln I_r(t)) >$
$\displaystyle\liminf_{t\to+\infty}(I_r(t)\ln I(t) - I_r(t)\ln I_r(t)) > \liminf_{t\to+\infty}(I_r(t)\ln I_r(t) - I_r(t)\ln I_r(t)) = 0$ so that $\displaystyle\liminf_{t\to+\infty}\ln\frac{I(t)^{I(t)}}{I_r(t)^{I_r(t)}} > 0$.
Hence, a contradiction to Barbalat´s lemma; or

(b) $\displaystyle\liminf_{t\to+\infty}(I_r(t) - I(t)) > 0$. Under a similar reasoning to that of a), one gets that $\displaystyle\liminf_{t\to+\infty}\ln\frac{I_r(t)^{I_r(t)}}{I(t)^{I(t)}} > 0$.
Again, a contradiction to Barbalat´s lemma.

The second result discusses the simplest case for $\eta \neq 1$. It is seen that the basic limit result $\displaystyle\lim_{t\to+\infty}(I(t) - I_r(t)) = 0$ of Proposition 1 is still kept under the reasonable assumption that the infection and reference infection functions are bounded. $\square$

Proposition 2. *Assume that $\eta \neq 1$, $I, I_r : \mathbf{R}_{0+} \to \mathbf{R}_{0+}$ are bounded and $\displaystyle\lim_{t\to+\infty}\left|\int_0^t \ln\left(\frac{I(\tau)^{I(\tau)}}{I_r(\tau)^{I_r\tau}t^{(\eta-1)(I_r(\varsigma)-I(\tau))}}\right)d\tau\right| <$*
$+\infty$.

Then, $\widetilde{S}_I(\eta) < +\infty$ for all $t \in \mathbf{R}_{0+}$ and $\displaystyle\lim_{t\to+\infty}(I(t) - I_r(t)) = 0$.

Proof. Note that $\widetilde{S}_I(1) < +\infty$ and that, from the uniform continuity of $\frac{I(t)^{I(t)}}{I_r(t)^{I_r(t)}t^{(\eta-1)(I_r(t)-I(t))}}$ everywhere in $\mathbf{R}_{0+}$, the boundedness of its integral on $[0, \infty)$ and Barbalat´s lemma, it follows that $\frac{I(t)^{I(t)}}{I_r(t)^{I_r(t)}t^{(\eta-1)(I_r(t)-I(t))}} \to 1$ as $t \to +\infty$ what implies that:

$$\lim_{t\to+\infty}(I(t)\ln I(t) - I_r(t)\ln I_r(t) + (1-\eta)(I_r(t) - I(t))\ln t) = 0$$

If $\eta > 1$ and $\ln t \to \infty$ as $t \to \infty$ then there exists some strictly increasing real sequence $\{t_i\}_{i=0}^\infty$, such that $\displaystyle\lim_{k\to\infty}\left|(1-\eta)(I_r((t_k)) - I((t_k)))\ln t_k\right| = \infty$ with $t_k \in \{t_i\}_{i=0}^\infty$ if $\displaystyle\lim_{t\to+\infty}(I(t) - I_r(t)) \neq 0$. But this can hold only if $\displaystyle\lim_{k\to+\infty}\left|I(t_k)\ln I(t_k) - I_r(t_k)\ln I_r(t_k)\right| = +\infty$. But, since $I_r : \mathbf{R}_{0+} \to \mathbf{R}_{0+}$ is bounded for all time, this implies that $I(t_k) \to +\infty$ as $t_k\big(\in \{t_i\}_{i=0}^\infty\big) \to +\infty$ and $I : \mathbf{R}_{0+} \to \mathbf{R}_{0+}$ is unbounded. But then

$$\lim_{k\to+\infty}(I(t_k)\ln I(t_k) - I_r(t_k)\ln I_r(t_k) + (\eta-1)(I(t_k) - I_r(t_k))\ln t_k) = \infty + \infty = \infty$$

and a contradiction follows to the above limit to be zero. As a result, $\displaystyle\lim_{t\to+\infty}(I(t) - I_r(t)) = 0$ if $\eta > 1$.

Now, assume that $\eta < 1$. Since $\displaystyle\lim_{k\to\infty}\left|(1-\eta)(I_r(t_k) - I(t_k))\ln t_k\right| = \infty$ for $t_k\big(\in \{t_i\}_{i=0}^\infty\big) \to +\infty$ and some strictly increasing real sequence $\{t_i\}_{i=0}^\infty$, provided that $\displaystyle\lim_{t\to+\infty}(I(t) - I_r(t)) \neq 0$, then $I(t_k) \to +\infty$ as $t_k\big(\in \{t_i\}_{i=0}^\infty\big) \to +\infty$ since $I_r : \mathbf{R}_{0+} \to \mathbf{R}_{0+}$ is bounded. Since $I : \mathbf{R}_{0+} \to \mathbf{R}_{0+}$ is unbounded, because it has a divergent subsequence $\{I(t_k)\}_{k=0}^\infty$ and it is a solution of a unstable time-invariant linear differential system, it is of positive exponential order $\varsigma_0 > 0$ and there exists a real constant $\varsigma < \varsigma_0$ such that $I(t_k) \geq e^{\varsigma t_k}$; $\forall t_k \in \{t_i\}_{i=0}^{\infty}$ and $I(t_k)/lnt_k \left(\geq e^{\varsigma t_k}/lnt_k \right) \to \infty$ as $t_k \left(\in \{t_i\}_{i=0}^{\infty} \right) \to \infty$ and, furthermore,

$$\lim_{k\to+\infty} \left(I(t_k)lnI(t_k) - I_r(t_k)lnI_r(t_k) \right) = (1-\eta) \lim_{k\to+\infty} \left(I(t_k) - I_r(t_k) \right) \ln t_k = \infty$$

but the expression below is an infinity limit (and not a $\infty - \infty$ indetermination since $I(t_k)/lnt_k \to \infty$):

$$\lim_{k\to+\infty} \left(I(t_k)lnI(t_k) - I_r(t_k)lnI_r(t_k) - (1-\eta)(I(t_k) - I_r(t_k)) \ln t_k \right) = \infty$$

which contradicts:

$$\lim_{t\to+\infty} \left(I(t)lnI(t) - I_r(t)lnI_r(t) + (1-\eta)(I_r(t) - I(t)) \ln t \right) = 0$$

As a result, $\lim_{t\to+\infty} (I(t) - I_r(t)) = 0$ if $\eta \neq 1$.

It is now briefly discussed the fact that the boundedness hypothesis of Proposition 2 is not very restrictive for some of the given examples, like for instance, Examples 2,3, where the infectious subpopulation converges asymptotically to zero. For such a purpose, note from (35) that $I_r(t) \to 0$ exponentially fast as $t \to \infty$. In example 2, $I(t) \to 0$ exponentially as $t \to \infty$ so their difference function also converges to zero exponentially as $t \to \infty$. The integral boundedness invoked in the assumption of Proposition 2 is of the form $F = \left| \int_0^\infty \ln x(t)dt \right| < +\infty$, where $x(t) = \dfrac{I(t)^{I(t)}}{I_r(t)^{I_r(t)} t^{(\eta-1)(I_rt-I(t))}}$ is everywhere differentiable with respect to time. In order to convert the elevant Riemann-Stieljes integral into a standard Riemann one, take $dx = \dot{x}(t)dt$ and, later on, perform the change of variable $x \to u$ defined by $u = lnx$, $du = dx/x$ to yield:

$$F = \left| \int_{x_0}^1 \frac{\ln x}{\dot{x}} dx \right| = \left| \int_{r_0}^1 \frac{\ln x}{x} \frac{x}{\dot{x}} dx \right| \leq \left(\sup_{0<t\leq+\infty} \left| \frac{x(t)}{\dot{x}(t)} \right| \right) \left| \int_{x_0}^1 \frac{\ln x}{x} dx \right|$$

$$= M(\eta) \left| \int_{lnx_0}^0 udu \right| = M(\eta) \frac{lnx_0^2}{2}$$

where $x(0) = x_0$ and $M(\eta) = \sup_{0\leq t\leq+\infty} \left| \frac{x(t)}{\dot{x}(t)} \right| \leq +\infty$ for the given constant η. Note that $M(\eta) < +\infty$ if and only if the set of zeros of $\dot{x}(t)$ at any finite time instant is empty, that is, if and only if $Z_{xdot}(\eta) = \varnothing$, where $Z_{xdot}(\eta) = \left\{ t \geq 0 : \dot{x}(t) = 0 \right\} = \varnothing$ (equivalently, $M(\eta) = +\infty$ if and only if $Z_{xdot}(\eta) \neq \varnothing$. Rewriting $x(t) = \frac{y(t)}{t^{\eta-1}z(t)}$ it follows that $\dot{x}(t) = 0$ for any $t \geq 0$ if and only the following constraint holds $t = \frac{(\eta-1)z(t)y(t)}{z(t)\dot{y}(t)-y(t)\dot{z}(t)}$. Therefore, $Z_{xdot}(\eta) = \left\{ t \geq 0 : t = \frac{(\eta-1)z(t)y(t)}{z(t)\dot{y}(t)-y(t)\dot{z}(t)} \right\} \neq \varnothing$ is an event of zero probability. Thus, the boundedness hypothesis of Proposition 2 happens almost surely in the event that the infectious subpopulation converges asymptotically to zero as time tends to infinity. $\square$

Propositions 1 and 2 yield the direct joint result independently of the value of η:

Proposition 3. *Assume that* $\eta \in \mathbf{R}_{0+}$, $I, I_r : \mathbf{R}_{0+} \to \mathbf{R}_{0+}$ *are bounded and* $\lim_{t\to+\infty} \left| \int_0^t \ln\left(\dfrac{I(t)^{I(t)}}{I_r(t)^{I_r(t)} t^{(\eta-1)(I_r(t)-I(t))}} \right) dt \right| < +\infty.$ *Then,* $\widetilde{S}_I(\eta) < +\infty$ *for all* $t \in \mathbf{R}_{0+}$ *and* $\lim_{t\to+\infty} (I(t) - I_r(t)) = 0.$

Concerning Proposition 3, note that the boundedness of $\widetilde{S}_I(\eta)$ does not guarantee that the linear part and the remaining part of higher- order terms in the decomposition of (42), subject to (43) and (44), are both finite. It could "a priori" happen that they both tend to infinity with opposite signs. But if any of them is bounded, the other one should be bounded as well according to Proposition 3. Fortunately, this does not happen under weak extra assumptions. In particular, the following result holds:

Proposition 4. *Assume that $\eta \in R_{0+}$, $I, I_r : R_{0+} \to R_{0+}$ are bounded, and*

$$\int_0^\infty \ln\left(\frac{I(t)^{I(t)}}{I_r(t)^{I_r(t)} t^{(\eta-1)(I_r(t)-I(t))}} - 1\right) dt < +\infty.$$

Then, $\widetilde{S}_{IL}(\eta) < +\infty$; $t \in R_{0+}$.

If, in addition, $\int_0^\infty \ln\left(\frac{I(t)^{I(t)}}{I_r(t)^{I_r(t)} t^{(\eta-1)(I_r(t)-I(t))}} dt\right) < +\infty$ then $\left|\widetilde{\widetilde{S}}_I(\eta)\right| < +\infty$ and $\widetilde{S}_I(\eta) < +\infty$ for all $t \in R_{0+}$ and $\lim_{t\to+\infty} (I(t) - I_r(t)) = 0$.

Proof. It is direct to see that $\widetilde{S}_{IL}(\eta) < +\infty$. Also, and again from Barbalat´s lemma, $\frac{I(t)^{I(t)}}{I_r(t)^{I_r(t)} t^{(\eta-1)(I_r(t)-I(t))}} \to 1$ as $t \to +\infty$. Thus, from Proposition 3, $\lim_{t\to+\infty} (I(t) - I_r(t)) = 0$. If, furthermore, $\int_0^\infty \ln\left(\frac{I(t)^{I(t)}}{I_r(t)^{I_r(t)} t^{(\eta-1)(I_r(t)-I(t))}}\right) dt < +\infty$ then, again from Proposition 3, $\widetilde{S}_I(\eta) < +\infty$ and

$$-\infty < -\left|\widetilde{S}(\eta) - \widetilde{S}_{IL}(\eta)\right| \le \widetilde{S}_I(\eta) \le \left|\widetilde{S}(\eta) - \widetilde{S}_{IL}(\eta)\right| < +\infty. \quad \square$$

Note that the above results agree with the asymptotic results of Examples 1–4, where $I(t) \to 0$ as $t \to \infty$, and with Theorem 1, since the reference $I_r(t) \to 0$, jointly implying $(I(t) - I_r(t)) \to 0$ as $t \to \infty$.

Remark 4. *The rationale behind the definition of a time-varying coefficient in (1) is to reduce the higher-order epidemic model with two or more states to a single-order differential equation based on the assumption that the log-normal distribution is a sufficiently accurate model for the infectious evolution. It is apparent that the profile of the log-normal distribution remembers the behavior of the strong infections in their blowing–up evolution phase along time. However, it is obvious that the epidemic models have the concourse of several coupled subpopulations so that it the model is reduced to a first-order dynamics the influence of the remaining dynamics should be accounted for through a time-varying parameterization and dynamics uncertainty in (1) since the model order is reduced to unity. The accuracy of the modeling procedure is evaluated by means of the entropy through (37). Hence if the actual infectious population curve is close to the reference one, then we have $S_I(\eta) = S_{I_r}(\eta)$ which generates the dissipation rate of the model. On the other hand, if the current system differs from the reference model, then the entropy becomes corrected with the additional term $\widetilde{S}_I(\eta)$. Therefore, the contributing terms in (37) provide an estimation of the modeling uncertainty based on the assumed log- normal reference distribution. As a result, the best approximation of the current model to the reference one is that which minimizes the error entropy $\widetilde{S}_I(\eta)$, i.e., the one which reduces as much as possible the uncertainty introduced by the approximation.*

Remark 5. *Note that the entropy of the infection $I(t)$ for $\eta = 1$ is defined as $S_I(1) = -\int_0^\infty I(\tau)\ln I(\tau)d\tau$ The entropy of the truncated function $I_t(\tau) = I(\tau)$ for $\tau \in [0,t]$ and $I_t(\tau) = 0$ for $\tau \notin [0,t]$ is $S_{I_t}(1) = -\int_0^\infty I_t(\tau)\ln I_t(\tau)d\tau = -\int_0^t I(\tau)\ln I(\tau)d\tau$. Note also that $\dot{S}_{I_t}(1) = -I(t)\ln I(t)$ and $\ddot{S}_{I_t}(1) = -\dot{I}(t)(1 + \ln I(t)) = 0$ if $t = D$. That is, the inflection point of the truncated entropy occurs at the relative extreme values of $I(t)$. In particular, if the infection is in its first expanding phase, this occurs at its maximum $t = D$.*

4.2. Estimation of Errors of the Distribution Widths between the Log-Normal Reference and Current Model Information Entropies

One gets from (38) for the usual reference entropy definition based on the log-normal distribution of width $\sigma_r = \frac{1}{\sqrt{2\eta}}$, [11,33,37], that:

$$S_{I_r}(\sigma_r, D_r, \eta) = \eta\left(\ln\left(\sqrt{2\pi}\sigma_r\right) + \eta\left(\ln D_r + \sigma_r^2\right) + \frac{1}{2}\right) \tag{45}$$

and the particular value:

$$\sigma_r = \sigma_r(\eta) = \arg\left(r \in \mathbf{R}_{0+} : \frac{d^2 S(\sigma_r, D_r, \eta)}{dr^2} = 0\right) \tag{46}$$

is the width distribution maximum value which makes the reference entropy to cease to increase while giving the maximum dissipation rate which leads to:

$$S_{I_r}\left(\sqrt{\frac{1}{2\eta}}, D_r, \eta\right) = \eta\left(\ln\sqrt{\frac{\pi}{\eta}} + \eta\left(\ln D_r + \sigma_r^2\right) + \frac{1}{2}\right) = \arg\left(S_{I_r}(\sigma_r, \eta) : \frac{d^2 S(\sigma_r, \eta)}{d\sigma_r^2} = 0\right) \tag{47}$$

Note that the above reference description is easily associated to an epidemic model given by a first-order differential equation involving only the infection evolution. Note also, in particular, that the infection curve solution is of exponential order as it is the log-normal function. Such an order is negative if the disease-free equilibrium point is globally asymptotically stable (that is, the reproduction number is less than one) so that the infection converges exponentially to zero. In other words, the curves (43) and (44) can be reasonably identified with each other as it has been made in the above subsection by considering the influence of the initial conditions. In more sophisticated models involving the concourse of more subpopulations (say susceptible, immune, etc.), like those discussed in the above section, the differential equation is of higher-order than one so that the $\alpha(t)$ -function describing the time evolution of $I(t)$ depends on the remaining subpopulations. This translates into the following facts:

(1) *Fact 1*: It is known that, for $\eta = 3$, $\sigma_r = \sqrt{\frac{1}{6}} \approx 0.408$; $\frac{D_r}{L_r} = 1.649$; $\frac{I(D_r)}{I(L_r)} = \frac{f(D_r)}{f(L_r)} = 2.120$, [11].

(2) *Fact 2*: A modification of the relevant time instants D and L of maximum infection and previous inflection point with respect to D_r and L_r, and the corresponding entropies as it has been discussed analytically in Section 4.1. Those parameters depend on each particular model. This also will translate, as a result, into a change of the distribution width σ related to the reference width σ_r for the maximum dissipation concerns.

(3) *Fact 3*: Although the above reference values σ_r and ratio D_r/L_r are independent on I_0, since they are got from the log-normal distribution function, the current ones are, in general, dependent on I_0. The entropy of the current given multi-subpopulation model is given explicitly by (37), subject to (38)–(39). The time instants D and L of the respective maximum and inflection infection time instants and their values $I(D)$ and $I(L)$ are calculated from the first zeros of the curves $\dot{I}(t)$ and $\ddot{I}(t)$, respectively which also lead directly to their corresponding rates.

(4) *Fact 4*: The entropy of the current model might be interpreted in terms of the maximum dissipation rate by assuming a description via a log-normal distribution. However, it is easy to verify that the log-normal function is zeroed as its argument is either zero or $+\infty$, although its profile is close, but not identical, to the solution of a first-order differential equation describing a decaying exponential infection evolution towards a disease- free equilibrium point. For this reason, and having in mind the comparison of the solution of models with more than one subpopulation (with associated differential system of order larger than one) to the log-normal distribution $f(t)$ which is zero at zero and at infinity and which satisfies $\int_0^\infty f(t)d(t) = 1$, we first normalize the infectious

subpopulation of the current model in order to get a comparable entropy to the reference one associated with the log-normal function, that is, we define:

$$I_n(t) = \frac{I(t)}{\int_0^\infty I(\tau)d\tau}; \; S_{I_n}(\eta) = -\int_0^\infty I_n(\tau)ln\left(I_n(\tau)/t^{\eta-1}\right)d\tau \tag{48}$$

4.3. Some Numerical Tests on Reference and Current Model Entropies

Now Example 2 and Example 4 are compared to the infection study of [11], by introducing the appropriate tools of normalized infection entropy (48) associated with the maximum dissipation rate for the choice $\eta = 1$. Recall the basic notation D_r, L_r, D and L being the first time instants such that $\dot{I}(D_r) = 0$, $\ddot{I}(L_r) = 0$, $\dot{I}(D) = 0$, $\ddot{I}(L) = 0$ (Examples 2 and 4). One gets from (45) for $\eta = 1$ and, correspondingly, $\sigma_r = \sqrt{1/2}$ that the parameterized reference entropy is:

$$S_{I_r}\left(\sqrt{\frac{1}{2}}, D_r, 1\right) = \left(ln\sqrt{\pi} + lnD_r + 1\right) \tag{49}$$

and one gets for Example 2 that its associated normalized entropy for $\eta = 1$ being un-parameterized in (D, σ) becomes from (48):

$$S_{I_n}(1) = -\int_0^\infty I_n(t)lnI_n(t)dt \tag{50}$$

Numerical experimentation with Example 2: Note that D is the first time instant such that $\dot{I}(D) = 0$ and $I(D)$ is a relative maximum, which in practice, gives the maximum expected infectious numbers. Also, L is the first time instant such that $\ddot{I}(L) = 0$. Note also that the basic model, of response being close to a log- normal function, has only an infectious subpopulation while the examples of Section 3 have more subpopulations integrated in the models. Therefore, the reasonable condition that the initial conditions of the infectious subpopulation are the one percent of the total population, we consider a total population of $N_0 = S_0 + I_0 + R_0 = 1$ for Example 2 in order to get a feasible comparison.

Thus, we perform several alternative experiments as follows:

(a) We get the values of the time instants D and L and the corresponding infection numbers $I(D)$ and $I(L)$, from the solution trajectory of Example 2 and its first two-time derivatives trajectories through time, as well as the normalized entropy $S_{I_n}(1)$ from (50). Later on, by equalizing (50) to (49), one then gets the value of D_{rm} which specifies the time instant given a maximum infectious subpopulation with a maximum dissipation rate in a log normal distribution. This equalization yields:
$$D_{rm} = e^{S_{I_n}(1)-1}/\sqrt{\pi} \tag{51}$$

(b) We equalize again (49) by fixing $D_r = D$ in (50). Then, we get the necessary value σ_{rm} for such an equality to hold.

(c) We define the variance with distribution function $I_n(t)$ and log- normal distribution resulting to be:
$$var(\omega) = \int_0^\infty \omega(t)\left(t - \int_0^\infty \omega(t)tdt\right)^2 dt \tag{52}$$

where $\omega(t) = I_n(t)$ or $\omega(t) = x(t, D_r, \sigma_r)$, the log-normal distribution. Then, we obtain the necessary $\sigma_{rmv} = \sigma_{rmv}(var(I_n), D)$ got from

$$var(I_n) = var(x(D_r = D, \sigma_{rmv})) \tag{53}$$

One observes that, in general, $\sigma_{rmv} \neq \sigma_r = \frac{1}{\sqrt{2}}$ which ensures that the variance of log- normal distribution is equal to $var(I_n)$ for such a value of σ_{rmv}. Some numerical data on Example 2 are now compared with the log-normal distribution function. The model parameters are $\beta = 13{,}065$ and $\gamma = 50.1$ year^{-1} what means that the average infectious period is $T_\gamma = 1/\gamma = 365/50.1 = 7.29$ days. The initial infectious subpopulation is the one percent of the normalized total one $N_0 = 1$. For those initialization, the quotient S_0/I_0 (percentage of initial susceptible subpopulation versus recovered subpopulation) is used to plot Figures 5–8 whose time scale are in years. Figure 5 displays the time instants of maximum infection and inflection point versus different values of S_0/R_0. The values of D_{rm} from (51) is also plotted. The corresponding infectious subpopulations are displayed in Figure 6. Figure 7 gives the entropies of (50) and (49). On the other hand, Figure 8 displays σ_{rm}, σ_{rmv} and the variance of the normalized infectious $I_n(t)$ of (52). It is basically concluded that for the model of example 2 which has three subpopulations, the results are distinct from to those obtained from the log-normal distribution which we can recall that behave closely to the solution of a first-order differential equation involving the infectious only for initial infection being close to zero and small susceptible amounts. The above discrepancy increases as the quotient S_0/I_0 increases. The reason of the approximation discrepancy is that the couplings of the infectious subpopulations with the remaining ones becomes increasingly relevant to the transient responses evolution as the proportion of susceptible to infectious increases.

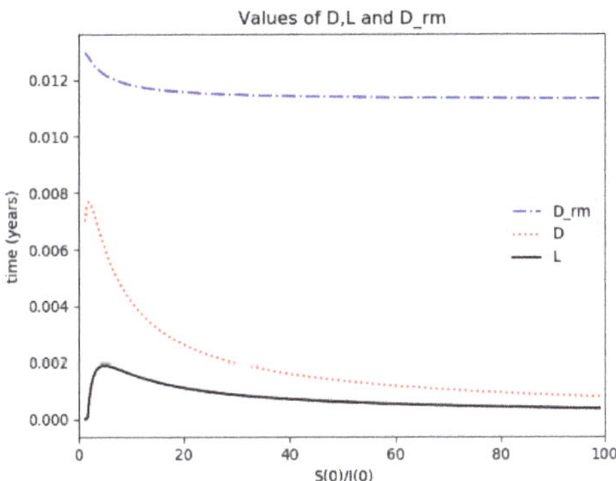

Figure 5. D (maximum infection time) and L (inflection point time) for Example 2.

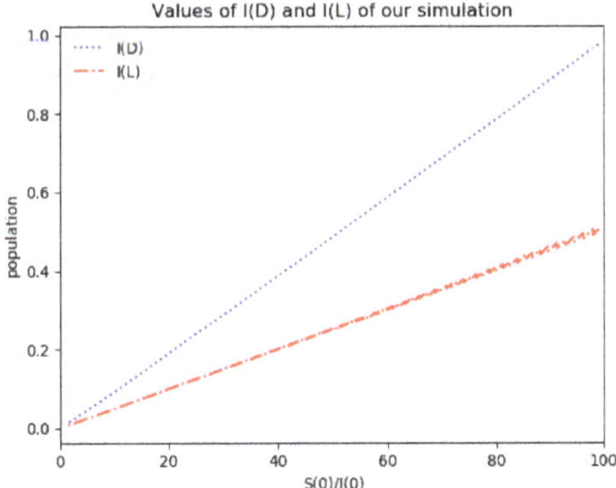

Figure 6. Maximum infection and inflection reached values I(D) and I(L) for Example 2.

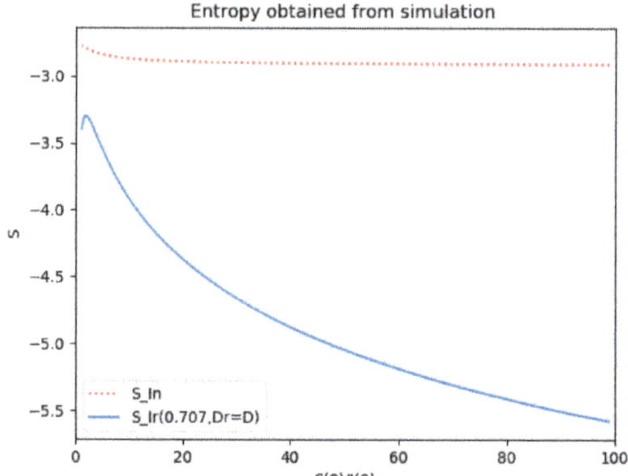

Figure 7. Reference and model entropies of Example 2.

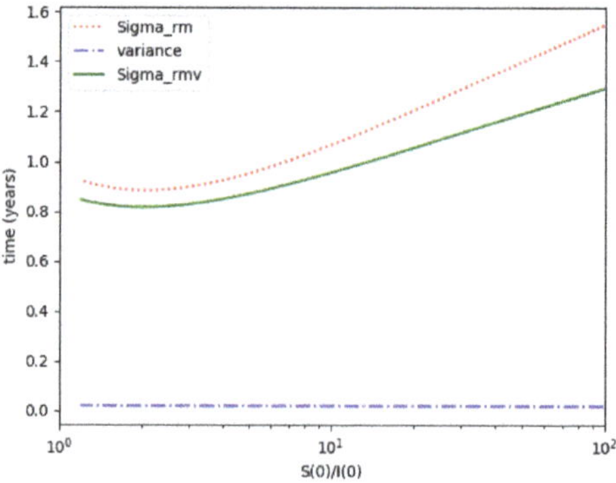

Figure 8. σ_{rm}, σ_{rmv} and variance for Example 2.

Numerical experimentation with Example 4: The initial values satisfy a normalization constraint $N_0 = S_0 + R_0 = 1$ with subpopulations $S_0 = 0.99$, $i_o = 0.01$ (that, is the initial infectious subpopulation is 1% of the total one) and $R_0 = 0$ since the recovered populations is compensatory in the model in order to take into account the effects of the intervention controls. The parameters β and γ are fixed as in Example 2. In particular, Figures 9 and 10 show the maximum infection and its previous value at the inflection time instant and the corresponding time instants without vaccination and with a vaccination effort rate of $k_T = 290$ for different values of the vaccination control gain. It is basically seen that the maximum and inflection amounts decrease as the treatment control gain gives a skip from zero to an important effort as that, in parallel, the above values also decrease as the vaccination control gain increases. Figures 10 and 11 describe parallel experiments where the roles of the vaccination and treatment control gains are reversed with respect to the data of Figures 9 and 10. The obtained conclusions are similar. The time instants of maximum infection and the inflection value are reached without and with vaccination control as the treatment control effort increases for Example 4 are plotted in Figure 12. The corresponding entropies for those to experiments compared to the reference entropy are displayed in Figures 13 and 14. Note that the entropies (48) and (50) reach negative values because of the normalization of the infection by the total infection integral contribution (48) used to evaluate the normalized entropy (50). Note that the vaccination control does not affect to the entropy as significantly as the treatment control gains since it influences less significantly to the model dynamics.

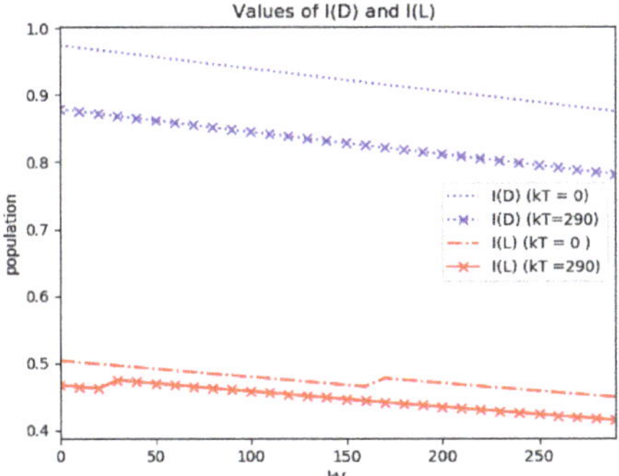

Figure 9. Maximum infection and its values at the inflection time instants without and with treatment control as the vaccination control effort increases for Example 4.

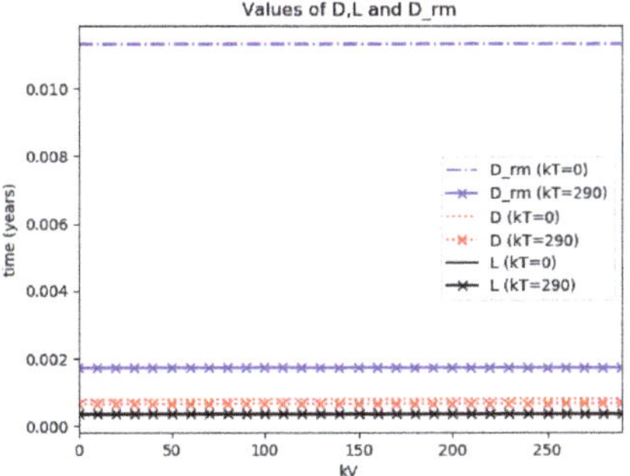

Figure 10. Time instants at which the maximum infection and the inflection value are reached without and with treatment control as the vaccination control effort increases for Example 4.

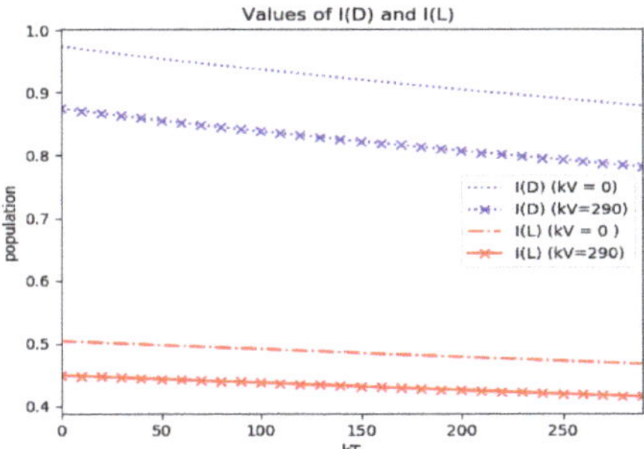

Figure 11. Maximum infection and its values at the inflection time instants without and with vaccination control as the treatment control gain increases for Example 4.

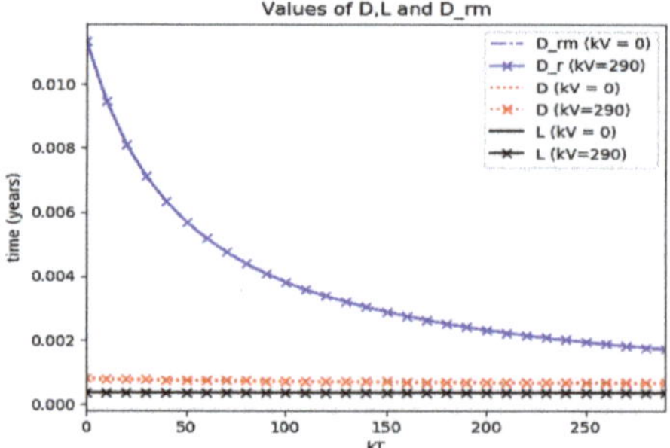

Figure 12. Time instants at which the maximum infection and the inflection value are reached without and with vaccination control as the treatment control effort increases for Example 4.

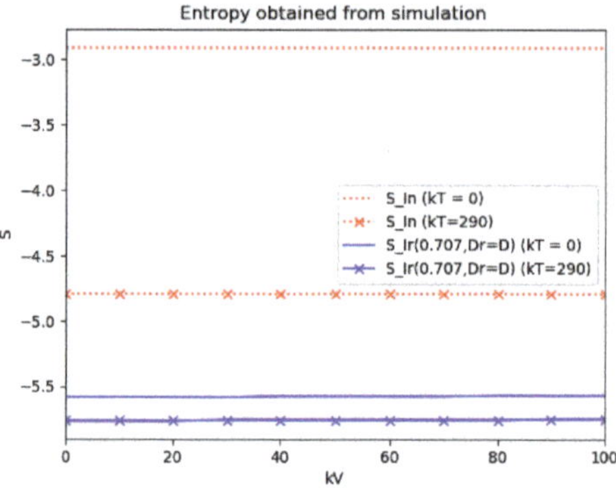

Figure 13. Entropies of the reference and the normalized model of Example 4 without and with treatment control as the vaccination control effort increases.

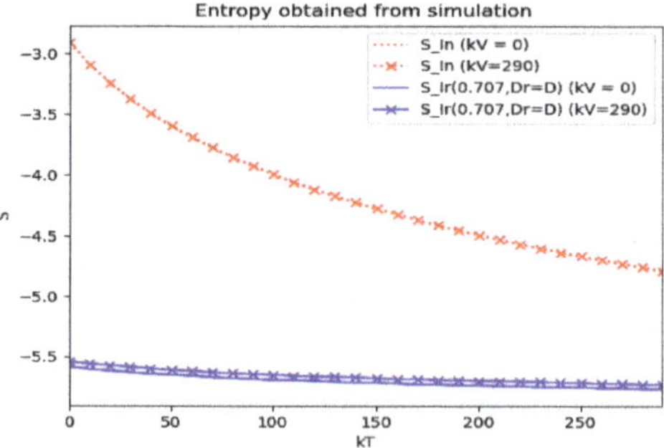

Figure 14. Entropies of the reference and the normalized model of Example 4 without and with vaccination control as the treatment control effort increases.

5. Conclusions

This paper has investigated the extensions of a first-order differential system describing the infection propagation through time to epidemic models integrating more than one subpopulation. The main involved tool has been the consideration of the coupling of inter-populations dynamics and the control intervention information through the structure of the time-varying coefficient which drives the basic differential equation model of first-order. The control of the infection along its transient to fight more efficiently against a potential initial exploding transmission from a high initial growth rate is considered relevant. Special attention has been paid throughout the manuscript to the discussion of the profiles of the transients of the infection curve in terms of the time instants of its first relative maximum towards its previous inflection time instant, so the study is mainly focused on the transient behavior characterization rather than on the steady-state equilibrium points. The time instants leading to the maximum infection and inflection numbers have been investigated via the Shannon´s information entropy for the maximum dissipation rate linked to a previous background study for a first-order differential equation describing the infection propagation. Since it is relevant to know the time instants of maximum infection and inflection as well as its numbers in order to monitor the availability of hospitalization resources, some examples related to existing epidemic models integrated by more than a subpopulation have been studied. The obtained results have been compared, both via theoretical work and also by numerical experimentation, to the background results obtained from a reference model, just involving a single infectious population, which is based on a description via a log-normal distribution which has a close profile to the solution response of a first-order differential equation. In those examples, special attention is paid to the comparisons of the maximum infection and inflection time dates for different values of initial conditions and to the entropy discrepancies related to the reference one. It can be concluded that the influence of the couplings of the dynamics of other subpopulations in the model to the infectious one is relevant to the infection evolution, especially, in the cases when the initial amounts of the susceptible are significantly large compared to the initial amounts of the infectious.

Author Contributions: Conceptualization, M.D.l.S. and R.N.; methodology, M.D.l.S. and R.N.; software, R.N.; validation, R.N., A.I. and A.J.G.; formal analysis, M.D.l.S.; investigation, M.D.l.S. and R.N.; resources, M.D.l.S. and A.I.; data curation, R.N. and A.I.; writing—original draft preparation, M.D.l.S.; writing—review and editing, M.D.l.S., R.N. and A.I.; visualization, R.N. and A.J.G.; supervision, M.D.l.S. and A.I.; project administration, M.D.l.S.; funding acquisition, M.D.l.S. and A.J.G. All authors have read and agreed to the published version of the manuscript.

Funding: This research was funded by MCIU/AEI/FEDER, UE, grant number RTI2018-094902-B-C22 and the APC was funded by RTI2018-094902-B-C22.

Acknowledgments: The authors are grateful to the Spanish Government for Grants RTI2018-094336-B-I00 and RTI2018-094902-B-C22 (MCIU/AEI/FEDER, UE) and to the Basque Government for Grant IT1207-19. They also thank the Instituto de Salud Carlos III and the Spanish Ministry of Science and Innovation for Grant COV20/01213 of the Program: "Expressions of interest for the support on SARS-COV-2 and COVID 19". The authors also thank the referees for their useful comments.

Conflicts of Interest: The authors declare no conflict of interest.

References

1. Khinchin, A.I. *Mathematical Foundations of Information Theory*; Dover Publications Inc.: New York, NY, USA, 1957.
2. Aczel, J.D.; Daroczy, Z. *On Measures of Information and Their Generalizations*; Academic Press: New York, NY, USA, 1975.
3. Ash, R.B. *Information Theory*; Interscience Publishers, John Wiley and Sons: New York, NY, USA, 1965.
4. Feynman, R.P. Simulating Physics and Computers. *Int. J. Theor. Phys.* **1982**, *21*, 467–488. [CrossRef]
5. Burgin, M.; Meissner, G. Larger than one probabilities in mathematical and practical finance. *Rev. Econ. Finance* **2012**, *4*, 1–13.
6. Tenreiro-Machado, J. Fractional derivatives and negative probabilities. *Commun. Nonlinear Sci. Numer. Simul.* **2019**, *79*, 104913. [CrossRef]
7. Baez, J.C.; Fritz, T.; Leinster, T. A characterization of entropy in terms of information loss. *Entropy* **2011**, *2011*, 1945–1957. [CrossRef]
8. Delyon, F.; Foulon, P. Complex entropy for dynamic systems. *Ann. Inst. Henry Poincarè Physique Théorique* **1991**, *55*, 891–902.
9. Nalewajski, R.F. Complex entropy and resultant information measures. *J. Math. Chem.* **2016**, *54*, 1777–2782. [CrossRef]
10. Goh, S.; Choi, J.; Choi, M.Y.; Yoon, B.G. Time evolution of entropy in a growth model: Dependence on the description. *J. Korean Phys. Soc.* **2017**, *70*, 12–21. [CrossRef]

11. Wang, W.B.; Wu, Z.N.; Wang, C.F.; Hu, R.F. Modelling the spreading rate of controlled communicable epidemics through an entropy-based thermodynamic model. *Sci. China Phys. Mech. Astron.* **2013**, *56*, 2143–2150. [CrossRef]

12. Tiwary, S. The evolution of entropy in various scenarios. *Eur. J. Phys.* **2020**, *41*, 025101. [CrossRef]

13. Koivu-Jolma, M.; Annila, A. Epidemic as a natural process. *Math. Biosci.* **2018**, *299*, 97–102. [CrossRef]

14. Artalejo, J.R.; Lopez-Herrero, M.J. The SIR and SIS epidemic models. A maximum entropy approach. *Theor. Popul. Biol.* **2011**, *80*, 256–264. [CrossRef] [PubMed]

15. Erten, E.Y.; Lizier, J.T.; Piraveenan, M.; Prokopenko, M. Criticality and Information Dynamics in Epidemiological Models. *Entropy* **2017**, *19*, 194. [CrossRef]

16. De la Sen, M. On the approximated reachability of a class of time-varying systems based on their linearized behaviour about the equilibria: Applications to epidemic models. *Entropy* **2019**, *21*, 1045.

17. Li, K.; Small, M.; Zhang, H.; Fu, X. Epidemic outbreaks on networks with effective contacts. *Nonlinear Anal. Real World Appl.* **2010**, *11*, 1017–1025. [CrossRef]

18. Cui, Q.; Qiu, Z.; Liu, W.; Hu, Z. Complex Dynamics of an SIR Epidemic Model with Nonlinear Saturate Incidence and Recovery Rate. *Entropy* **2017**, *19*, 305. [CrossRef]

19. Nistal, R.; de la Sen, M.; Alonso-Quesada, S.; Ibeas, A. Supervising the vaccinations and treatment control gains in a discrete SEIADR epidemic model. *Int. J. Innov. Comput. Inf. Control* **2019**, *15*, 2053–2067.

20. Verma, R.; Sehgal, V.K.; Nitin, V. Computational Stochastic Modelling to Handle the Crisis Occurred During Community Epidemic. *Ann. Data Sci.* **2016**, *3*, 119–133. [CrossRef]

21. Iggidr, A.; Souza, M.O. State estimators for some epidemiological systems. *J. Math. Boil.* **2018**, *78*, 225–256. [CrossRef]

22. Yang, H.M.; Ribas-Freitas, A.R. Biological view of vaccination described by mathematical modellings: From rubella to dengue vaccines. *Math. Biosci. Eng.* **2019**, *16*, 3195–3214. [CrossRef]

23. De la Sen, M. On the Design of Hyperstable Feedback Controllers for a Class of Parameterized Nonlinearities. Two Application Examples for Controlling Epidemic Models. *Int. J. Environ. Res. Public Health* **2019**, *16*, 2689. [CrossRef]

24. De la Sen, M. Parametrical non-complex tests to evaluate partial decentralized linear-output feedback control stabilization conditions for their centralized stabilization counterparts. *Appl. Sci.* **2019**, *9*, 1739. [CrossRef]

25. Meyers, L. Contact network epidemiology: Bond percolation applied to infectious disease prediction and control. *Bull. Am. Math. Soc.* **2006**, *44*, 63–87. [CrossRef]

26. De la Sen, M.; Alonso-Quesada, S. Control issues for the Beverton–Holt equation in ecology by locally monitoring the environment carrying capacity: Non-adaptive and adaptive cases. *Appl. Math. Comput.* **2009**, *215*, 2616–2633. [CrossRef]

27. De la Sen, M.; Ibeas, A.; Alonso-Quesada, S.; Nistal, R. On a SIR Model in a Patchy Environment Under Constant and Feedback Decentralized Controls with Asymmetric Parameterizations. *Symmetry* **2019**, *11*, 430. [CrossRef]

28. Herrmann-Pillath, C.; Salthe, S.N. Triadic conceptual structure of the maximum entropy approach to evolution. *Biosyst.* **2011**, *103*, 315–330. [CrossRef] [PubMed]

29. Ulanowicz, R.E. The balance between adaptability and adaptation. *Biosystems* **2002**, *64*, 13–22. [CrossRef]

30. Keeling, M.; Rohani, P. *Modeling Infectious Diseases in Humans and Animals*; Princeton University Press: Princeton, NJ, USA, 2008.

31. Hartonen, T.; Annila, A. Natural networks as thermodynamic systems. *Complexity* **2012**, *18*, 53–62. [CrossRef]

32. Moreno, Y.; Pastor-Satorras, R.; Vespignani, A. Epidemic outbreaks in complex heterogeneous networks. *Phys. Condens. Matter* **2002**, *26*, 521–529. [CrossRef]

33. Ziegler, H. *An introduction to Thermomechanics*; North-Holland Publishing Company: Amsterdam, The Netherlands, 1983.

34. Prigogine, I. *Introduction to Thermodynamics of Irreversible Processes*, 3rd ed.; John Wiley & Sons: New York, NY, USA, 1967.

35. De Groot, S.R.; Mazur, P. *Non-Equilibrium Thermodynamics*; Dover Publications, Inc.: New York, NY, USA, 1984.

36. Stein, S. *Calculus in the First Three Dimensions*; Dover: New York, NY, USA, 2016.

37. On the Generalized Lognormal Distribution. *J. Probab. Stat.* **2013**, *2013*, 432642.

38. Khan, H.; Mohapatra, R.N.; Vajravelu, K.; Liao, S.J. The explicit series solution of SIR and SIS epidemic models. *Appl. Math. Comput.* **2009**, *215*, 653–669. [CrossRef]
39. Harko, T.; Lobo, F.S.N.; Mak, M.K. Exact analytical solutions of the susceptible-infected-recovered (SIR) epidemic model and on SIR model with equal death and birth rates. *Appl. Math. Comput.* **2014**, *236*, 184–194. [CrossRef]
40. Hethcote, H.W. The Mathematics of Infectious Diseases. *SIAM Rev.* **2000**, *42*, 599–653. [CrossRef]

Article

Analysis of HIV/AIDS Epidemic and Socioeconomic Factors in Sub-Saharan Africa

Shuman Sun, Zhiming Li, Huiguo Zhang , Haijun Jiang and Xijian Hu *

College of Mathematics and System Science, Xinjiang University, Urumqi 830046, China; Sunshuman2222@163.com (S.S.); zmli@xju.edu.cn (Z.L.); zhang_huiguo@163.com (H.Z.); jianghaixju@163.com (H.J.)
* Correspondence: xijianhu@xju.edu.cn

Received: 21 September 2020; Accepted: 26 October 2020; Published: 29 October 2020

Abstract: Sub-Saharan Africa has been the epicenter of the outbreak since the spread of acquired immunodeficiency syndrome (AIDS) began to be prevalent. This article proposes several regression models to investigate the relationships between the HIV/AIDS epidemic and socioeconomic factors (the gross domestic product per capita, and population density) in ten countries of Sub-Saharan Africa, for 2011–2016. The maximum likelihood method was used to estimate the unknown parameters of these models along with the Newton–Raphson procedure and Fisher scoring algorithm. Comparing these regression models, there exist significant spatiotemporal non-stationarity and auto-correlations between the HIV/AIDS epidemic and two socioeconomic factors. Based on the empirical results, we suggest that the geographically and temporally weighted Poisson autoregressive (GTWPAR) model is more suitable than other models, and has the better fitting results.

Keywords: HIV/AIDS epidemic; regression model; Newton–Raphson procedure; Fisher scoring algorithm; time series

---

1. Introduction

Acquired immunodeficiency syndrome (AIDS) is a malignant infectious disease with a high fatality rate caused by human immunodeficiency virus (HIV). The HIV/AIDS epidemic has been one of the greatest global public health and social development problems since 1981, particularly in Sub-Saharan Africa. As of 31 December 2016, over 30 million people had died from the disease [1]. More than 70% of the 35 million people are infected with the HIV/AIDS disease in Sub-Saharan Africa. Thus, the HIV/AIDS epidemic of Sub-Saharan Africa has attracted extensive attention from researchers around the world [2–4].

In earlier studies, Janet et al. [5] and Hallman et al. [6] demonstrated the relationship between the disease and socioeconomic status. Chris et al. [7] indicated socioeconomic factors to explain this disease outperformed cultural ones in South Africa. Mathematical models always play an important role in evaluating the trends of the HIV/AIDS epidemic [8]. For example, regression models have been widely used in the study of the relationship between this disease and influencing factors. Shiboski et al. [9] considered a generalized linear model to obtain the statistical analysis of the HIV/AIDS disease. A mixed-effects linear regression model was used to analyze the correlation between national population and antenatal care [10]. Laurence et al. [11] applied a spatial regression model to show that the epidemic had substantial geographic variance across Sub-Saharan Africa.

This paper proposes several regressive models to investigate the relationships between the HIV/AIDS epidemic, the gross domestic product (GDP) per capita and the population density in ten countries of Sub Saharan Africa. The Poisson regression model is introduced in Section 2.1. Sections 2.2 and 2.3

describe two spatial models, respectively. A spatiotemporal autoregressive model is proposed in Section 2.4. The maximum likelihood method is used to obtain the iterative formulas of coefficient estimations in Section 3. The main results are shown in Section 4, followed by discussion in Section 5.

2. Methodologies

2.1. Poisson Regression Model

Regression models are a set of statistical processes for estimating the relationships between response and explanatory variables. The classical model is a linear regression. Nelder and Wedderburn [12] extended the linear model to a generalized linear regression for solving the discrete data problem. This kind of models are very important in ecology, medicine and economics [13–15]. Suppose that $Y = (Y_1, Y_2, \ldots, Y_n)$ is the response variable, where $Y_i (i = 1, \ldots, n)$ are independent. The density function is

$$f(y_i; \theta_i, \phi_i) = \exp\left(\frac{y_i\theta_i - b(\theta_i)}{a(\phi_i)} + c(y_i, \phi_i)\right),$$

where $a(\cdot), b(\cdot), c(\cdot, \cdot)$ are known functions, and θ_i, ϕ_i are unknown parameters for $i = 1, 2, \ldots, n$. Denote $\mu_i = E(Y_i)$, and $g(\mu_i) = \ln(\mu_i)$ is a link function. Let X_{ij} be explanatory variables for the ith observation in the jth variable. Then, the Poisson regression (PR) model is given by

$$g(\mu_i) \triangleq \eta_i = \sum_{j=1}^{p} \beta_j X_{ij}, \tag{1}$$

where $i = 1, 2, \ldots, n$, and $\beta_j (j = 1, 2, \ldots, p)$ are unknown parameters.

2.2. Geographically Weighted Poisson Regression Model

With in-depth study, regression models have been frequently applied in epidemiology and health geography for trying to investigate the persistent geographical variations in disease [16]. Based on the generalized linear regression, Brunsdon et al. [17] proposed the geographically weighted regression model to analyze the spatial non-stationary processes of discrete data. The disease maps arising from this process are considered through the establishment of the geographically weighted Poisson regression (GWPR) model [18–20] below

$$g(\mu_i) \triangleq \eta_i = \sum_{j=1}^{p} \beta_j(u_i, v_i) X_{ij}, \tag{2}$$

where $(u_i, v_i)(i = 1, 2, \ldots, n)$ are the geographical locations, and $\beta_j(u_i, v_i)(j = 1, 2, \ldots, p)$ are unknown parameters at the position (u_i, v_i).

2.3. Geographically Weighted Poisson Autoregressive Model

Another issue deserving of special attention is whether there exists an interaction between different regions in terms of spatial data. Previous studies [21–24] showed that spatial data has not only spatial non-stationarity but also correlation. Zhang [25] proposed the geographically weighted Poisson autoregressive (GWPAR) model as follows:

$$g(\mu_i) \triangleq \eta_i = \rho \sum_{k=1}^{n} c_{ik}\eta_k + \sum_{j=1}^{p} \beta_j(u_i, v_i) X_{ij}, \tag{3}$$

where ρ is a scalar autoregressive parameter, and $c_{ik}(i,k=1,2,\ldots,n)$ is the adjacency relation between the ith and kth locations. Let c^i be the number of regions adjacent to the ith position. If the kth position is next to the ith's, then $c_{ik}=1/c^i$. Otherwise, $c_{ik}=0$.

2.4. Geographically and Temporally Weighted Poisson Autoregressive Model

Recently, many spatiotemporal models have been proposed to describe the spatiotemporal variations in the relationships of response and explanatory variables [26,27]. Concerning the modeling of spatiotemporal data, there are two important properties: non-stationarity and auto-correlation. The non-stationarity indicates that there exists more than one linear relation between response and explanatory variables. It can be used to identify where interesting relationships are likely to occur or where detailed investigation is necessary in the study areas [28]. Spatiotemporal auto-correlation is an important factor to determine the temporal correlations of observations [29]. These two problems always appeared together [30]. A geographically and temporally weighted autoregressive (GTWPAR) model can be applied to account for non-stationary and auto-correlated effects simultaneously.

Let Y be the response variable, and $Y_{ik}(i=1,2,\ldots,n_k,k=1,2,\ldots,T)$ be the independent variables of Y in the ith position and the kth time. The density function can be defined as follows:

$$f(y_{ik};\theta_{ik},\phi_{ik}) = \exp\left(\frac{y_{ik}\theta_{ik}-b(\theta_{ik})}{a(\phi_{ik})}+c(y_{ik},\phi_{ik})\right),$$

where the parameters are similar to Section 2.1. Denote $\mu_{ik}=E(Y_{ik})$, and $g(\mu_{ik})=\ln(\mu_{ik})$. Let $X_{ijk}(j=1,2,\ldots,p)$ be the jth explanatory variable. The GTWPAR model is expressed by

$$g(\mu_{ik}) \triangleq \eta_{ik} = \rho \sum_{m=1}^{T}\sum_{l=1}^{n_k} c_{lm}^{(ik)}\eta_{lm} + \sum_{j=1}^{p}\sum_{k=1}^{T}\beta_{jk}(u_{ik},v_{ik},t_k)X_{ijk}, \tag{4}$$

where $\{\beta_{jk}(u_{ik},v_{ik},t_k)\}$ is a set of unknown parameters at the ith position in the kth time, and $c_{lm}^{(ik)}$ is the adjacent relation between the location (u_{ik},v_{ik},t_k) and (u_{lm},v_{lm},t_m). Following the work of [31], the spatiotemporal distance between the locations (u_{ik},v_{ik},t_k) and (u_{lm},v_{lm},t_m) can be defined as

$$d_{lm}^{(ik)} = \sqrt{\lambda[(u_{ik}-u_{lm})^2+(v_{ik}-v_{lm})^2]+\mu(t_k-t_m)^2},$$

where μ and λ are used to balance spatiotemporal distances. Suppose that

$$c_{lm}^{(ik)} = \begin{cases} 1/c^{ik}, & 0 < d_{lm}^{(ik)} < d, \\ 0, & \text{otherwise}, \end{cases}$$

where d is a constant and satisfies $\min\{d_{lm}^{(ik)}\} < d < \max\{d_{lm}^{(ik)}\}$.

Next, we rewrite the model (4) in a matrix form

$$\eta = \rho C\eta + \mathbf{B}'X',$$

where $\eta = (\eta_{11},\cdots,\eta_{n_11},\eta_{12},\cdots,\eta_{n_22},\cdots,\eta_{1T},\cdots,\eta_{n_TT})'$, $C = (c_{lm}^{(ik)})$, $X = (X_{ijk})$ and $\mathbf{B} = (\beta_{jk}(u_{ik},v_{ik},t_k))$. For convenience, define η_K as the Kth element of η; C_{IK} and X_{IK} are the Ith row and the Kth column of the matrices C and X, respectively. The detailed expressions of C, X and $\mathbf{B}$ are given in Appendix A.1.

Remark 1. *For the GTWPAR model (4), if $\rho = 0$ and $\beta_{jk}(u_{ik}, v_{ik}, t_k)$ is independent of the spatiotemporal effect, the model is a PR model. If $\rho = 0$ and $\beta_{jk}(u_{ik}, v_{ik}, t_k)$ is dependent on spatial effect but independent of temporal effect, the model becomes GWPR model. If $\rho \neq 0$ and $\beta_{jk}(u_{ik}, v_{ik}, t_k)$ is independent of temporal effect, it is the GWPAR model. Thus, PR, GWPR and GWPAR models are the special cases of the GTWPAR model.*

3. Coefficient Estimation

In this section, we only provide the estimation method of the GTWPAR model since the PR, GWPR and GWPAR models are its special cases (Remark 1). Let $(u_{ik}, v_{ik}, t_k)(i = 1, 2, \ldots, n_k, k = 1, 2, \ldots, T)$ be any point in the studied spatiotemporal region. We fix a point (u_{00}, v_{00}, t_0) and assume that $\beta_{jk}(u_{ik}, v_{ik}, t_k) \approx \beta_{j0}(u_{00}, v_{00}, t_0)(j = 1, 2, \ldots, p)$. Then, the model (4) can be rewritten by

$$\eta_{ik} = g(\mu_{ik}) = \rho \sum_{m=1}^{T} \sum_{l=1}^{n_k} c_{lm}^{(ik)} \eta_{lm} + \sum_{j=1}^{p} \sum_{k=1}^{T} \beta_{j0}(u_{00}, v_{00}, t_0) X_{ijk}. \tag{5}$$

Denote $\beta(u_{00}, v_{00}, t_0) = (\beta_{10}, \ldots, \beta_{p0})'$, $\mathbf{X} = \mathrm{diag}(X_{i\cdot})$ and $X_{i\cdot} = (X_{i1}, \ldots, X_{ip})$. The corresponding matrix form can be represented as $\eta = \rho C \eta + \beta'(u_{00}, v_{00}, t_0) \mathbf{X}'$.

3.1. Estimation of Parameter Vector β

For the fixed point (u_{00}, v_{00}, t_0), we define a spatiotemporal distance $d_{ik}^{(0)}$ from this point to (u_{ik}, v_{ik}, t_k) as $d_{ik}^{(0)} = \sqrt{\lambda[(u_{00} - u_{ik})^2 + (v_{00} - v_{ik})^2] + \mu(t_0 - t_k)^2}$. The Gauss kernel function of these two points can be written by

$$
\begin{aligned}
w_{ik}(u_{00}, v_{00}, t_0) &= \frac{1}{\sqrt{2\pi}} \exp\left\{ -\frac{1}{2}\left(\frac{d_{ik}^{(0)}}{h_{ST}}\right)^2 \right\} \\
&= \frac{1}{\sqrt{2\pi}} \exp\left\{ -\frac{1}{2} \frac{\lambda[(u_{00} - u_{ik})^2 + (v_{00} - v_{ik})^2] + \mu(t_0 - t_k)^2}{h_{ST}^2} \right\} \\
&= \frac{1}{\sqrt{2\pi}} \exp\left\{ -\frac{1}{2}\left(\frac{(u_{00} - u_{ik})^2 + (v_{00} - v_{ik})^2}{h_S^2} + \frac{(t_0 - t_k)^2}{\tau h_S^2}\right) \right\},
\end{aligned}
$$

where h_{ST} and h_S are the space-time bandwidth and space bandwidth, respectively. Meanwhile, we have $h_{ST}^2 = \lambda h_S^2$, and $\tau = \lambda/\mu$ is a spatiotemporal factor. Without loss of generality, let $\lambda = 1$. Then, the weighted maximum likelihood of $Y_{ik}(i = 1, 2, \ldots, n_k, k = 1, 2, \ldots, T)$ at the point (u_{00}, v_{00}, t_0) is

$$L(\beta_{10}, \beta_{20}, \cdots, \beta_{p0}) = \prod_{k=1}^{T} \prod_{i=1}^{n_k} f(y_{ik}; \theta_{ik}, \phi_{ik}) w_{ik}(u_{00}, v_{00}, t_0),$$

where $f(y_{ik}; \theta_{ik}, \phi_{ik})$ is the density function. The log-likelihood can be obtained as follows:

$$\mathbf{L}_1(\beta(u_{00}, v_{00}, t_0)) = \sum_{k=1}^{T} \sum_{i=1}^{n_k} \left(\frac{y_{ik}\theta_{ik} - b(\theta_{ik})}{a(\phi_{ik})} + c(y_{ik}, \phi_{ik})\right) w_{ik}(u_{00}, v_{00}, t_0).$$

Note that $c(y_{ik}, \phi_{ik}) = -\ln(y_{ik}!)$, $b(\theta_{ik}) = \mu_{ik} = \exp(\theta_{ik})$, and $a(\phi_{ik}) = \phi_{ik} = 1$. Thus, $E(Y_{ik}) = b'(\theta_{ik}) = \exp(\theta_{ik}) = \mu_{ik}$, $Var(Y_{ik}) = b''(\theta_{ik})a(\phi_{ik}) = \exp(\theta_{ik}) = \mu_{ik}$. Differentiating $\mathbf{L}_1$ with respect to $\beta(u_{00}, v_{00}, t_0)$ yields

$$\frac{\partial \mathbf{L}_1}{\partial \beta_{r0}} = \sum_{k=1}^{T} \sum_{i=1}^{n_k} \left(\frac{y_{ik} - \mu_{ik}}{a_{ik}\phi} \frac{\partial \theta_{ik}}{\partial \beta_{r0}} \right) w_{ik}(u_{00}, v_{00}, t_0) = 0, \tag{6}$$

where $\beta_{r0} = \beta_r(u_{00}, v_{00}, t_0)(r = 1, 2, \cdots, p)$, and

$$\frac{\partial \theta_{ik}}{\partial \beta_{r0}} = \left(\frac{\partial \mu_{ik}}{\partial \theta_{ik}} \right)^{-1} \frac{\partial \mu_{ik}}{\partial g(\mu_{ik})} \frac{\partial g(\mu_{ik})}{\partial \beta_{r0}} = \frac{1}{b''(\theta_{ik})} \frac{1}{g'(\mu_{ik})} \frac{\partial \eta_{ik}}{\partial \beta_{r0}}.$$

For convenience, let $N = \sum_{k=1}^{N} n_k$ and $W = (w_{ik}(u_{00}, v_{00}, t_0))_{N \times N}$. Denote $A = (I_N - \rho C)^{-1}$, $Y = (Y_{11}, \cdots, Y_{n_1 1}, \cdots, Y_{1T}, \cdots, Y_{n_T T})'$, $\mu = (\mu_{11}, \cdots, \mu_{n_1 1}, \cdots, \mu_{1T}, \cdots, \mu_{n_T T})'$, $\theta = (\theta_{11}, \cdots, \theta_{n_1 1}, \cdots, \theta_{1T}, \cdots, \theta_{n_T T})'$, $\phi = (\phi_{11}, \cdots, \phi_{n_1 1}, \cdots, \phi_{1T}, \cdots, \phi_{n_T T})'$. Suppose that Y_K, μ_K, θ_K and ϕ_K are the Kth elements of Y, μ, θ and ϕ, respectively. Then, we take the derivative of the model (5) with respect to β_{r0}, and obtain

$$\frac{\partial \eta_l}{\partial \beta_{r0}} = \sum_{h=1}^{N} A_{lh} X_{hr} = A_l. X._r, \quad l = 1, 2, \ldots, N.$$

The calculation process is given in Appendix A.2. Thus, the Equation (6) can be rewritten as

$$\frac{\partial \mathbf{L}_1}{\partial \beta_{r0}} = \frac{1}{\phi} \sum_{l=1}^{N} T_l A_l. X._r (Y_l - \mu_l) g'(\mu_l) W_l(u_{00}, v_{00}, t_0) = 0.$$

However, there is not a close-form solution for $\beta(u_{00}, v_{00}, t_0)$. The Newton–Raphson procedure and Fisher scoring algorithm are used to get the estimation of β. The iterative formula is expressed as

$$
\begin{aligned}
\hat{\beta}^{(m+1)}(u_{00}, v_{00}, t_0) &= \hat{\beta}^{(m)}(u_{00}, v_{00}, t_0) + \mathbf{I}^{-1}(\hat{\beta}^{(m)}(u_{00}, v_{00}, t_0)) S(\hat{\beta}^{(m)}(u_{00}, v_{00}, t_0)) \\
&= ((A^{(m)} X)' T^{(m)} W(u_{00}, v_{00}, t_0)(A^{(m)} X))^{-1} \\
&\quad \times (A^{(m)} X)' T^{(m)} W(u_{00}, v_{00}, t_0) Z^{(m)},
\end{aligned}
\tag{7}
$$

where the Fisher information matrix $\mathbf{I}(\beta) = E(I(\beta))$, and

$$S(\hat{\beta}^{(m)}(u_{00}, v_{00}, t_0)) = \left(\frac{\partial \mathbf{L}_1}{\partial \beta_{10}}, \frac{\partial \mathbf{L}_1}{\partial \beta_{20}}, \cdots, \frac{\partial \mathbf{L}_1}{\partial \beta_{p0}} \right)'$$

is the scalar vector. The detail process is provided in Appendix A.2. For the fixed point $(u_{ik}, v_{ik}, t_k)(i = 1, 2, \ldots, n_k; k = 1, 2, \ldots, T)$, $\hat{\beta}_{jk}(u_{ik}, v_{ik}, t_k)$ can be obtained by (7).

Remark 2. *The estimations $\hat{\beta}(u_{ik}, v_{ik}, t_k)(i, l = 1, 2, \ldots, n_k, k, m = 1, 2, \ldots, T)$ are related to the temporal and spatial effects in the GTWPAR model. If $m \neq k$, $w_{ik}(u_{lm}, v_{lm}, t_m) = 0$ and $c_{lm}^{(ik)} = 0$, then $\hat{\beta}(u_{ik}, v_{ik}, t_k) = \hat{\beta}(u_i, v_i)$ correspond to the parameter estimations of the GWPAR model. If $w_{ik}(u_{lm}, v_{lm}, t_m) = 0(m \neq k)$ and $C = 0$, they are the estimations of the GWPR model. If $W = 0$ and $C = 0$, then $\hat{\beta}(u_{ik}, v_{ik}, t_k) = \hat{\beta}$ are the global estimation values of the PR model.*

3.2. Estimation of Parameter ρ

Based on the density function, the log-likelihood function of ρ is

$$\mathbf{L}_2(\rho) = \sum_{k=1}^{T} \sum_{i=1}^{n_k} \left(\frac{y_{ik}\theta_{ik} - b(\theta_{ik})}{a(\phi_{ik})} + c(y_{ik}, \phi_{ik}) \right).$$

Differentiating $\mathbf{L}_2(\rho)$ with respect to ρ, we have

$$\frac{\partial \mathbf{L}_2}{\partial \rho} = \sum_{k=1}^{T} \sum_{i=1}^{n_k} \left(\frac{y_{ik} - \mu_{ik}}{a_{ik}\phi} \frac{\partial \theta_{ik}}{\partial \rho} \right) = 0, \tag{8}$$

where $\frac{d\theta_{ik}}{d\rho} = \frac{1}{b''(\theta_{ik})g'(\mu_{ik})} \frac{d\eta_{ik}}{d\rho}$. Then, we take the derivative of the model (5) with respect to ρ as follows:

$$\frac{d\eta_l}{d\rho} = \frac{dg(\mu_l)}{d\rho} = \sum_{h=1}^{N} A_l.C_{\cdot h}\eta_h.$$

The detail calculation is given in Appendix A.3. Then, the Equation (8) can be rewritten in the following nonlinear form

$$\frac{d\mathbf{L}_2}{d\rho} = \sum_{l=1}^{N} \frac{(Y_l - \mu_l) \sum_{h=1}^{N} A_l.C_{\cdot h}\eta_h}{a_l \phi V(\mu_l)g'(\mu_l)} = 0.$$

According to the Newton–Raphson procedure and Fisher scoring algorithm, the iterative formula of $\hat{\rho}^{(m+1)}$ is

$$\begin{aligned} \hat{\rho}^{(m+1)} &= \hat{\rho}^{(m)} + \mathcal{I}^{-1}(\hat{\rho}^{(m)})S(\hat{\rho}^{(m)}) \\ &= \hat{\rho}^{(m)} + ((A^{(m)}C\eta^{(m)})'T^{(m)}(A^{(m)}C\eta^{(m)}))^{-1} \\ &\quad \times (A^{(m)}C\eta^{(m)})'T^{(m)}(Z^{(m)} - \eta^{(m)}), \end{aligned} \tag{9}$$

where the scalar vector $S(\hat{\rho}^{(m)}) = \frac{1}{\phi}(AC\eta)'T(Z - \eta)$ and the Fisher information matrix $\mathcal{I}(\rho) = \frac{1}{\phi}(AC\eta)'T(AC\eta)$. The calculation process of the scalar vector $S(\hat{\rho}^{(m)})$ and the information matrix $\mathcal{I}$ is given in Appendix A.3.

4. Main Results

In this section, we apply the PR, GWPR, GWPAR and GTWPAR models to analyze the relationships between the HIV/AIDS epidemic, the GDP per capita and population density in ten countries of Sub-Saharan Africa from 2011 to 2016. The ten countries are Angola, Botswana, Lesotho, Malawi, Mozambique, Namibia, South Africa, Swaziland, Zimbabwe and Zambia. The parameters of these four models are estimated by the Newton–Raphson procedure and Fisher scoring algorithm. The coefficient of determination R^2, the corrected Akaike information criterion (AICc), the deviation (D) and mean-square error (MSE) are used to compare the performances of the four models [18].

4.1. The HIV/AIDS Epidemic Models

The data of HIV/AIDS incidence, GDP per capita and population density were derived from http://data.cnki.net/InternationalData/Report. Readers should note that authorization is required to access the database on this website. Figure 1 describes the HIV/AIDS incidence in ten countries from 2011 to 2016. It shows that the incidence varies significantly in different regions. Angola has a minimum incidence of less than 5%, while Botswana and Swaziland have higher incidences of more than 20% every years. Therefore, it may be necessary to consider the temporal and spatial factors in analyzing the HIV/AIDS epidemic.

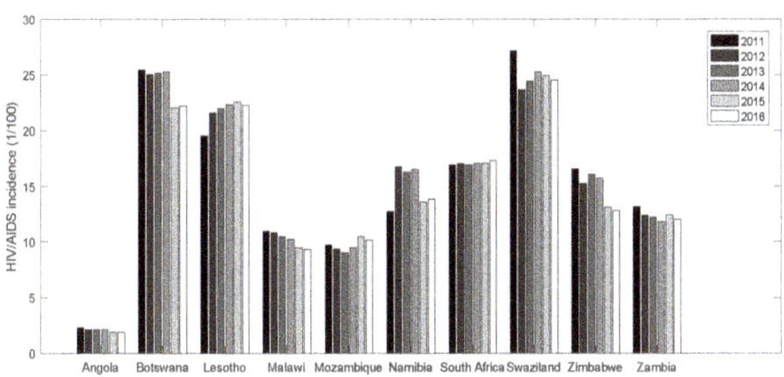

Figure 1. Spatiotemporal HIV/AIDS incidence of ten countries, 2011–2016.

The distributions of HIV/AIDS cases, GDP per capita and population density are displayed in Figure 2. The Pearson correlation coefficients between these cases and GDP per capita and population density are 0.2739 and −0.1179, respectively. Meanwhile, the two socioeconomic factors have different effects on the HIV/AIDS cases at the spatiotemporal locations. These reflect a spatiotemporal non-stationarity between the cases and two factors in ten countries from 2011 to 2016. Table 1 lists the p-values of the first-order autocorrelation of HIV/AIDS cases in the different years of the same region or the different regions of the same year. Each region has a significant spatial autocorrelation (p-value < 0.01) each year. Lesotho and South Africa had temporal autocorrelation during 2011 to 2016. Thus, the spatial and temporal autocorrelation should not be ignored.

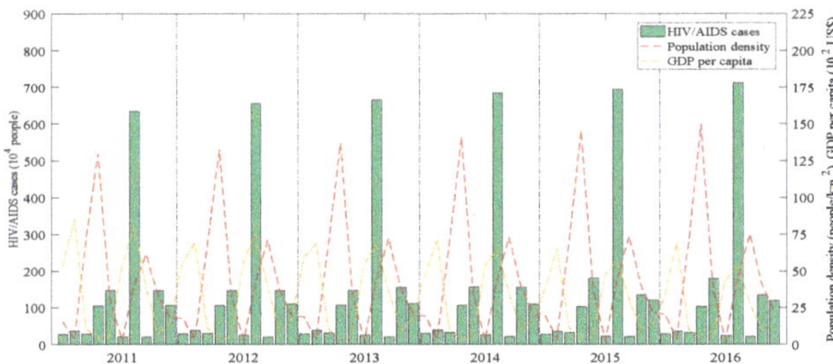

Figure 2. Distributions of HIV/AIDS cases, GDP per capita and population density of ten countries, 2011–2016.

Table 1. *p*-values of the spatial and temporal autocorrelation analysis.

Time	2011	2012	2013	2014	2015	2016
p-value	0.0024	0.0028	0.0021	0.0019	0.0017	0.0022
Regions	**Angola**	**Botswana**	**Lesotho**	**Malawi**	**Mozambique**	**Namibia**
p-value	0.9094	0.8807	0.0092	0.5300	0.1289	0.8267
	South Africa	**Swaziland**	**Zimbabwe**	**Zambia**		
	0.0045	0.1491	0.5688	0.2231		

Next, we standardized the two socioeconomic factors. The multiplex collinear test [32] was performed by the condition number $k = \sqrt{\lambda_{\max}/\lambda_{\min}} = 1.804 (\leq 15)$ (λ is the eigenvalue of explanatory variable matrix). If $k > 15$, then the data have collinearity. Otherwise, there is no collinearity. Thus, there is no collinearity between the two factors. Let μ_{ik}, r_{ik} and P_{ik} be the annual HIV/AIDS cases (Unit: 1/1000 people), incidence (Unit: 1/100) and total population (unit: 100,000 people) in the kth year of the ith region, respectively. Denote $g(\mu_{ik}) = \eta_{ik} = \ln \mu_{ik} = \ln r_{ik} + \ln P_{ik} (\mu_{ik} = r_{ik}P_{ik}, i = 1, 2, \ldots, 10$ and $k = 1, 2, \ldots, 6)$. Let X_{i1k} and X_{i2k} be the GDP per capita and population density in the ith region at the kth year, respectively. The PR model is written by

$$g(\mu_{ik}) = \beta_0 + \beta_1 X_{i1k} + \beta_2 X_{i2k}, \quad i = 1, 2, \ldots, 10, k = 1, 2, \ldots, 6, \tag{10}$$

where $\beta_j (j = 0, 1, 2)$ are unknown constants. The GWPR model is introduced as

$$g(\mu_{ik}) = \beta_0(u_{ik}, v_{ik}) + \beta_1(u_{ik}, v_{ik}) X_{i1k} + \beta_2(u_{ik}, v_{ik}) X_{i2k}, \quad i = 1, 2, \ldots, 10, \tag{11}$$

where k is a fixed constant taken from $\{1, 2, \ldots, 6\}$, and $\beta_j(u_{ik}, v_{ik})$ are unknown spatial parameters for the ith country (u_{ik}, v_{ik}) in the kth year. Let ρ be a scalar autoregressive parameter, and c_{il} be a constant that represents an adjacency relation. The GWPAR model is

$$g(\mu_{ik}) = \rho \sum_{l=1}^{n} c_{il}\eta_{ik} + \beta_0(u_{ik}, v_{ik}) + \beta_1(u_{ik}, v_{ik}) X_{i1k} + \beta_2(u_{ik}, v_{ik}) X_{i2k}, \tag{12}$$

where $n = 10$, k is a fixed constant, and $\beta_j(u_{ik}, v_{ik})$ are defined as above. Let $c_{lm}^{(ik)}$ be a spatiotemporal adjacency relation, and $\beta_{jk}(u_{ik}, v_{ik}, t_k)(k = 1, 2, \ldots, 6)$ be unknown spatiotemporal parameters in the ith country (u_{ik}, v_{ik}) in the kth year. The GTWPAR model is established as follows:

$$\begin{aligned} g(\mu_{ik}) &= \rho \sum_{m=1}^{T} \sum_{l=1}^{n_k} c_{lm}^{(ik)} \eta_{lm} + \beta_{0k}(u_{ik}, v_{ik}, t_k) \\ &\quad + \beta_{1k}(u_{ik}, v_{ik}, t_k) X_{i1k} + \beta_{2k}(u_{ik}, v_{ik}, t_k) X_{i2k}, \end{aligned} \tag{13}$$

where $T = 6$; $n_k = 10$ for every k years; and ρ is defined as above.

Algorithms I, II, III and IV of PR, GWPR, GWPAR and GTWPAR models are provided in Appendix A.4, respectively.

4.2. Statistical Analysis

For the PR model, we get the estimated values of unknown parameters by Algorithm I. Then, the best space bandwidth is chosen by the cross-validation method. Following Huang et al. [28], the range [0.09, 2.49] of the space bandwidth is selected according to the minimum and maximum distance of the geographical positions. In the GWPR model, the best space bandwidth is $h = 0.62, 0.59, 0.62, 0.61, 0.60, 0.60$, and the estimations of coefficient functions are given by Algorithm II. The optimal space bandwidth of the GWPAR model is selected as $h = 1.2895, 1.1316, 1.1316, 1.0526, 1.0526, 1.0526$. Based on Algorithm III, we can get the estimations of coefficient functions and the scalar autoregressive parameter $\hat{\rho} = 0.267, 0.269, 0.263, 0.264, 0.264, 0.264$. For the GTWPAR model, we chose $h_s = 1.1316, 0.9737, 0.8947, 0.9211, 0.7842, 0.8789$ and $\tau = 0.1$, where $\tau(> 0)$ is a balanced parameter. The coefficient estimations and scalar autoregressive parameter $\hat{\rho} = 0.126$ can be obtained by Algorithm IV. The quantile and mean values of coefficient estimations and response variables are shown in Table 2. We note that the GWPR, GWPAR and GTWPAR models can reflect the non-stationarity property of the influencing factors; the PR model cannot. Moreover, the GTWPAR model has a better performance than other models by comparing the true and fitted values.

Table 2. The quantile and mean values of coefficient estimations and response variables.

Model	Coefficient	Min	1st Qu	Median	3rd Qu	Max	Mean
True	η	5.293	5.644	6.455	7.288	8.871	6.559
PR	$\hat{\beta}_1$	0.581	0.581	0.581	0.581	0.581	0.581
	$\hat{\beta}_2$	0.385	0.385	0.385	0.385	0.385	0.385
	$\hat{\eta}$	6.344	6.527	7.185	7.447	8.162	7.078
GWPR	$\hat{\beta}_1$	−0.519	−0.249	0.009	0.641	1.774	0.288
	$\hat{\beta}_2$	−0.672	−0.139	0.068	2.044	3.803	0.747
	$\hat{\eta}$	5.618	6.422	7.022	7.202	8.813	6.932
GWPAR	$\hat{\beta}_1$	−6.286	−1.258	−0.910	0.240	3.036	−0.556
	$\hat{\beta}_2$	−2.089	−0.669	0.019	2.281	9.152	0.912
	$\hat{\eta}$	5.176	5.625	6.435	7.219	8.940	6.539
GTWPAR	$\hat{\beta}_1$	−5.642	−0.956	−0.677	0.186	2.992	−0.443
	$\hat{\beta}_2$	−1.865	−0.464	0.215	2.134	9.104	0.988
	$\hat{\eta}$	5.200	5.589	6.387	7.215	8.803	6.489

The average estimated coefficients are visualized in Figure 3. For the PR model, the GDP per capita and population density had the same effect on the HIV/AIDS epidemic for ten countries in six years. However, there exist significant spatial non-stationarity and auto-correlation for different countries under the GWPR, GWPAR and GTWPAR models. Figure 4 shows the spatial distribution of the average MSE of their response variables. The lighter the color, the smaller the average error is. Thus, the GWPAR and GTWPAR models have the better fitting results.

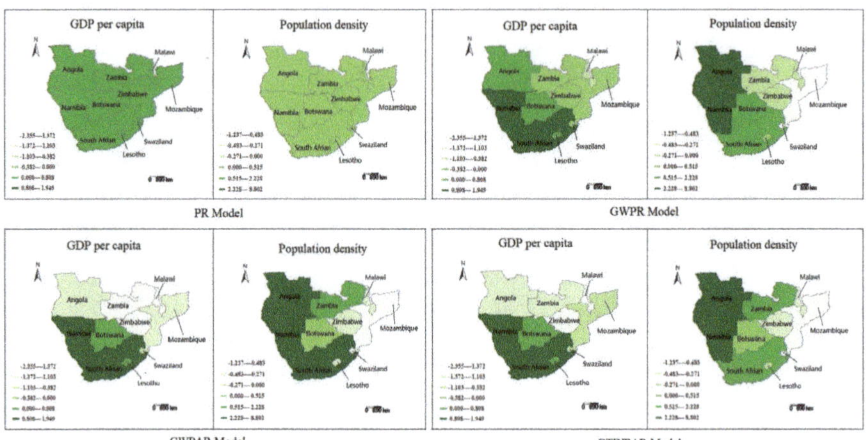

Figure 3. The spatial distribution of the average coefficient estimations in four models.

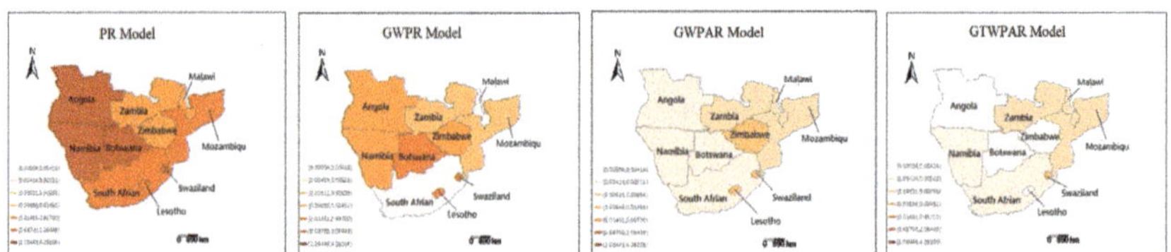

Figure 4. The average MSE of response variables.

These four indicators can effectively compare the performances of the proposed models (Table 3). The calculation formulas of R^2, AICc, D and MSE are given in Appendix A.5. The coefficient of determination R^2 gradually increases from 12.91% of the PR model to 99.57% of the GTWPAR model. The MSE, AICc and D values of the GTWPAR model are smaller than those of other models. Therefore, the GTWPAR model is more suitable to investigate the spatiotemporal HIV/AIDS epidemic.

Table 3. The comparison of the four models.

Model	R^2	AICc	D	MSE
PR	0.1291	42,504.40	42,624.07	1.6488
GWPR	0.6139	6495.12	6613.02	0.4326
GWPAR	0.9940	155.46	236.02	0.0067
GTWPAR	0.9957	115.25	190.05	0.0048

Based on the GTWPAR model, the mean values and 95% confidence intervals of the coefficient estimations are shown in Figure 5. The mean estimations are represented by the dot, and the 95% confidence intervals are given by the upper and lower lines. Note that the GDP per capita in Botswana, Namibia and South Africa has a positive effect on the HIV/AIDS cases. Six other countries (except Lesotho) had the opposite results. The population density for five countries had a positive effect on the HIV/AIDS cases—Angola, Botswana, Namibia, South Africa and Zambia. The population density of other five countries had the negative effect. Moreover, the impact of the GDP per capita on HIV/AIDS epidemic had a strong spatiotemporal non-stationarity in Lesotho, Malawi and Zimbabwe, while the population density had a strong spatiotemporal non-stationarity in Angola.

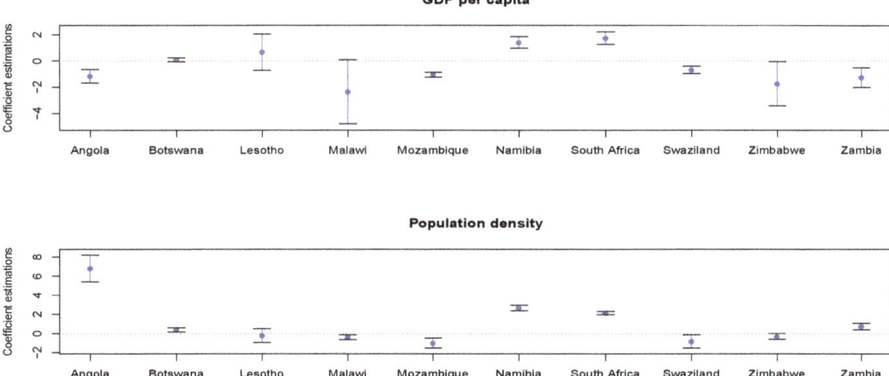

Figure 5. The mean values and 95% confidence intervals of coefficient estimations.

5. Conclusions

In this paper, we propose four regression models, including the PR, GWPR, GWPAR and GTWPAR, to investigate the non-stationary and auto-correlation properties. The relationships between the HIV/AIDS epidemic, GDP per capita and population density were analyzed in ten countries of Sub-Saharan Africa from 2011 to 2016. The unknown parameters of these models can be estimated by the Newton–Raphson procedure and Fisher scoring algorithm.

The PR model is a classical generalized model, which considers the global relationships between the response and explanatory variables. The GWPR and GWPAR models have been introduced to determine the spatial non-stationarity or auto-correlation. The GTWPAR model proposed by this article can be used to investigate not only spatiotemporal non-stationary but also auto-correlation. Thus, the PR, GWPR and GWPAR models are several special cases of the GTWPAR model (see Remark 1 and Remark 2). The performances of these models were evaluated by analyzing the correlations between the HIV/AIDS epidemic and two socioeconomic factors. The parameter estimations of the models can be obtained by Algorithms I, II, III and IV in Appendix A.4.

The results show that the impacts of GDP per capita and population density on HIV /AIDS cases had significant spatiotemporal non-stationarity and auto-correlation. The GWPR, GWPAR and GTWPAR models can reflect the strong spatial or spatiotemporal non-stationarity. The auto-correlation can be reflected in the GWPAR and GTWPAR models. Compared with other models, the GTWPAR model is more effective in terms of four comparison indicators. Thus, we suggest that the GTWPAR model can be used to analyze the spatiotemporal characteristics of the HIV/AIDS epidemic and the influences of the GDP per capita and population density.

Further work also exists in our study. For example, we observed that the effects of the GDP per capita for Lesotho, Malawi and Zimbabwe and the population density for Angola on HIV/AIDS had strong spatiotemporal non-stationarity. These may be the result of local environmental or political factors. Whether the fitting results of these regions will perform better if explanatory variables such as local unique environmental or political factors are added needs to be further investigated.

Author Contributions: Conceptualization, Z.L., H.Z. and X.H.; methodology, S.S., H.Z. and X.H.; software, S.S.; supervision, Z.L. and H.J.; visualization, Z.L.; writing—original draft, S.S.; writing—review & editing, S.S., Z.L. and H.J. All authors have read and agreed to the published version of the manuscript.

Funding: This work was supported by the Natural Science Foundation of Xinjiang Uygur Autonomous Region (XJEDU2017M001, 2018Q011), and the National Natural Science Foundation of China (U1703237, 11661076, 12061070).

Conflicts of Interest: The authors declare no conflict of interest.

Abbreviations

The following abbreviations are used in this manuscript:

HIV/AIDS	Human immunodeficiency virus/Acquired immunodeficiency syndrome
PR	Poisson regression model
GWPR	Geographically weighted Poisson regression model
GWPAR	Geographically weighted Poisson autoregressive model
GTWPAR	Geographically and temporally weighted Poisson autoregressive model
GDP	Gross domestic product

Appendix A. Detailed Processes

Appendix A.1. The Expressions of C, X and B

In model $\eta = \rho C \eta + \mathbf{B}'X'$, the expressions of C, X and B are

$$
C = \begin{pmatrix}
c_{11}^{(11)} & \cdots & c_{11}^{(n_1 1)} & \cdots & c_{11}^{(1T)} & \cdots & c_{11}^{(n_T T)} \\
\vdots & \vdots & \vdots & \vdots & \vdots & \vdots & \vdots \\
c_{(n_1 1)}^{(11)} & \cdots & c_{(n_1 1)}^{(n_1 1)} & \cdots & c_{(n_1 1)}^{(1T)} & \cdots & c_{(n_1 1)}^{(n_T T)} \\
\vdots & \vdots & \vdots & \vdots & \vdots & \vdots & \vdots \\
c_{1T}^{(11)} & \cdots & c_{1T}^{(n_1 1)} & \cdots & c_{1T}^{(1T)} & \cdots & c_{1T}^{(n_T T)} \\
\vdots & \vdots & \vdots & \vdots & \vdots & \vdots & \vdots \\
c_{(n_T T)}^{(11)} & \cdots & c_{(n_T T)}^{(n_1 1)} & \cdots & c_{(n_T T)}^{(1T)} & \cdots & c_{(n_T T)}^{(n_T T)}
\end{pmatrix}, X = \begin{pmatrix}
X_{111} & X_{121} & \cdots & X_{1p1} \\
\vdots & \vdots & \vdots & \vdots \\
X_{n_1 11} & X_{n_1 21} & \cdots & X_{n_1 p1} \\
\vdots & \vdots & \vdots & \vdots \\
X_{11T} & X_{12T} & \cdots & X_{1pT} \\
\vdots & \vdots & \vdots & \vdots \\
X_{n_T 1T} & X_{n_T 2T} & \cdots & X_{n_T pT}
\end{pmatrix},
$$

where C_{IK}, X_{IK} are respectively the *I*th row and the *K*th column of the matrices C and X. Moreover,

$$
\begin{aligned}
\mathbf{B} = {} & (\beta_{11}(u_{11}, v_{11}, t_1), \cdots, \beta_{p1}(u_{11}, v_{11}, t_1), \cdots, \beta_{11}(u_{n_1 1}, v_{n_1 1}, t_1), \cdots, \beta_{p1}(u_{n_1 1}, v_{n_1 1}, t_1), \cdots, \\
& \beta_{1T}(u_{1T}, v_{1T}, t_T), \cdots, \beta_{pT}(u_{1T}, v_{1T}, t_T), \cdots, \beta_{1T}(u_{n_T 1}, v_{n_T 1}, t_1), \cdots, \beta_{pT}(u_{n_T 1}, v_{n_T 1}, t_1))'.
\end{aligned}
$$

Appendix A.2. Formula and Information Matrix of $\beta(u, v, t)$

(1) For the matrix form of model (5), we obtain

$$
\begin{cases}
(1 - \rho C_{11})\eta_1 - \rho C_{12}\eta_2 - \cdots - \rho C_{1N}\eta_N = \displaystyle\sum_{j=1}^{p} \beta_{j0} X_{1j}, \\[2ex]
-\rho C_{21}\eta_1 + (1 - \rho C_{22})\eta_2 - \cdots - \rho C_{2N}\eta_N = \displaystyle\sum_{j=1}^{p} \beta_{j0} X_{2j}, \\[2ex]
\quad\vdots \\[1ex]
-\rho C_{N1}\eta_1 - \rho C_{N2}\eta_2 - \cdots + (1 - \rho C_{NN})\eta_N = \displaystyle\sum_{j=1}^{p} \beta_{j0} X_{Nj}.
\end{cases}
$$

Differentiating the above equations with β_{r0} yields

$$
\begin{cases}
(1 - \rho C_{11})\dfrac{\partial \eta_1}{\partial \beta_{r0}} - \rho C_{12}\dfrac{\partial \eta_2}{\partial \beta_{r0}} - \cdots - \rho C_{1N}\dfrac{\partial \eta_N}{\partial \beta_{r0}} = X_{1r}, \\[2ex]
- \rho C_{21}\dfrac{\partial \eta_1}{\partial \beta_{r0}} + (1 - \rho C_{22})\dfrac{\partial \eta_2}{\partial \beta_{r0}} - \cdots - \rho C_{2N}\dfrac{\partial \eta_N}{\partial \beta_{r0}} = X_{2r}, \\[2ex]
\vdots \\[2ex]
- \rho C_{N1}\dfrac{\partial \eta_1}{\partial \beta_{r0}} - \rho C_{N2}\dfrac{\partial \eta_2}{\partial \beta_{r0}} - \cdots + (1 - \rho C_{NN})\dfrac{\partial \eta_N}{\partial \beta_{r0}} = X_{Nr}.
\end{cases}
$$

Then,

$$
\begin{pmatrix}
1 - \rho C_{11} & -\rho C_{12} & \cdots & -\rho C_{1N} \\
-\rho C_{21} & 1 - \rho C_{22} & \cdots & -\rho C_{2N} \\
\vdots & \vdots & \vdots & \vdots \\
-\rho C_{N1} & -\rho C_{N2} & \cdots & 1 - \rho C_{NN}
\end{pmatrix}
\begin{pmatrix}
\frac{\partial \eta_1}{\partial \beta_{r0}} \\
\frac{\partial \eta_2}{\partial \beta_{r0}} \\
\vdots \\
\frac{\partial \eta_N}{\partial \beta_{r0}}
\end{pmatrix}
=
\begin{pmatrix}
X_{1r} \\
X_{2r} \\
\vdots \\
X_{Nr}
\end{pmatrix}.
$$

Denote $\dfrac{\partial \eta}{\partial \beta_{r0}} = \left(\dfrac{\partial \eta_1}{\partial \beta_{r0}}, \dfrac{\partial \eta_2}{\partial \beta_{r0}}, \ldots, \dfrac{\partial \eta_N}{\partial \beta_{r0}}\right)'$ and

$$
A =
\begin{pmatrix}
1 - \rho C_{11} & -\rho C_{12} & \cdots & -\rho C_{1N} \\
-\rho C_{21} & 1 - \rho C_{22} & \cdots & -\rho C_{2N} \\
\vdots & \vdots & \vdots & \vdots \\
-\rho C_{N1} & -\rho C_{N2} & \cdots & 1 - \rho C_{NN}
\end{pmatrix}^{-1},
$$

$$
X_{\cdot r} = (X_{1r}, X_{2r}, \ldots, X_{Nr})', \quad A_{l\cdot} = (A_{l1}, A_{l2}, \ldots, A_{lN})'.
$$

Thus, $\dfrac{\partial \eta}{\partial \beta_{r0}} = A X_{\cdot r}$, that is

$$
\frac{\partial \eta_l}{\partial \beta_{r0}} = \sum_{h=1}^{N} A_{lh} X_{hr} = A_{l\cdot} X_{\cdot r}, \quad l = 1, 2, \ldots, N.
$$

(2) The element I_{rb} of $I(\beta)$ satisfies

$$
\begin{aligned}
I_{rb}(\beta) &= -\frac{\partial^2 \mathbf{L}_1}{\partial \beta_{b0} \partial \beta_{r0}} \\[1ex]
&= -\frac{\partial}{\partial \beta_{b0}}\left(\sum_{l=1}^{N}\left(\frac{Y_l - \mu_l}{a_l \phi}\right)\left(\frac{A_{l\cdot} X_{\cdot r}}{V(\mu_l) g'(\mu_l)}\right) W_l(u_{00}, v_{00}, t_0) \right) \\[1ex]
&= -\sum_{l=1}^{N}\left(\frac{Y_l - \mu_l}{a_l \phi}\right)\frac{\partial}{\partial \beta_{b0}}\left(\frac{A_{l\cdot} X_{\cdot r}}{V(\mu_l) g'(\mu_l)}\right) W_l(u_{00}, v_{00}, t_0) \\[1ex]
&\quad - \sum_{l=1}^{N}\left(\frac{A_{l\cdot} X_{\cdot r}}{V(\mu_l) g'(\mu_l)}\right)\frac{\partial}{\partial \beta_{b0}}\left(\frac{Y_l - \mu_l}{a_l \phi}\right) W_l(u_{00}, v_{00}, t_0) \\[1ex]
&= -\sum_{l=1}^{N}\left(\frac{Y_l - \mu_l}{a_l \phi}\right)\frac{\partial}{\partial \beta_{b0}}\left(\frac{A_{l\cdot} X_{\cdot r}}{V(\mu_l) g'(\mu_l)}\right) W_l(u_{00}, v_{00}, t_0) \\[1ex]
&\quad + \sum_{l=1}^{N}\frac{A_{l\cdot} X_{\cdot r} A_{l\cdot} X_{\cdot b}}{a_l \phi V(\mu_l)(g'(\mu_l))^2} W_l(u_{00}, v_{00}, t_0).
\end{aligned}
$$

The Fisher information matrix is

$$
\mathbf{I}(\beta) = E(I(\beta)) = E((I_{rb}(\beta))_{p\times p}) \;=\; \sum_{l=1}^{N} \frac{A_l.X_{\cdot r}A_l.X_{\cdot b}}{a_l\phi V(\mu_l)(g'(\mu_l))^2} W_l(u_{00}, v_{00}, t_0)
$$

$$
=\; \frac{1}{\phi}\sum_{l=1}^{N} T_l A_l.X_{\cdot r}A_l.X_{\cdot b}W_l(u_{00}, v_{00}, t_0).
$$

Appendix A.3. Formula and Information Matrix of ρ

(1) Differentiating (4) with ρ yields

$$
\begin{cases}
(1-\rho C_{11})\dfrac{d\eta_1}{d\rho} - \rho C_{12}\dfrac{d\eta_2}{d\rho} - \cdots - \rho C_{1N}\dfrac{d\eta_N}{d\rho} = C_{11}\eta_1 + C_{12}\eta_2 + \cdots + C_{1N}\eta_N, \\[2mm]
-\rho C_{21}\dfrac{d\eta_1}{d\rho} + (1-\rho C_{22})\dfrac{d\eta_2}{d\rho} - \cdots - \rho C_{2N}\dfrac{d\eta_N}{d\rho} = C_{21}\eta_1 + C_{22}\eta_2 + \cdots + C_{2N}\eta_N, \\[2mm]
\vdots \\[2mm]
-\rho C_{N1}\dfrac{d\eta_1}{d\rho} - \rho C_{N2}\dfrac{d\eta_2}{d\rho} - \cdots + (1-\rho C_{NN})\dfrac{d\eta_N}{d\rho} = C_{N1}\eta_1 + C_{N2}\eta_2 + \cdots + C_{NN}\eta_N.
\end{cases}
$$

That is to say

$$
\begin{pmatrix}
1-\rho C_{11} & -\rho C_{12} & \cdots & -\rho C_{1N} \\
-\rho C_{21} & 1-\rho C_{22} & \cdots & -\rho C_{2N} \\
\vdots & \vdots & \vdots & \vdots \\
-\rho C_{N1} & -\rho C_{N2} & \cdots & 1-\rho C_{NN}
\end{pmatrix}
\begin{pmatrix}
\frac{d\eta_1}{d\rho} \\
\frac{d\eta_2}{d\rho} \\
\vdots \\
\frac{d\eta_N}{d\rho}
\end{pmatrix}
=
\begin{pmatrix}
C_{11} & C_{12} & \cdots & C_{1N} \\
C_{21} & C_{22} & \cdots & C_{2N} \\
\vdots & \vdots & \vdots & \vdots \\
C_{N1} & C_{N2} & \cdots & C_{NN}
\end{pmatrix}
\begin{pmatrix}
\eta_1 \\
\eta_2 \\
\vdots \\
\eta_N
\end{pmatrix}.
$$

Let $\frac{d\eta}{d\rho} = (\frac{d\eta_1}{d\rho}, \ldots, \frac{d\eta_N}{d\rho})'$. Then, $\frac{d\eta}{d\rho} = AC\eta$, where

$$
\frac{d\eta_l}{d\rho} = \frac{dg(\mu_l)}{d\rho} = \sum_{h=1}^{N} A_l.C_{\cdot h}\eta_h, \quad l = 1, 2, \cdots, N.
$$

(2) The scalar vector of ρ is

$$
S(\rho) \;=\; \frac{d\mathbf{L}_2}{d\rho} = \frac{1}{\phi}\sum_{l=1}^{N} T_l\Big(\sum_{h=1}^{N} A_l.C_{\cdot h}\eta_h\Big)(Y_l - \mu_l)g'(\mu_l)
$$

$$
=\; \frac{1}{\phi}(AC\eta)'T(Z - \eta).
$$

The information matrix is

$$
I(\rho) \;=\; -\frac{\partial^2 \mathbf{L}_2}{\partial\rho^2} = -\frac{d}{d\rho}\sum_{l=1}^{N}\Big(\frac{Y_l - \mu_l}{a_l\phi}\Big)\Big(\frac{\sum_{h=1}^{N} A_l.C_{\cdot h}\eta_h}{V(\mu_l)g'(\mu_l)}\Big)
$$

$$
=\; -\sum_{l=1}^{N}\Big(\frac{Y_l - \mu_l}{a_l\phi}\Big)\frac{d}{d\rho}\Big(\frac{\sum_{h=1}^{N} A_l.C_{\cdot h}\eta_h}{V(\mu_l)g'(\mu_l)}\Big)
$$

$$
-\sum_{l=1}^{N}\Big(\frac{\sum_{h=1}^{N} A_l.C_{\cdot h}\eta_h}{V(\mu_l)g'(\mu_l)}\Big)\frac{d}{d\rho}\Big(\frac{Y_l - \mu_l}{a_l\phi}\Big),
$$

where

$$\frac{d\mu_l}{d\rho} = \frac{d\mu_l}{dg(\mu_l)}\frac{dg(\mu_l)}{d\rho} = \frac{1}{g'(\mu_l)}\sum_{h=1}^{N} A_{l\cdot}C_{\cdot h}\eta_h.$$

Thus,

$$I(\rho) = -\sum_{l=1}^{N}\left(\frac{Y_l - \mu_l}{a_l\phi}\right)\frac{d}{d\rho}\left(\frac{\sum_{h=1}^{N}A_{l\cdot}C_{\cdot h}\eta_h}{V(\mu_l)g'(\mu_l)}\right) + \sum_{l=1}^{N}\frac{(\sum_{h=1}^{N}A_{l\cdot}C_{\cdot h}\eta_h)^2}{a_l\phi V(\mu_l)(g'(\mu_l))^2}.$$

The Fisher information matrix is

$$\begin{aligned}
\mathcal{I}(\rho) &= E(I(\rho)) = \sum_{l=1}^{N}\frac{(\sum_{h=1}^{N}A_{l\cdot}C_{\cdot h}\eta_h)^2}{a_l\phi V(\mu_l)(g'(\mu_l))^2}\\
&= \frac{1}{\phi}\sum_{l=1}^{N}T_l\left(\sum_{h=1}^{N}A_{l\cdot}C_{\cdot h}\eta_h\right)^2 = \frac{1}{\phi}(AC\eta)'T(AC\eta).
\end{aligned}$$

Appendix A.4. Algorithms of Coefficient Estimation

We provide four algorithms to estimate the unknown coefficients of PR, GWPR, GWPAR and GTWPAR models in Section 4.

Algorithm I: Estimate the unknown parameters in the PR model. Take the initial values $g(\mu_{ik}^{(0)}) = \eta_{ik}^{(0)} = \ln\mu_{ik}^{(0)}$, $y_{ik} = \mu_{ik}^{(0)}$, $Z_{ik}^{(0)} = \eta_{ik}^{(0)} + g'(\mu_{ik}^{(0)})(y_{ik} - \mu_{ik}^{(0)})$, and

$$w_{ik}^{(0)} = \frac{1}{a_{ik}V(\mu_{ik}^{(0)})(g'(\mu_{ik}^{(0)}))^2}, i = 1,2,\ldots,10, k = 1,2,\ldots,6.$$

The iterative formula of $\hat{\beta}^{(m+1)}$ is

$$\hat{\beta}^{(m+1)} = \left(X'W^{(m)}X\right)^{-1}X'W^{(m)}Z^{(m)}.$$

Repeat the above step until convergence yields. The estimated value $\hat{\beta} = \hat{\beta}^{(m)}$ can be obtained.

Algorithm II: Estimate the unknown coefficients in the GWPR model. (Note that k is a fixed constant taken from $\{1,2,\ldots,6\}$ and the following steps should be repeated six times independently). Take the initial values $\eta_{ik}^{(0)} = g(\mu_{ik}^{(0)})$, $y_{ik} = \mu_{ik}^{(0)}$, $Z_{ik}^{(0)} = \eta_{ik}^{(0)} + g'(\mu_{ik}^{(0)})(y_{ik} - \mu_{ik}^{(0)})$, and

$$t_{ik}^{(0)} = \frac{1}{a_{ik}V(\mu_{ik}^{(0)})(g'(\mu_{ik}^{(0)}))^2}, i = 1,2,\ldots,10.$$

Let $Z^{(0)} = (Z_{1k}^{(0)}, Z_{2k}^{(0)}, \cdots, Z_{10k}^{(0)})'$ and $T^{(0)} = \operatorname{diag}(T_{1k}^{(0)}, T_{2k}^{(0)}, \cdots, T_{10k}^{(0)})$. The iterative formula of $\hat{\beta}^{(m+1)}$ at the location (u_{0k}, v_{0k}) is

$$\hat{\beta}^{(m+1)}(u_{0k}, v_{0k}) = (A'(u_{0k}, v_{0k})T^{(m)}W(u_{0k}, v_{0k})A(u_{0k}, v_{0k}))^{-1}A'(u_{0k}, v_{0k})T^{(m)}W(u_{0k}, v_{0k})Z^{(m)},$$

where $W(u_{0k}, v_{0k}) = \operatorname{diag}(w_1(u_{0k}, v_{0k}), w_2(u_{0k}, v_{0k}), \cdots, w_{10}(u_{0k}, v_{0k}))$ and

$$w_i(u_{0k}, v_{0k}) = \frac{1}{\sqrt{2\pi}}\exp\left(-\frac{1}{2}\left(\frac{d_{ik}^{(0)}}{h}\right)^2\right).$$

Repeat the above step until convergence. When (u_{0k}, v_{0k}) takes all the locations (u_{ik}, v_{ik}), we will get the estimated value $\hat{\beta} = \hat{\beta}^{(m)}$ in a fixed the kth year.

Algorithm III: Estimate unknown coefficients in the GWPAR model. (Note that k is a fixed constant and the following steps should be repeated six times as in Algorithm II). Take the initial value $\beta_1^{(0)}$, $\beta_2^{(0)}$ from Algorithm II, and $\rho^{(0)}$ is the absolute value of spatial Moran's I = 0.4480. The initial values $\eta^{(0)} = (I - \rho^{(0)}C)^{-1}X\beta^{(0)}$, $\mu_{ik}^{(0)} = g^{-1}(\eta_{ik}^{(0)})$, and

$$
Z_{ik}^{(0)} = g(\mu_{ik}^{(0)}) + g'(\mu_{ik}^{(0)})(y_{ik} - \mu_{ik}^{(0)}),\ t_{ik}^{(0)} = \frac{1}{a_{ik}V(\mu_{ik}^{(0)})(g'(\mu_{ik}^{(0)}))^2}, i = 1, 2, \cdots, 10.
$$

The $(m+1)$th iterative estimation $\hat{\beta}^{(m+1)}(u_{0k}, v_{0k})$ and $\hat{\rho}^{(m+1)}$ is

$$
\begin{aligned}
\hat{\beta}^{(m+1)}(u_{0k}, v_{0k}) &= ((A^{(m)}X)'T^{(m)}W(A^{(m)}X))^{-1}(A^{(m)}X)'T^{(m)}WZ^{(m)}, \\
\hat{\rho}^{(m+1)} &= \rho^{(m)} + ((A^{(m)}C\eta^{(m)})'T^{(m)}(A^{(m)}C\eta^{(m)}))^{-1}(A^{(m)}C\eta^{(m)})'T^{(m)}(Z^{(m)} - \eta^{(m)}).
\end{aligned}
$$

If (u_{0k}, v_{0k}) takes all the locations (u_{ik}, v_{ik}), the estimate $\hat{\beta}^{(m+1)}$ can be given. When all estimated values arrive to converge, we will get $\hat{\beta} = \hat{\beta}^{(m)}(u_{ik}, v_{ik})$ and $\hat{\rho} = \hat{\rho}^{(m)}$ in a fixed the kth year.

Algorithm IV: Estimate the unknown coefficients in the GTWPAR model. Take the initial values $\beta_{1k}^{(0)}(u_{ik}, v_{ik}, t_k), \beta_{2k}^{(0)}(u_{ik}, v_{ik}, t_k), i = 1, 2, \ldots, 10, k = 1, 2, \ldots, 6$ from Algorithm III, and $\rho^{(0)}$ is the absolute value of spatiotemporal Moran's I = 0.2143. The initial value vector $\eta^{(0)} = (I - \rho^{(0)}C)^{-1}X\beta^{(0)}$, $\mu_{ik}^{(0)} = g^{-1}(\eta_{ik}^{(0)})$, $Z^{(0)} = (Z_1^{(0)}, Z_2^{(0)}, \cdots, Z_{60}^{(0)})'$, $T^{(0)} = \text{diag}(T_1^{(0)}, T_2^{(0)}, \ldots, T_{60}^{(0)})$ and

$$
Z_{ik}^{(0)} = g(\mu_{ik}^{(0)}) + g'(\mu_{ik}^{(0)})(y_{ik} - \mu_{ik}^{(0)}),\ T_{ik}^{(0)} = \frac{1}{a_{ik}V(\mu_{ik}^{(0)})(g'(\mu_{ik}^{(0)}))^2}.
$$

The $(m+1)$th iterative estimations $\hat{\beta}(u_{00}, v_{00}, t_0)$ and $\hat{\rho}$ are

$$
\begin{aligned}
\hat{\beta}^{(m+1)}(u_{00}, v_{00}, t_0) &= ((A^{(m)}X)'T^{(0)}W(A^{(m)}X))^{-1} \\
&\quad \times (A^{(m)}X)'T^{(m)}WZ^{(m)}, \\
\hat{\rho}^{(m+1)} &= \rho^{(m)} + ((A^{(m)}C\eta^{(m)})'T^{(m)}(A^{(m)}C\eta^{(m)}))^{-1} \\
&\quad \times (A^{(m)}C\eta^{(m)})'T^{(m)}(Z^{(m)} - \eta^{(m)}).
\end{aligned}
$$

where $W = \{w_{ik}(u_{00}, v_{00}, t_0)\}$ and

$$
w_{ik}(u_{00}, v_{00}, t_0) = \frac{1}{\sqrt{2\pi}}\exp\left\{-\frac{1}{2}\left(\frac{(u_{00} - u_{ik})^2 + (v_{00} - v_{ik})^2}{h_S^2} + \frac{(t_0 - t_k)^2}{\tau h_S^2}\right)\right\}.
$$

A detailed definition is given in Section 3.1. If (u_{00}, v_{00}, t_0) takes all the locations (u_{ik}, v_{ik}, t_k) and all estimations converge, we will get $\hat{\beta} = \hat{\beta}^{(m)}(u_{ik}, v_{ik}, t_k)$ and $\hat{\rho} = \hat{\rho}^{(m)}$.

It is worth noting that we use the parameter estimates of the previous model as the initial values of the next model to reduce the number of iterations and improve the operational efficiency. For example, the estimations of the GWPR model are selected as the initial values of the GWPAR model.

Appendix A.5. The Model Comparison Indicators (See Table 3)

(1) The coefficient of determination is defined by

$$R^2 = 1 - \frac{\sum(\eta - \hat{\eta})^2}{\sum(\eta - \bar{\eta})^2},$$

where η is a set of vectors $\{\eta_{ik}\}$, $\hat{\eta}$ and $\bar{\eta}$ are the parameter estimate and the mean value of η, respectively.

(2) Deviation can be defined as

$$D = \sum\left(y ln(\frac{\hat{\mu}}{y}) + (y - \hat{\mu})\right),$$

where y is a set of response variables, and $\hat{\mu}$ is the estimation of μ=E(y).

(3) The corrected Akaike information criterion is

$$AICc = D + 2P + 2\frac{P(P+1)}{N-P-1},$$

where D, P and N are the deviation, the number of parameters and the number of samples, respectively.

(4) Mean-square error is given by

$$MSE = \frac{1}{N-P}\sum(\hat{\eta} - \eta)^2,$$

where the parameter settings are the same as above.

References

1. Hlongwa, P. Current ethical issues in HIV/AIDS research and HIV/AIDS care. *Oral Dis.* **2016**, *22*, 61–65. [CrossRef] [PubMed]
2. Buve, A.; Bishikwabo-Nsarhaza, K.; Mutangadura, G. The spread and effect of HIV-1 infection in sub-Saharan Africa. *Lancet* **2002**, *359*, 2011–2017. [CrossRef]
3. Vandormael, A.; Oliveira, T.D.; Tanser, F.; Till, B.; Herbeck, J.T. High percentage of undiagnosed HIV cases within a hyperendemic South African community: A population-based study. *J. Epidemiol Community Health* **2018**, *72*, 168–172. [CrossRef] [PubMed]
4. Abbas, U.L.; Robert, G.; Anuj, M.; Gregory, H.; Mellors, J.W. Antiretroviral therapy and pre-exposure prophylaxis: Combined impact on HIV transmission and drug resistance in South Africa. *J. Infect. Dis.* **2013**, *208*, 224–234. [CrossRef] [PubMed]
5. Wojcicki, J.M. Socioeconomic status as a risk factor for HIV infection in women in East, Central and Southern Africa: A systematic review. *J. Biosoc. Sci.* **2005**, *37*, 1–9. [CrossRef] [PubMed]
6. Hallman, K. Gendered socioeconomic conditions and HIV risk behaviours among young people in South Africa. *Afr. J. AIDS Res.* **2005**, *4*, 37–50. [CrossRef]
7. Chris, K.; Sizwe, Z.; Robert, C.; Sipho, D. Why have socioeconomic explanations been favoured over cultural ones in explaining the extensive spread of hiv in south africa? *S. Afr. J. HIV Med.* **2012**, *13*, 14–16.
8. Stover, J. Projecting the demographic consequences of adult hiv prevalence trends: The spectrum projection package. *Sex. Transm. Infect.* **2004**, *80*, 14–18. [CrossRef]
9. Shiboski, S.C.; Jewell, N.P. Statistical analysis of the time dependence of HIV infectivity based on partner study data. *J. Am. Stat. Assoc.* **1992**, *87*, 360–372. [CrossRef]
10. Marsh, K.; Mahy, M.; Salomon, J.A.; Hogan, D.R. Assessing and adjusting for differences between hiv prevalence estimates derived from national population-based surveys and antenatal care surveillance, with applications for spectrum 2013. *AIDS* **2014**, *28*, S497–S505. [CrossRef]

11. Palk, L.; Blower, S. Geographic variation in sexual behavior can explain geospatial heterogeneity in the severity of the HIV epidemic in Malawi. *BMC Med.* **2018**, *16*, 22–31. [CrossRef] [PubMed]

12. Nelder, J.A.; Wedderburn, R.W.M. Generalized linear models. *J. R. Stat. Soc.* **1972**, *135*, 370–384.

13. Myers, R.H.; Montgomery, D.C. A tutorial on generalized linear models. *J. Qual. Technol.* **1997**, *29*, 274–275. [CrossRef]

14. Madden, L.V.; Boudreau, M.A. Effect of strawberry density on the spread of anthracnose caused by colletotrichum acutatum. *Phytopathology* **1997**, *87*, 828–838. [CrossRef]

15. Guisan, A.; Edwards, T.C.; Hastie, T. Generalized linear and generalized additive models in studies of species distributions: Setting the scene. *Ecol. Model.* **2002**, *157*, 89–100. [CrossRef]

16. Marshall, R.J. A review of methods for the statistical analysis of spatial patterns of disease. *J. R. Stat. Soc.* **1991**, *154*, 421–441. [CrossRef]

17. Brunsdon, C.; Fotheringham, A.S.; Charlton, M.E. Geographically weighted regression: A method for exploring spatial nonstationarity. *Geogr. Anal.* **1996**, *28*, 281–298. [CrossRef]

18. Nakaya, T.; Fotheringham, A.S.; Brunsdon, C.; Charlton, M.E. Geographically weighted Poisson regression for disease association mapping. *Stat. Med.* **2005**, *24*, 2695–2717. [CrossRef]

19. Hadayeghi, A.; Shalaby, A.S.; Persaud, B.N. Development of planning level transportation safety tools using geographically weighted Poisson regression. *Accid. Anal. Prev.* **2010**, *42*, 676–688. [CrossRef]

20. Li, Y.; Jiao, Y.; Browder, J.A. Modeling spatially-varying ecological relationships using geographically weighted generalized linear model: A simulation study based on longline seabird bycatch. *Fish. Res.* **2016**, *181*, 14–24. [CrossRef]

21. Anselin, L. Spatial econometrics: Methods and models. *J. Am. Stat. Assoc.* **1990**, *85*, 160–161.

22. Lichstein, J.W.; Simons, T.R.; Franzreb, S.K.E. Spatial Autocorrelation and Autoregressive Models in Ecology. *Ecol. Monogr.* **2002**, *72*, 445–463. [CrossRef]

23. Lee, L.F. Asymptotic distributions of quasi-maximum likelihood estimators for spatial autoregressive models. *Econometrica* **2004**, *72*, 1899–1925. [CrossRef]

24. Brunsdon, C.; Fotheringham, A.S.; Charlton, M.E. Some notes on parametric significance tests for geographically weighted regression. *J. Reg. Sci.* **2010**, *39*, 497–524. [CrossRef]

25. Zhang, Y.H. Generalized Space Variable Coefficient Autoregressive Model and Its Application in the Impact of Macroscopic-Factors on the HIV/AIDS Incidence. Master's Dissertation, Xinjiang University, Urumqi, China, 2017.

26. Yan, N.; Mei, C.L. A two-step local smoothing approach for exploring spatio-temporal patterns with application to the analysis of precipitation in the mainland of China during 1986–2005. *Environ. Ecol. Stat.* **2014**, *21*, 373–390. [CrossRef]

27. Fotheringham, A.S.; Crespo, R.; Yao, J. Geographical and temporal weighted regression (GTWR). *Geogr. Anal.* **2015**, *47*, 431–452. [CrossRef]

28. Huang, B.; Wu, B.; Michael, B. Geographically and temporally weighted regression for modeling spatio-temporal variation in house prices. *Int. J. Geogr. Inf. Sci.* **2010**, *24*, 383–401. [CrossRef]

29. Pace, R.K.; Barry, R.; Clapp, J.M.; Rodriquez, M. Spatiotemporal autoregressive models of neighborhood effects. *J. R. Estate Financ. Econ.* **1998**, *17*, 15–33.

30. Wu, B.; Li, R.; Huang, B. A geographically and temporally weighted autoregressive model with application to housing prices. *Int. J. Geogr. Inf. Sci.* **2014**, *28*, 1186–1204. [CrossRef]

31. Yan, N.; Mei, C.L.; Wang, N. A unified bootstrap test for local patterns of spatio-temporal association. *Environ. Plan. A* **2015**, *47*, 227–242.

32. Wu, X.Z. Multivariate analysis. In *Complex Data Statistical Methods Application Based on R*; Wu, X.Z., Ed.; China Renmin University Press: Beijing, China, 2013; pp. 116–144.

Article

Characterization of Pathogen Airborne Inoculum Density by Information Theoretic Analysis of Spore Trap Time Series Data

Robin A. Choudhury [1] and Neil McRoberts [2,*]

[1] School of Earth, Environmental, and Marine Sciences, University of Texas, Rio Grande Valley, Edinburg, TX 78541, USA; robchoudhury@gmail.com
[2] Quantitative Biology and Epidemiology Group, Plant Pathology Department, University of California, Davis, Davis, CA 95616, USA
*   Correspondence: nmcroberts@ucdavis.edu

Received: 10 October 2020; Accepted: 23 November 2020; Published: 27 November 2020

Abstract: In a previous study, air sampling using vortex air samplers combined with species-specific amplification of pathogen DNA was carried out over two years in four or five locations in the Salinas Valley of California. The resulting time series data for the abundance of pathogen DNA trapped per day displayed complex dynamics with features of both deterministic (chaotic) and stochastic uncertainty. Methods of nonlinear time series analysis developed for the reconstruction of low dimensional attractors provided new insights into the complexity of pathogen abundance data. In particular, the analyses suggested that the length of time series data that it is practical or cost-effective to collect may limit the ability to definitively classify the uncertainty in the data. Over the two years of the study, five location/year combinations were classified as having stochastic linear dynamics and four were not. Calculation of entropy values for either the number of pathogen DNA copies or for a binary string indicating whether the pathogen abundance data were increasing revealed (1) some robust differences in the dynamics between seasons that were not obvious in the time series data themselves and (2) that the series were almost all at their theoretical maximum entropy value when considered from the simple perspective of whether instantaneous change along the sequence was positive.

Keywords: time series; entropy; average mutual information; stochastic processes; deterministic dynamics

1. Introduction

> *"We now have to look at apparently random time series of data, be they from the stock market, or currency exchanges, or in ecology and ask are we seeing "random walks down Wall street" or deterministic chaos, or, often more likely, some mixture of the two."*
>
> —Sir Robert May [1]

The study of disease dynamics in plant pathology has been dominated by analysis of situations where disease increases monotonically within single growing seasons or over several seasons [2]. Reflecting this focus, the literature on the use of monotonic growth curve models or, more recently, compartment models consisting of linked differential equations is extensive and the methodology is well developed. In contrast, the literature on how to handle long, oscillating data series for plant pathogen populations is rather thin, with only isolated case studies [3–7] employing a range of statistical approaches. To date, there has been no concerted effort in botanical epidemiology to establish general properties of time series data associated with pathogen populations or disease intensity. This is due in part, no doubt, to the fact that time series methods have been considered relevant mostly for multi-season contexts, and multi-season datasets are scarce in plant pathology. However, with the advent of molecular probes for studying the airborne inoculum of plant pathogens, it has become much easier to capture time series data within single growing seasons [5,6,8].

Developments in technology for monitoring airborne inoculum of target species offer a promise of methodological advance to epidemiologists with an interest in creating evidence-based, within-season decision rules for disease management in crops. Given such potential applications for spore traps and quantification of target nucleic acid sequences, it is important that efforts are made to develop an analytical approach which takes into account the relevant statistical properties of the data that these monitoring methods generate. What can experimenters expect to see when they collect such data? What types of dynamical behavior are likely to be apparent, and how should the results be interpreted in relation to the use of the data in disease management?

The work we report here falls into the broad theme on decision-making that runs through several of the contributions to this Special Issue of Entropy. In the case of the current work, our effort is aimed more at understanding the basic properties of the data than in deriving decision rules from them. The work is motivated by our belief that it is important to be aware of any informational limitations inherent in the data, so that efforts to use air sampling as a means of forecasting interventions occur with realistic expectations. The work is intended to be an initial contribution to the literature; one from which we hope a range of further investigations covering a wider range of pathogen systems will develop.

As already noted, airborne concentrations of pathogen inoculum have been monitored using vortex (spinning rod) air samplers combined with species-specific quantitative polymerase chain reaction (qPCR) in a number of situations. In some cases, the approach has already been used commercially for disease management. Carisse and colleagues were pioneers, developing one of the first examples in commercial agriculture; in their case, to manage fungicide applications to control Botrytis leaf blight in onion in Quebec, Canada [5,9,10]. Their work (along with characterization of effective fungicide regimes and conducive weather conditions) helped to improve monitoring and to reduce disease outbreaks.

The use of spore traps linked with qPCR assays has been developed successfully for disease monitoring in several other pathosystems, including monitoring for early season inoculum for grape powdery mildew [11], where mitigating early season inoculum can reduce yield losses in susceptible varieties. These studies show that managing disease based on the binary presence or absence of pathogen primary inoculum can be quite successful, since what is needed in that situation is to detect the first occurrence of pathogen activity at the start of the growing season. The use of these systems for mitigating the impacts from secondary inoculum is more challenging.

Spinach downy mildew, caused by the obligate oomycete pathogen *Peronospora effusa*, is the most important threat to spinach production worldwide. Choudhury et al. [6] examined several sets of qPCR-based spore trap data collected from the Salinas Valley in California. The resulting time series were analyzed by fitting statistical models to characterize both trend and periodicity. While the approach was successful in producing a description of the observed dynamics, and in linking important statistical features to plausible biological mechanisms, it offered little in the way of general understanding of inoculum dynamics. Analyses of the coefficients of prediction and the Lyapunov exponents of the time series suggested that the datasets were quasi-chaotic. Further analyses of this example dataset could reveal general dynamics of airborne inoculum for plant pathogens.

Recent developments in time series analysis [12] based on information-theoretic quantities offer some promise in being able to extract more generic properties from the available data. Our objectives in this paper are to revisit the data originally studied by Choudhury et al. [6] and to apply the methods suggested by Huffaker et al. [12] in order to describe the dynamics of pathogen airborne inoculum in information theoretic terms. The analyses also place our data from botanical epidemiology in the wider context of the analysis of dynamical systems allowing interdisciplinary comparison. Our primary intended audience is plant pathologists and epidemiologists who might be interested in an introduction to these topics. For that reason, our approach is somewhat pedagogical and does not delve deeply into the underlying technical details.

2. Materials and Methods

2.1. Data Collection

Airborne inoculum of *P. effusa* was sampled at four locations in the Salinas Valley of California in 2013 and 2014 using vortex air samplers constructed by Dr. Walt Mahaffee (USDA-ARS Corvallis, Corvallis, OR, USA) and operated by Dr. Steven Klosterman (USDA-ARS Salinas, Salinas, CA, USA). The presence of the inoculum and quantification were achieved using qPCR amplification of a species-specific DNA sequence in the total DNA extract from the sampler rods. Details of the sampling procedure, qPCR primers, reaction conditions, and translation of the qPCR cycle threshold number to daily pathogen DNA copy number are described in Klosterman et al. [8].

2.2. Data Preparation

Samples were recovered from the air samplers on an irregular sampling interval of two or three days depending on the availability of technical staff. In the original 2016 study [6], we accommodated the irregular sampling interval by fitting a flexible sine function to the observations, having first removed any temporal trend by linear regression. In the current work, in order to utilize nonlinear methods incorporating information quantities, we interpolated the raw data to produce time series with a regular time step of one day. All nine data series were processed in the same way so that we could compare their statistical properties directly. The interpolation was achieved by linear averaging between the measured data points. The interpolation method will have the effect of smoothing the data to some extent, and the interpretation of the results takes that into account. We avoid overinterpretation of fine-grain aspects of the analyzed series and focus on the major dynamic features that are unlikely to be strongly influenced by the interpolation.

2.3. Basic Time Series Analysis

After interpolation of the data to a daily time step, each of the nine time series consisted of 129 observations of the estimated target DNA copy number of *P. effusa* trapped over the preceding 24 h period. The nine time series were first inspected for evidence of an overall trend in copy number with time. Increasing trends were detected in 7 of the 9 series, and the series were tagged accordingly to indicate their status. Irrespective of whether or not the initial inspection suggested a trend to be present, in order to standardize the pretreatment of the data, a simple linear regression with time (i.e., data point in sequence, $t = 1, 2, 3,... 129$) was fitted to the natural logarithm of the estimated copy number. The residuals from the regression were then exponentiated to produce the detrended series that were subsequently used analysis. In what follows, we refer to these series as N_t, indicating the (detrended) copy number on day t. When corresponding log-transformed values are analyzed, they are denoted n_t.

For each series, we obtained the autocorrelation function (ACF), the partial autocorrelation function (PACF), and the phase plot of the log-transformed series with $n_{t+1} = \ln(N_{t+1})$ on the ordinate and $n_t = \ln(N_t)$ on the abscissa. The PACF differs from the standard autocorrelation function in that it considers only the direct effect of observations at one point in the series on observations separated by lag τ, indirect effects, operating through the interposing points in the series that are removed.

2.4. Nonlinear Time Series Analysis

To characterize the time series in terms of nonlinear dynamics, we followed an approach suggested by Huffaker et al. [12] and by Kantz and Schreiber [13]. The various quantities estimated for each series were obtained using functions provided in the R packages *"nonlinearTseries"* [14], *"TseriesChaos"* [15], or *"TseriesEntropy"* [16]. Additional calculations to obtain empirical entropy values used the package *"entropy"* [17] or were coded directly in R. As with many other aspects of applied data analysis, for several of the steps in a nonlinear time series analysis, there is no single method that is guaranteed to provide optimal results under every circumstance. For many of the procedures, there are no formal test statistics to indicate that a "significant" result has been obtained; we followed the approaches suggested in the references. We provide R code and data necessary to replicate a full set of analyses for one of the 9 time series analyzed in the repository at the following URL: https://github.com/robchoudhury/spore_trap_information_theory. The R code is provided as is, and we offer no guarantee that it will work when adapted to other data sets.

2.4.1. Surrogate Testing for Nonlinear Dependence

Since nonlinear analysis (NLTS) can be time-consuming, an initial step should be to test for lack of linear dependence in the observed data. An agreed approach for performing this is to perform surrogate tests [12,14]. Different versions of the surrogate test are implemented in *nonlinearTseries* and *TseriesChaos*. The basic idea in both cases is to construct an empirical hypothesis test by resampling from the observed data, with the test statistic being a suitable property of the data that will hold under linear dependence but not otherwise. One of the simplest approaches, the one implemented in *nonlinearTseries*, relies on the idea that a Gaussian linear process will show time reversibility. Randomized permutations are obtained using a method in which the phases of the Fourier transform of the observed data are randomized. A two-sided hypothesis test is implemented to examine whether there is evidence that the value calculated from the observed data differs from the set of surrogates generated in the data resampling routine. We set the "significance level" option at 0.02, which results in the observed data being treated as one observation in a set of 100, with the two-sided test examining whether the observed data are in the $p = 0.02$ upper or lower tail of the sample. The supplied function includes a built-in diagnostic plot of the resampling test, but we implemented our own diagnostic graphical representation of the outcome for the test.

TseriesEntropy implements a more complex surrogate testing procedure. First, the best-fitting linear autoregressive (AR) model is selected on the basis of the Akaike Information Criterion (AIC). The residuals of the best AR model are resampled (with replacement). For each resampled series, a metric entropy measure (the Bhattacharya–Matusita–Hellinger measure, S_p) [18] is calculated at different lags. Based on the relevant properties of the resampled data, the 95% confidence band for S_p can be calculated and the values for the observed series are compared with the confidence band. If S_p for the observed series falls outside the band, the series can be considered to show nonlinear as opposed to linear dependence at the relevant lags. The entropy-based approach in *TseriesEntropy* is computationally more demanding than the expectation-based approach in *nonlinearTseries*. In the initial work, we examined both approaches. The results reported here are for the time-reversibility approach implemented in *nonlinearTseries*. The code supplied in the Supplemental Materials includes an example of the regression-based approach, deactivated by comment markers.

2.4.2. Characterizing Nonlinear Properties

Assuming that the surrogate tests indicate sufficient reason to proceed with NLTS, characterization of the dynamics in terms of their tendency to chaotic versus stochastic uncertainty is an important component of the ensuing effort. Following the pioneering work of Takens [19], one widely accepted approach to NLTS proceeds by attempting to reconstruct important features of the complete (and only partially observed) phase space of the whole system, using the methods of time delay embedding to characterize the time series of a single observed component of the system.

In the current context, where the ultimate motivation is the hope of using similar series in disease management, the capacity to reconstruct the phase portrait of the whole system is of secondary importance to characterizing the dynamics of the observed series. However in this initial study the focus is on understanding the dynamics rather than immediate practical application, and the time delay embedding approach may be valuable because the features of the dynamics it reveals are informative.

Three properties of the series are important in NLTS, these being (i) the average mutual information (AMI), $I(N_{t,}; N_{t-\tau})$, of the time series data at successive lags, $\tau = 0, 1, 2, \ldots \tau_{max}$; (ii) the Theiler Window, tw; and (iii) the embedding dimension, m.

The AMI Function

The AMI function is calculated by binning observations and by calculating the mutual information obtained about observation N_t being in the ith bin from knowing observation $N_{t-\tau}$ is in the jth bin. The results are averaged over all of the available data to produce the average mutual information. A graphical plot of $I(N_{t,}; N_{t-\tau})$ against lag, $\tau = 0, 1, 2, \ldots \tau_{max}$ produces an information-theoretic analogue of the ACF plot, but one in which the AMI's general measure of lagged association, as opposed to the linear lagged dependence captured by ACF, is visualized. The first minimum, or the first occurrence of a value below an empirical threshold, of the AMI function is taken to be an indication of the embedding time delay, d, of the series, since this value indicates a time lag at which observations have, in a general sense, low correlation.

The Theiler Window (tw)

The Theiler Window [20,21] is used to define the minimum separation along the time series that two points must have in order to be included in procedures used to find the embedding dimension, m (see below). Theiler's review [21] gives a detailed and technical account of the issues and the various approaches suggested (up to that time) for finding the embedding dimension.

For long time series, both *TseriesChaos* and *nonlinearTseries* offer functions to generate a space-time plot [22] from which tw can be selected by choosing a value at which there is a low probability of points being close in the phase space for a given time lag separation. For short time series, such as what we have in the present study, the space-time plot approach may not give usable results and other options may be needed; this was the case with our datasets which consist of 129 observations.

As an easily obtained first approximation, Huffaker et al. [12] suggested using the first minimum of the standard autocorrelation function (ACF). Since ACF is a linear function, there are risks in using it to estimate correlation structure of nonlinear data [23]; indeed, this issue was one of the motivations for Theiler's review [21] of methods for identifying the dimensionality of nonlinear attractors. The problem, in general, appears to be that nonlinear correlation may occur at a larger lag separation that would be suggested by the ACF.

In the current case, lacking a reasonable alternative, we opted for a trial-and-error approach. With both the AMI function and the ACF available, we had estimates of both general association and linear correlation with lag, while the original time series and the corresponding phase plots also help to indicate suitable values of tw. For each series, we started with the value suggested by the first minimum of the ACF, noting also whether this lag separation was longer or shorter than the value suggested by the AMI. Where the AMI reached its first minimum at longer lag than the ACF, we used a range of estimates for tw and examined the effect of changing tw on the estimated embedding dimension, m.

The Embedding Dimension, m

Options for estimating, m, are either the method of False Nearest Neighbors (FNN) offered in *TseriesChaos* (Huffaker et al. [12] pp. 67–69) or Cao's [24] algorithm implemented in *nonlinearTseries*. Briefly, the motivation for the FNN approach comes from the idea that (in the current case), the observed time series of pathogen DNA copies represent only one dimension of a higher-order dynamical system. We can think of the observed series as representing the whole higher-order dynamical system projected onto a single dimension. With this perspective, points that appear close to one another may actually be widely separated in the full dimensional space of the dynamical system. The idea of FNN computation is to select a subset of points within a given "radius" of each other but separated by at least the value of tw and to track whether they remain as neighbors as the dimensionality of the assumed attractor is incrementally increased. If the proportion of FNN is plotted against the number of dimensions,

m, the first value of m at which the proportion of FNN is minimized provides an estimate of the embedding dimension.

In the approach suggested by Cao [24], the embedding dimension is identified by calculating a pair of functions, referred to as $E1(m)$ and $E2(m)$, of putative values for the embedding dimension, m. Note that Cao's original notation used d in place of m. Cao's method starts by calculating an overall Euclidean distance measure between pairs of points on time delay vectors for successively larger assumed values of m. Function $E1(m)$ calculates the ratio of the distance measure at successive pairs of values, $(m+1, m)$. Cao's insight was that this ratio stabilizes close to 1 if the data are generated by an attractor. The second function, $E2(m)$, focuses on the distance between only the nearest neighbors in the time delay vectors and operates on the distance measure based only on those. As with $E1(m)$, the function returns the ratio between successive pairs $(m+1, m)$. If the data are generated by a deterministic attractor, $E2(m)$ has the property that, at some value m^*, $E2(m^*)! = 1$, whereas if the data are generated by a process dominated by stochastic noise, $E2(m) \cong 1, \forall m$. Thus, in addition to providing an estimate of the relevant embedding dimension, Cao's method offers the advantage over the FNN approach of providing an indication of whether the data-generating process is characterized by deterministic or stochastic uncertainty.

2.5. Additional Entropy Measures

In addition to the characterization of the dynamics provided by the time-delay-embedding approach, we calculated two empirical entropy values to help in understanding the uncertainty in the data for airborne pathogen DNA. The first approach worked directly on the DNA copy number time series (following detrending if necessary, see above). The entropy () function from the R package *entropy* was used to calculate empirical estimates of the entropy in the data at each time point by iteratively adding the datum for each time point to the entropy calculation. Calculation using this approach starts by constructing a binning structure for the data and then by estimating the entropy based on the frequencies of observation in each bin. We started the iterative process at the 10th time point, so that the first estimate of entropy was based on the first 10 observations of each series. The calculation then proceeded as just outlined, with the second estimate being based on the first 11 data points and so on. The maximum likelihood option for the entropy function was used throughout.

As a second approach to characterize uncertainty in the time series data in relation to decision making, we first transformed each series into a binary string of length (t_{max}-1). First differences between successive pairs of values were calculated, and if the resulting difference was greater than 0 (indicating $N_{t+1} > N_t$), then 1 was entered for the corresponding value of the string; $N_{t+1} \leq N_t$ resulted in 0. The calculation then proceeded along similar lines to those outlined for the entropy of the copy number, iteratively increasing the size of the dataset by one time point and calculating a new entropy value. In the current case, at each time point, we calculated the proportion of the data that were 1s and then used Shannon's equation for expected information to give an entropy value in bits for the string at each time point (including all data up to that time). The calculation was coded directly in R. We initiated the calculation with the first two observations and then iterated the calculation one time point at a time.

2.6. Linear Autoregressive Models

In discussing the analysis of time series data for biological populations, Royama [25] noted that for autoregressive (AR) models where an instantaneous growth rate is modeled as a function of lagged population sizes, there is a qualitative difference in the types of behavior that a second-order lag model can display compared with a first-order model. Further, given the capacity for second-order linear models to generate quite complex oscillatory patterns, even when completely deterministic, Royama [25] suggested they could be expected to approximate the behavior of simple nonlinear models. Since the main aim of our investigation is to look at the utility of nonlinear methods, the linear AR models included here were fitted for the purpose of illustrating the extent to which a linear model can account for the observed behavior of the data collected from air samplers.

We followed a conceptual approach that draws on the work of Royama [25] and Turchin [26] in fitting the AR linear models. The process starts with the log-transformed (log_e) time series, denoted n_t. The instantaneous log growth rate R_t is defined as $n_{t+1} - n_t$, and the estimated linear AR model is then

$$R_t = a_0 + a_1 n_t + a_2 n_{t-\tau} + \varepsilon, \tau = 1, 2, \ldots \tau_{max} \tag{1}$$

in which a_0, a_1, etc. are parameters to be estimated; ε is an error term; and τ is an index indicating lag dependence. Selection of the order of lag dependence (i.e., the value τ_{max}) to use in fitting the AR models in each case was guided by the estimates of ACF and AMI functions (see Section 2.4.2 above). Parameter estimation was achieved by the standard least-squares approach implemented in the lm () function in the R base statistics package. For the selected model in each case, we noted the percent variance accounted for by the model in form of the standard adjusted-R^2 and a coefficient of prediction similar to the one proposed by Turchin [26]. The coefficient was obtained as follows: We fitted a model consisting of only the mean value of the dependent variable and captured the residual sum of squares, (RSS_{mn}). Next, we calculated $1-(RSS_{mod}/RSS_{mn})$, in which RSS_{mod} is the residual sum of squares from the selected model. When $RSS_{mod} > RSS_{mn}$, the coefficient has a negative value and indicates that the model fits noise. Values approaching 1 occur when the observed series has a pattern of oscillations that can be captured reasonably well in simple autoregressive models. Finally, values in the region of 0 indicate that the series is dominated by noise and, possibly, too short and complex to be characterized well.

3. Results

3.1. Time Series Properties and Nonlinearity

Time series graphs for the nine series of spore trap DNA copy number data are shown in Figure 1. Two of the nine series did not require detrending prior to analysis, these being King City South, 2014 and Gonzales, 2014. The results of testing for evidence of nonlinear dependence using Cao's method are shown in Table 1 along with other summary parameters of interest for the nine series. For four series—Salinas 2013, Soledad 2013, King City North 2013, and Gonzales 2014—the surrogate (bootstrap) test led to rejection of the null hypothesis that the data were compatible with a stochastic linear (i.e., time-reversible) process. The output from the bootstrap analysis for each series is shown in Figure 2.

The results of using Cao's [24] method to test for deterministic versus stochastic dynamics indicated that all 9 series had a stochastic nature; the value of function E2(m) stayed close to the value 1 for all values of m tested. Graphical output from the R function is given in Appendix A in Figure A1. Note that the R function uses the symbol d in the place of m.

The phase plots (Figure 3) for the detrended series show a strong tendency for the points to lie along the diagonal on which $n_t = n_{t+1}$, with short orbits away from this line, typically lasting no more than three to four time steps. These features are indicative of stochastic variation around a fixed value with a mixture of immediate and time-delayed feedback Turchin [26].

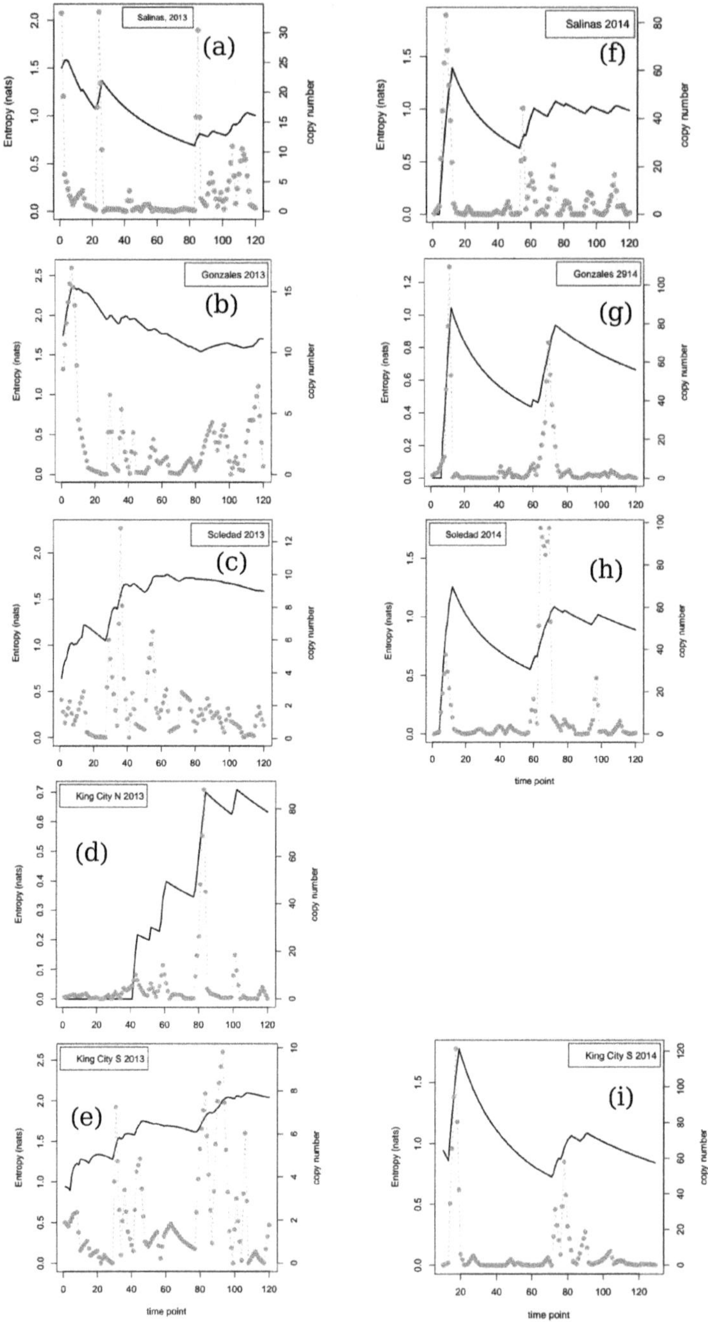

Figure 1. Detrended daily pathogen DNA copy number trapped (right axis scale) and cumulative entropy (nats, left axis scale) in the copy number series for 9 location/year combinations in which vortex air samplers were used to sample for the presence of DNA from the downy mildew pathogen of spinach, *P. effuse*, in the Salinas Valley of California: (**a–e**), 2013; (**f–i**), 2014. The King City, North location was sampled only in 2013.

Table 1. Summary statistics for the 9 time series of pathogen DNA copy number.

	Sal13	Sal14	Gon13	Gon14	Sol13	Sol14	KcN13	KcS13	KcS14 [2]
ACF [1]	3	6	9	7	5	8	5	7	5
PACF	4	8	3	10	6	5	4	5	5
AMI	5	11	6	10	4	10	6	7	6
m	6	6	6	6	9	6	6	7	7
λ_1	0.05	0.16	0.09	0.06	0.13	0.18	0.04	0.04	0.05
Linear?	N	Y	Y	N	N	Y	N	Y	Y
Entropy, nats (copy no.)	1.00	0.98	1.58	0.88	1.70	0.88	0.63	2.04	1.14
Entropy, bits (binary)	0.99	0.99	1.00	1.00	0.98	0.98	1.00	1.00	1.00
%VAF	11.4	6.6	3.6	4.2	7.1	17.0	26.4	20.1	7.5
pred Coeff	0.14	0.14	0.10	0.06	0.08	0.18	0.30	0.22	0.09

[1] ACF, lag at which series autocorrelation function has first minimum; PACF, lag at which the partial autocorrelation function has its first minimum; AMI, lag at which the series average mutual information function has its first minimum; m, estimated embedding dimension; λ_1, the maximum Lyapunov exponent; Linear?, outcome of surrogate test for compatibility of the series with stochastic linearity; Entropy copy no., estimated entropy (nats) of the copy number time series; Entropy binary, entropy (bits) of the binary series indicating if the copy number increased between successive pairs of observations; %VAF, percent variance accounted for in the best autoregressive linear model for the series of instantaneous rates of change in the log copy number data; pred Coeff, prediction coefficient for the autoregressive linear model (see text for details). [2] Location/year combination: Sal, Salinas; Gon, Gonzales; Sol, Soledad; KcN, King City, North; KcS, King City, South; 13, 2013; 14, 2014.

For all 9 series, the value of the dominant Lyapunov exponent (λ_1) was greater than 0, indicating that chaotic divergence would occur in independent realizations generated by the same data generating process. Although positive, the values of λ_1 were small, ranging from 0.04 to 0.17 (Table 1). Across the nine series, the value of the Lyapunov exponent was negatively correlated with the percent variance accounted for in fitting linear regression models to the series of instantaneous log growth rates and the coefficient of prediction for the linear fits; series with a higher Lyapunov exponent gave rise to poorer linear autoregressive models.

The best fitting autoregressive models for the time series of instantaneous rates generally captured only a low proportion of the variance in the series (Table 1). In general, the fitted values from the models showed less variability than the observed data, although in some cases, the qualitative fit to the series, in tracking the direction of the oscillations, was reasonably good. The main result from these analyses was that, while the data neither exhibited oscillations that could be easily attributed to a low-dimensional nonlinear attractor nor were they easily described by autoregressive linear models. The fitted autoregressive models and series of R_t for each location/year combination are shown in Figure A3 in Appendix A.

The AMI and ACF functions were correlated, but there was no consistent tendency for the AMI to reach its first minimum at higher lag than ACF. The AMI function minimized at higher lag than the ACF in 6 of the 9 cases; the functions minimized at the same lag in one case; and in the remaining two cases, the AMI minimized at lower lag than the ACF. In general, the estimated embedding dimension, m, was similar to the value suggested by the first minimum of the AMI and ACF functions; across the nine series, m was negatively correlated with both AMI and ACF. The relatively large estimated values for m are indicative of complex dynamics in the observed data, but we note, again, that the data series are relatively short, which may affect the accuracy of the estimated parameter.

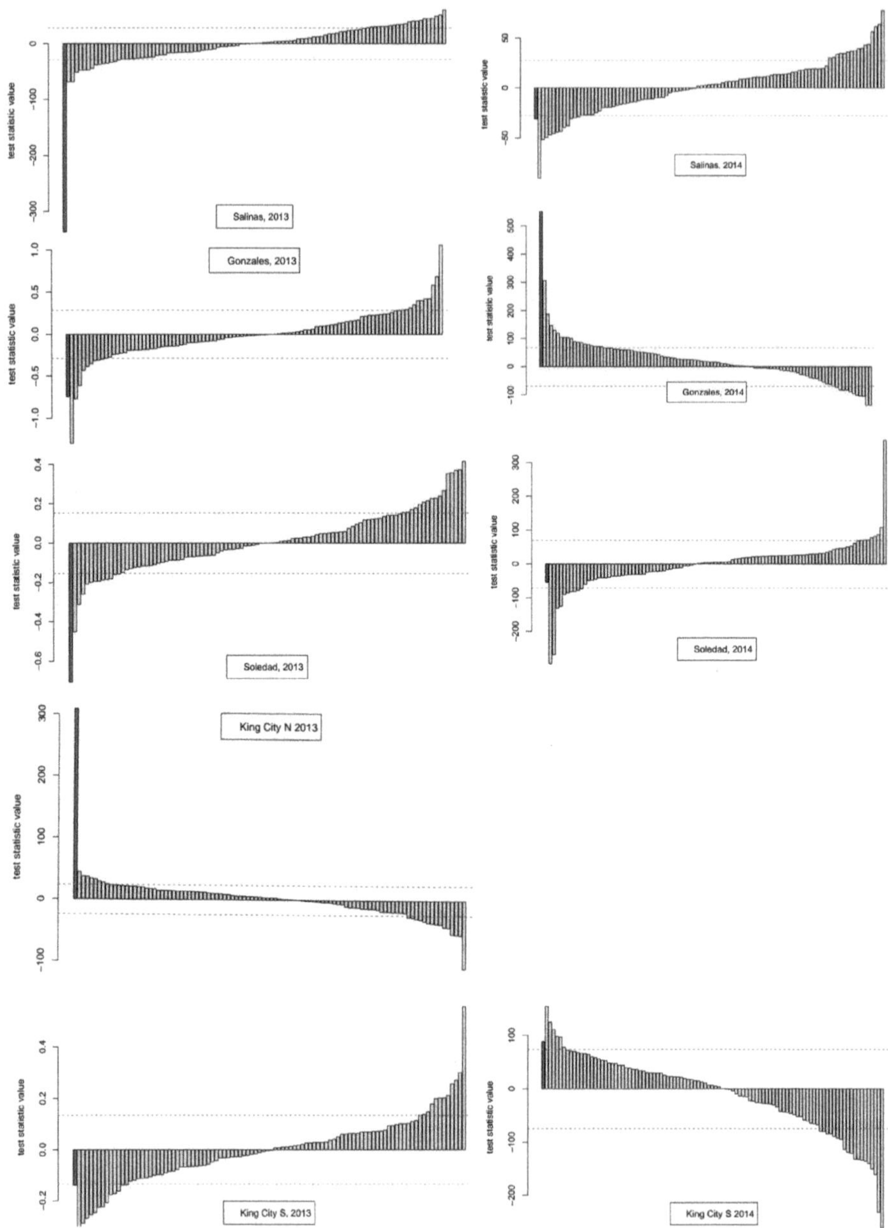

Figure 2. Results from surrogate tests (i.e., bootstrap data resampling) to assess the compatibility of spore trap data giving daily pathogen DNA copy numbers with time reversibility: the initial bar (in red) in each graph is the test statistic calculated for the original data. The remaining bars are the values calculated for bootstrap resamples of the data constructed in such a way to break any temporal autocorrelation in the original data. The dashed horizontal lines show the standard deviation of the surrogates above and below zero. Four of the nine series fail the two-sided hypothesis test for compatibility with time reversibility (i.e., stochastic linearity). Further details are given in the main text.

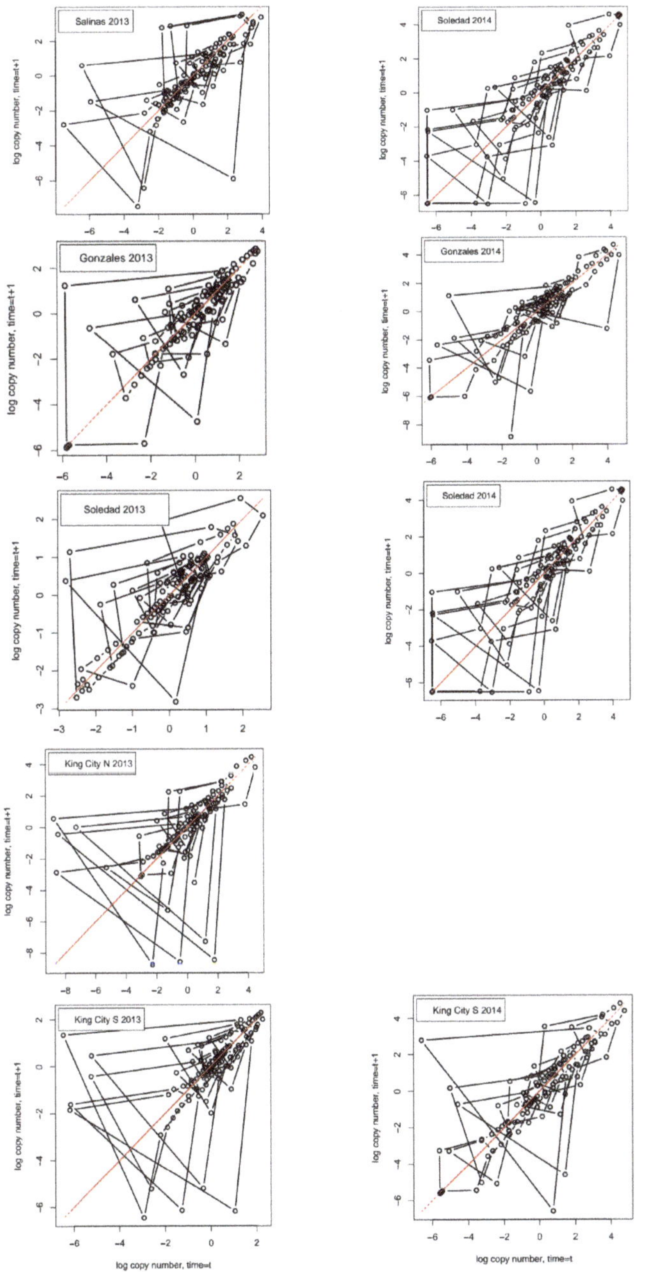

Figure 3. Phase space plots for the 9 series of detrended daily pathogen DNA copy numbers: the data are $\log_e$ values of the detrended data. Series orbiting a fixed attractor or a limit cycle show clockwise orbits. The obvious tendency for the phase portraits to lie along the diagonal for which $n_t = n_{t+1}$ is partly an artifact of detrending and partly a result of the fact that the series all contain sequences of observations that are very close to the mean value of the detrended series.

3.2. Entropy Measures of Time Series Uncertainty

We calculated entropy values along the time series for each location/year combination in two different ways. For the detrended copy number data, the entropy was calculated (in nats) using an automated binning procedure. The resulting series of entropy values are shown together with the data in Figure 1.

In the second year of observations (2014), the detection of pathogen DNA on the traps was sporadic. All four series showed an early peak in copy numbers around day 10 and then a long period of low-to-no detection until around day 80, when all locations experienced another peak in detection. Apart from these two shared features, the time series of trap counts were superficially dissimilar across the four locations sampled in 2014, but the series of cumulative entropy values showed a similar pattern in all four cases, with an initial peak corresponding to the trap data at day 10 followed by a long reduction as successive, similar trap results resulted in a reduction in heterogeneity in the data. The peak in trap counts caused a further peak in entropy around day 80, followed by a second period of decline. In general, the cumulative entropies in 2014 did not exceed 1.5 nats except in the case of King City, South, for which the initial peak was 1.78 nats. The final values for the entropy of the four series in 2014 are given in Table 1 and range from 0.88 to 1.14 nats.

In contrast to the more or less consistent pattern revealed by the 2014 data, the cumulative entropy values for the 2013 data sets were more variable. The final values for the five series tended to be higher than those in 2014 ranging from 1.00 to 2.04 nats with the exception of the King City North location, which had a final entropy value of 0.63 nats. In Salinas and Gonzales, the entropy value peaked early at over 2 nats and declined somewhat over the course of the season, although still finishing at or above 1.00 nats. In contrast to this early peak and decline pattern, at the remaining three sites in 2013, uncertainty increased through much of the season, in association with repeated oscillations in the trap copy number data.

In addition to characterizing uncertainty in the daily trap data directly, we also assessed the uncertainty in the simpler issue of whether the observed series increased between each successive pair of days. Figure 4 shows the time series for the entropy of the cumulative binary series together with the corresponding series of R_t, the instantaneous change in the log copy numbers between pairs of observations. The analysis showed that, in all nine series, the entropy remained close to its theoretical maximum value (i.e., 1 bit) over much of the season following an initial transient period lasting approximately 30 days. In three of the series (King City, North 2013; Soledad, 2013; and Salinas, 2014), the entropy did not settle close to its maximum until later in the season, but even in these cases, the final entropy value was close to the theoretical maximum of 1 bit. Note that, in Figure 4, the entropy values are shown on a log scale to allow detail of the changes over time to be visible.

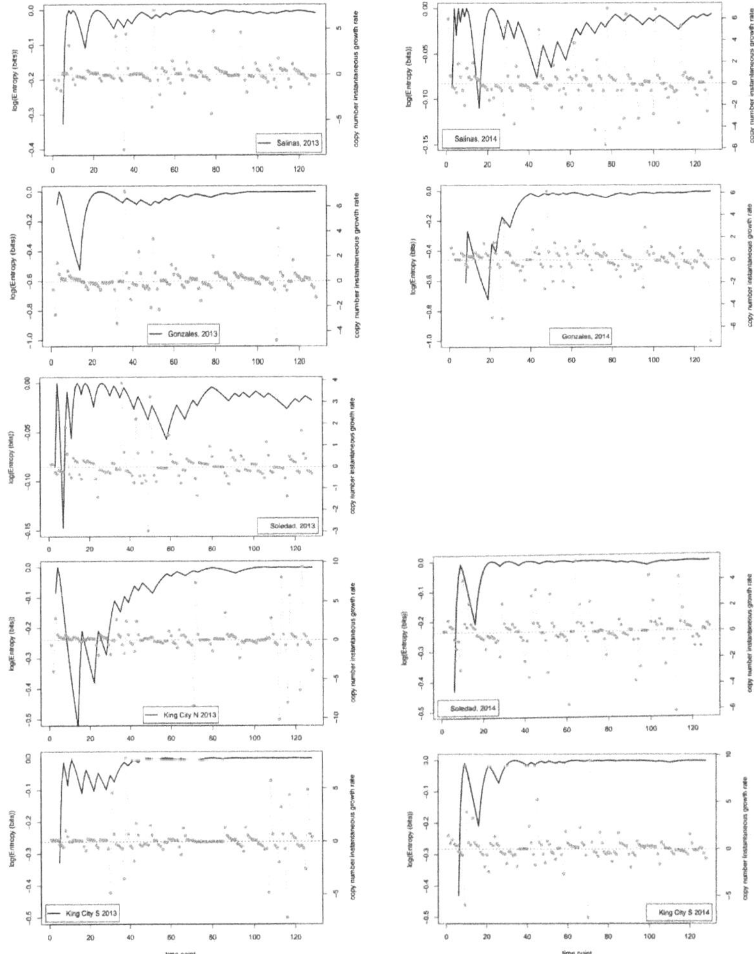

Figure 4. Graphs for 9 location years showing the series of instantaneous growth rates between successive time points (right axis scale) and the cumulative entropy (bits) of the series of binary values indicating whether the growth rate series is positive (left axis scale): the left axis is shown on a $\log_e$ scale to allow the variation in the entropy values to be visible. Note that, on this scale, the theoretical maximum value is 0.

4. Discussion

The quotation from the late Sir Robert May's introduction to the Landmark edition [1] of his monograph *Stability and Complexity in Model Ecosystems* was chosen deliberately and for more than one reason. First, May's point that the dynamics of real systems are likely to be a mixture of stochastic and deterministic processes applies directly to our observations on the time series of spore trap DNA copy numbers for *P. effusa* in the Salinas Valley of California. Secondly, May was an advocate of the idea that models can and should be used in biology in a strategic way to try to understand broad types of behavior without necessarily considering immediate questions of application or numerical accuracy in any specific case, while our analyses are predominantly statistical in nature, they are nonetheless carried out from a strategic perspective. Our aim in this study was not so much to produce accurate predictive models of any of the series as it was to use the tools of nonlinear time series analysis, together with some linear methods, to investigate the broad properties of pathogen DNA copy data collected from vortex air samplers.

A correlation matrix plot for the numerical data in Table 1 is given as Figure A2 in Appendix A. Summarizing the results for the diagnosis of time series properties, a mixture of findings resulted. In some cases, there were indications of deterministic chaos—i.e., positive estimated values for the Lyapunov coefficient, failure of time reversibility test in surrogates—while others were indicative of stochastic noise, i.e., in 5 out of 9 cases, the surrogate test failed to reject the hypothesis of time reversibility, and the first minimum values of the ACF and AMI functions were generally similar, indicating that the more general information-theoretic test of association based on average mutual information did not detect dependence in the series beyond the linear association measured by the ACF.

All of the series had positive Lyapunov exponents, indicating a tendency for deterministic sensitivity to initial conditions [26]. On the other hand, application of Cao's approach [24] indicated that the series were stochastic. The surrogate (bootstrap) test of time reversibility indicated that five series were compatible with the hypothesis that they were generated by a stochastic linear process while four were not. Relatively low values for the coefficient of prediction and adjusted-R^2 calculated from linear autoregressive models, ranging from a minimum of 3.6% (Gonzales, 2013) to a maximum of 26.4% (King City, N, 2013), also suggested that the series were strongly influenced by stochastic noise. Taken together, these results indicated that the series lie in the transition between stochastic and deterministic uncertainty in what Turchin [26] refers to as *quasi-chaotic* territory at the boundary between the two types of dynamics.

It seems reasonable, based on the dependence of oomycete pathogens such as *P. effusa* on suitable weather for spore production and release, that the copy number on air sampler traps would show appreciable stochasticity. Not only is the number of DNA copies detected dependent on the response of the pathogen to uncertain weather conditions, the physical processes of dispersal, and transport in air, together with the vortex sampling process itself, meaning that there are multiple sources of stochasticity between the release of spores and subsequent trapping events. However, at the same time, crop management practices such as planting and harvesting salad spinach happen on cycles of between 21 and 45 days, and may be a source of deterministic forcing in the data complicating the dynamics. If the data are predominantly stochastic in nature, then traditional statistical models should be able to describe the pattern and to characterize the uncertainty. Similarly, Turchin [26] argues that dynamic patterns generated by low-dimensional attractors can also successfully be described by relatively simple models. Our analyses indicate that, at least in the case of *P. effusa* in the Salinas Valley in California, the observed dynamics may fall between these two preferred situations, making characterization of the dynamics difficult and leading to low overall predictability. The estimated embedding dimension for the series (after detrending) ranged from 6 to 9, indicating that they did not have dynamics compatible with a low-dimensional attractor.

If we consider the data in relation to variability in time and space, there are clear implications for making robust inferences about the quantity of pathogen inoculum in the air. For example, at three of the four locations where samplers were deployed in two successive years, the dynamics were classed as linear in one year and not linear in the other. The four locations sampled span a linear distance of approximately 80 km from Salinas in the north to King City in the south. In 2014, a year with relatively little pathogen activity, peaks in trap counts, and corresponding time series of entropy values showed relatively good agreement. In contrast, in 2013, when inoculum pressure was higher, generally, there was much less agreement between locations, and extrapolation from one location to another would not necessarily have yielded robust conclusions about the dynamics of the pathogen. The most striking example is the contrast between Salinas and King City S. In Salinas between day 20 and day 80, trap catches were relatively low and the cumulative value of the entropy showed a steady decline from approximately 1.5 nats to under 1 nat. In contrast, over the same period in King City, multiple peaks in trap catches were noted and entropy in the catch data rose from approximately 1 nat to approximately 2 nats.

Inevitably, in a first use of a new methodology in a specific field of application, there are numerous things that could have been done that were not. The focus of our analyses was on the dynamics and properties of the times series when analyzed on a daily time step. It is probably not surprising that the binary series indicating the direction of change was close to its maximum entropy value over much of the season in both years at most locations. This result suggests that, on average on any given day, a sample from the next day is as likely to be higher as it is to be lower than the sample from the current day. Technical advances in sample preparation are reducing the time it takes to process nucleic acid samples from spore traps. As a consequence, the apparent possibility of real-time forecasting of disease risk on a daily basis is increasing. One possible interpretation of our findings is, however, that the usefulness of such forecasts may be limited by the inherent uncertainty of the data. Is there predictive value for decision making in knowing today's trap value if tomorrow's value may be higher or lower with equal probability? The results obtained here indicate that the binary strings derived from the time series data are close to being simple sequences of independent Bernoulli trials with a probability of 0.5 determining the outcome. As Grünwald [27] points out, model selection and fitting can be considered as analogous to data compression, and when a string of bits is essentially random, it is difficult to achieve an accurate description (i.e., compression) of the data that is more concise than simply writing the data out.

Aggregating results to form moving averages over longer runs of days would perhaps lead to information values that were more easily linked to disease outcomes, but the detailed work to examine this issue lies beyond the scope of this study. The objective of our work was not to explore whether entropy values can be used as a predictive indicator for disease risk but to characterize the uncertainty of time series data from spore traps to give practitioners a richer perspective on the level and nature of the uncertainty inherent in the data they collect.

Looking at the entropy values for the two seasons retrospectively, it is clear (Figure 1) that the cumulative entropy values have a qualitatively different nature in the two seasons. Moving from consideration of uncertainty within a season to differences in uncertainty between seasons, the results presented here suggest that information quantities might provide an alternative means of classifying growing seasons, but the extra value to epidemiology (compared with what can be learned from direct comparison of the trap data themselves, for example) that might be gained from long-term comparative analyses is not known at this time. We hope others will be encouraged to further analyze the open questions raised here; this is a new field for future research.

Our analyses suggest that there may be quite severe practical limitations to being able to characterize pathogen dynamics using the combination of vortex air sampling and DNA target amplification. There are clear cases where detection of primary inoculum helps to improve disease management [9,11], but the situation regarding season-long disease management based on measuring secondary inoculum is much less clear. There are few, if any, other published datasets for comparison so the potential remains uncertain. However, even with data series extending to over 100 data points, the fact that the coefficient of prediction for autoregressive models was close to zero is an indication that the time series may be so noisy that extracting a useful, succinct model of the dynamics may be difficult. While our results point to restrictions in the utility of dynamical analysis for helping with practical problems in disease risk forecasting, at the same time, they suggest a great deal of interesting investigative research on inoculum dynamics and their positions on the continuum form pure deterministic complexity to pure stochastic noise.

Supplementary Materials: The following are available online at https://github.com/robchoudhury/spore_trap_information_theory, R code to reproduce results for Salinas, 2013:Sal_13.R, Data required to run the analysis for Salinas 2013: spore_fill.csv, R Project file:spore_trap_information_theory.Rproj.

Author Contributions: Conceptualization, N.M. and R.A.C.; methodology, N.M. and R.A.C.; software, N.M. and R.A.C.; formal analysis, N.M.; resources, N.M. and R.A.C.; data curation, R.A.C.; writing—original draft preparation, N.M.; writing—review and editing, R.A.C.; visualization, N.M. and R.A.C.; funding acquisition, N.M. and R.A.C. All authors have read and agreed to the published version of the manuscript.

Entropy **2020**, 22, 1343

Funding: This research received no external funding.

Acknowledgments: Work by N.M. on this paper is aligned with USDA-NIFA Hatch project CA-D-PPA-2131-H. R.C. was supported by funds from the University of Texas, Rio Grande Valley.

Conflicts of Interest: The authors declare no conflict of interest.

Appendix A

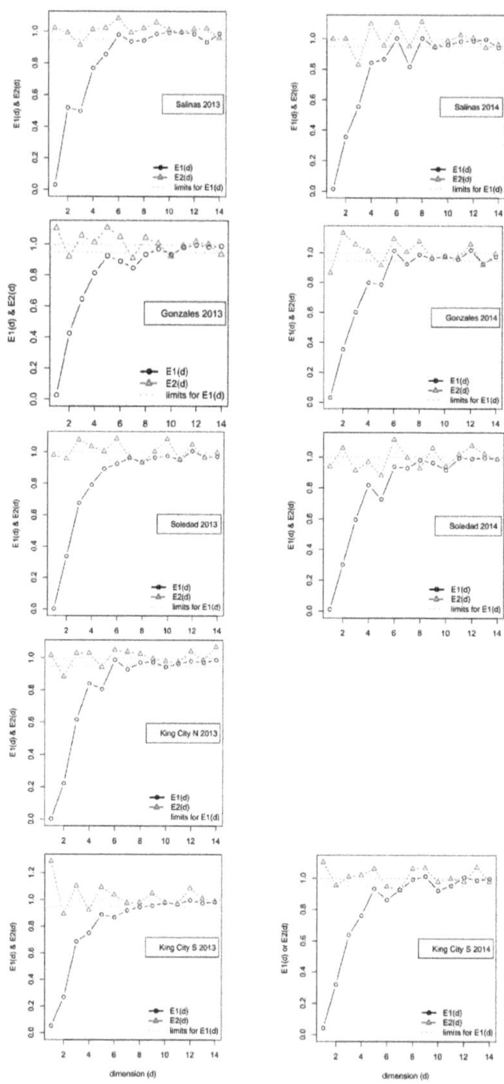

Figure A1. Output from the R function *estimateEmbeddingDim* for each series: the dimension, d, at which function E1 first falls within the critical region around 1 is taken to be the embedding dimension. The behavior of function E2 indicates the nature of the series. E2 is approximately equal to 1 for all values of d for all series, indicating that the series are stochastic in nature. Note that the embedding dimension is denoted with the letter *m* in the main text of the paper.

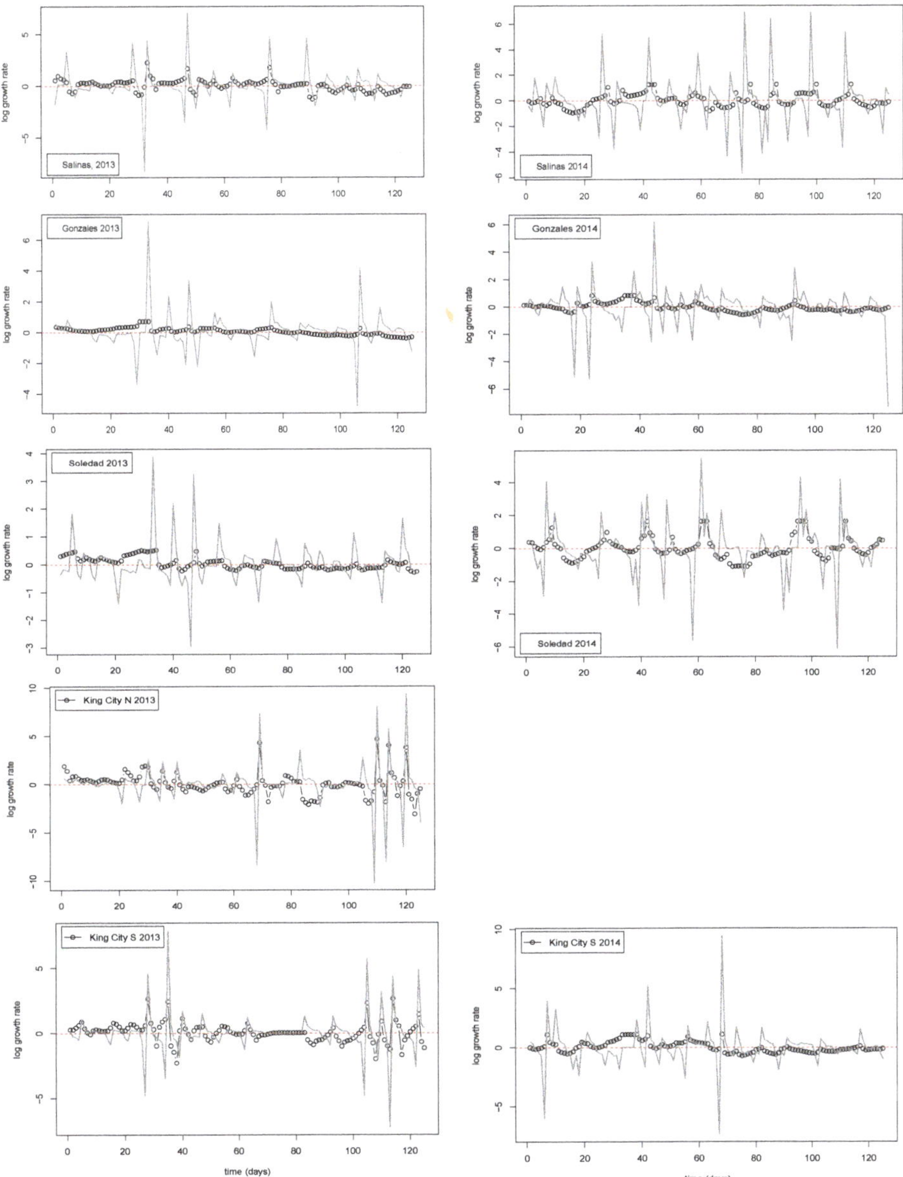

Figure A2. Correlation matrix plot indicating the numerical value and direction of correlations among the summary variables for time series properties across the 9 example times series. Abbreviations: ACF, autocorrelation function; PACF, partial autocorrelation function; AMI, average mutual information; m estimated embedding dimension; ly1, dominant Lyapunov exponent; Enctopy_CN, entropy (nats) for the pathogen DNA copy number; Entropy bin, entropy (bits) for the binary series indicating whether first differences in trap copy numbers are positive; VAF, adjusted percentage variance accounting for (adjusted R2) the autoregressive linear model, Coeff_pred, prediction coefficient comparing the autoregressive model with the mean of the series used as a simple predictor.

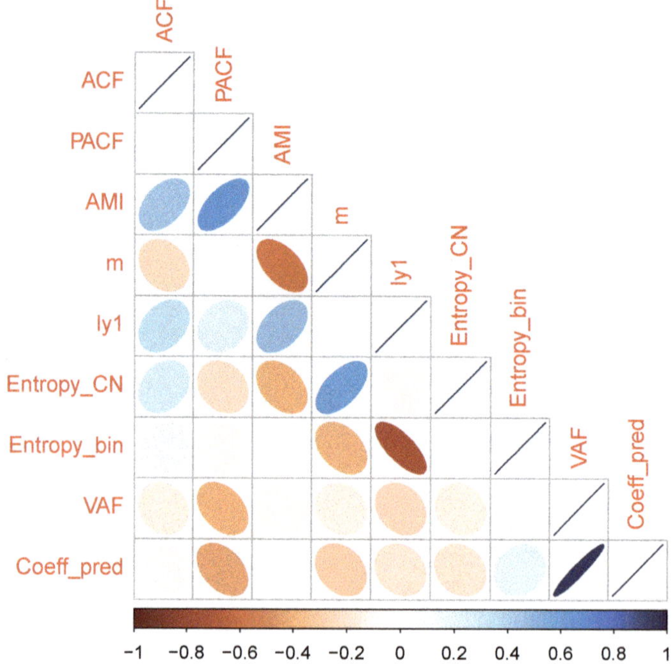

Figure A3. Observed data and fitted values for linear autoregressive models fitted to Equation (1) in the main text in the case of each of the nine series. Observed data: gray dashed line with open symbols; fitted values: black open symbols. The simple autoregressive models are capable of capturing some of the dynamic behavior in the series but generally lack the amplitude of the observed data and are poor at representing abrupt changes from large positive to large negative growth rates.

References

1. May, R.M. *Stability and Complexity in Model Ecosystems*; Princeton University Press: Princeton, NJ, USA, 2019; Volume 1.
2. Madden, L.V.; Hughes, G.; Van Den Bosch, F. *The Study of Plant Disease Epidemics*; American Phytopathological Society: St. Paul, MN, USA, 2007.
3. Zwankhuizen, M.; Zadoks, J. *Phytophthora infestans*'s 10-year truce with Holland: A long-term analysis of potato late-blight epidemics in the Netherlands. *Plant Pathol.* **2002**, *51*, 413–423. [CrossRef]
4. Kriss, A.; Paul, P.; Madden, L. Relationship between yearly fluctuations in Fusarium head blight intensity and environmental variables: A window-pane analysis. *Phytopathology* **2010**, *100*, 784–797. [CrossRef] [PubMed]
5. Carisse, O.; McRoberts, N.; Brodeur, L. Comparison of monitoring-and weather-based risk indicators of botrytis leaf blight of onion and determination of action thresholds. *Can. J. Plant Pathol.* **2008**, *30*, 442–456. [CrossRef]
6. Choudhury, R.; Koike, S.; Fox, A.; Anchieta, A.; Subbarao, K.; Klosterman, S.; McRoberts, N. Season-long dynamics of spinach downy mildew determined by spore trapping and disease incidence. *Phytopathology* **2016**, *106*, 1311–1318. [CrossRef] [PubMed]
7. Kasprzyk, I.; Worek, M. Airborne fungal spores in urban and rural environments in Poland. *Aerobiologia* **2006**, *22*, 169. [CrossRef]
8. Klosterman, S.J.; Anchieta, A.; McRoberts, N.; Koike, S.T.; Subbarao, K.V.; Voglmayr, H.; Choi, Y.-J.; Thines, M.; Martin, F.N. Coupling spore traps and quantitative PCR assays for detection of the downy mildew pathogens of spinach (*Peronospora effusa*) and beet (*P. schachtii*). *Phytopathology* **2014**, *104*, 1349–1359. [CrossRef] [PubMed]

9. Carisse, O.; Tremblay, D.; Lévesque, C.; Gindro, K.; Ward, P.; Houde, A. Development of a TaqMan real-time PCR assay for quantification of airborne conidia of Botrytis squamosa and management of Botrytis leaf blight of onion. *Phytopathology* **2009**, *99*, 1273–1280. [CrossRef] [PubMed]

10. Carisse, O.; Tremblay, D.; McDonald, M.; Brodeur, L.; McRoberts, N. Management of Botrytis leaf blight of onion: The Québec experience of 20 years of continual improvement. *Plant Dis.* **2011**, *95*, 504–514. [CrossRef] [PubMed]

11. Falacy, J.; Grove, G.; Mahaffee, W.; Galloway, H.; Glawe, D.; Larsen, R.; Vandemark, G. Detection of *Erysiphe necator* in air samples using the polymerase chain reaction and species-specific primers. *Phytopathology* **2007**, *97*, 1290–1297. [CrossRef] [PubMed]

12. Huffaker, R.; Bittelli, M.; Rosa, R. *Nonlinear Time Series Analysis with R*; Oxford University Press: Oxford, UK, 2017.

13. Kantz, H.; Schreiber, T. *Nonlinear Time Series Analysis*; Cambridge University Press: Cambridge, UK, 2004; Volume 7.

14. Garcia, C.; Sawitzki, G. Nonlineartseries: Nonlinear Time Series Analysis. In R Package Version 0.2. 2020, Volume 3. Available online: https://cran.r-project.org/web/packages/nonlinearTseries/index.html (accessed on 24 November 2020).

15. Fabio Di Narzo, A. TseriesChaos: Analysis of Nonlinear Time Series. R Package Version 0.1-13. 2019. Available online: https://cran.r-project.org/web/packages/tseriesChaos/tseriesChaos.pdf (accessed on 24 November 2020).

16. Giannerini, S. TseriesEntropy: Entropy Based Analysis and Tests for Time Series. R Package Version 0.5-13. 2017. Available online: https://cran.r-project.org/web/packages/tseriesEntropy/tseriesEntropy.pdf (accessed on 24 November 2020).

17. Hausser, J.; Strimmer, K. Entropy: Estimation of Entropy, Mutual Information and Related Quantities. R Package Version. 2015, Volume 1. Available online: https://cran.r-project.org/web/packages/entropy/entropy.pdf (accessed on 24 November 2020).

18. Granger, C.W.; Maasoumi, E.; Racine, J. A dependence metric for possibly nonlinear processes. *J. Time Ser. Anal.* **2004**, *25*, 649–669. [CrossRef]

19. Takens, F. Detecting strange attractors in turbulence. In *Dynamical Systems and Turbulence*; Springer: Berlin/Heidelberg, Germany, 1981; pp. 366–381.

20. Theiler, J. Spurious dimension from correlation algorithms applied to limited time series data. *Phys. Rev. A* **1986**, *34*, 2427. [CrossRef] [PubMed]

21. Theiler, J. Estimating fractal dimension. *JOSA A* **1990**, *7*, 1055–1073. [CrossRef]

22. Provenzale, A.; Smith, L.A.; Vio, R.; Murante, G. Distinguishing between low-dimensional dynamics and randomness in measured time series. *Phys. D Nonlinear Phenom.* **1992**, *58*, 31–49. [CrossRef]

23. Casdagli, M.; Eubank, S.; Farmer, J.; Gibson, J.; Desjardins, D.; Hunter, N.; Theiler, J. Nonlinear modeling of chaotic time series: Theory and applications. In Proceedings of the Electric Power Research Institute (EPRI) Workshop on Applications of Chaos, San Francisco, CA, USA, 4–7 December 1990.

24. Cao, L. Practical method for determining the minimum embedding dimension of a scalar time series. *Phys. D Nonlinear Phenom.* **1997**, *110*, 43–50. [CrossRef]

25. Royama, T. *Analytical Population Dynamics*; Springer Science & Business Media: Berlin/Heidelberg, Germany, 2012; Volume 10.

26. Turchin, P. *Complex. Population Dynamics: A Theoretical/Empirical Synthesis*; Princeton University Press: Princeton, NJ, USA, 2003; Volume 35.

27. Grünwald, P.G. *The Minimum Description Length Principle*; The MIT Press: Cambridge, MA, USA, 2007.

Publisher's Note: MDPI stays neutral with regard to jurisdictional claims in published maps and institutional affiliations.

Article

An Information-Theoretic Measure for Balance Assessment in Comparative Clinical Studies

Jarrod E. Dalton [1,*], **William A. Benish** [2] **and Nikolas I. Krieger** [1]

[1] Department of Quantitative Health Sciences, Cleveland Clinic, Cleveland Clinic Lerner College of Medicine at Case Western Reserve University, 9500 Euclid Avenue, Cleveland, OH 44126, USA; kriegen@ccf.org

[2] Department of Internal Medicine, Case Western Reserve University, Cleveland, OH 44106, USA; wab4@cwru.edu

* Correspondence: daltonj@ccf.org

Received: 4 December 2019; Accepted: 12 February 2020; Published: 15 February 2020

Abstract: Limitations of statistics currently used to assess balance in observation samples include their insensitivity to shape discrepancies and their dependence upon sample size. The Jensen–Shannon divergence (JSD) is an alternative approach to quantifying the lack of balance among treatment groups that does not have these limitations. The JSD is an information-theoretic statistic derived from relative entropy, with three specific advantages relative to using standardized difference scores. First, it is applicable to cases in which the covariate is categorical or continuous. Second, it generalizes to studies in which there are more than two exposure or treatment groups. Third, it is decomposable, allowing for the identification of specific covariate values, treatment groups or combinations thereof that are responsible for any observed imbalance.

Keywords: balance; Jensen–Shannon divergence; observational study; relative entropy; selection bias

1. Introduction

The goal of comparative studies is to measure the effect of two or more treatment (or exposure) groups on an outcome. A potential source of bias in these studies is the association between the treatment groups and one or more confounding variables. Randomized clinical trials mitigate this risk through randomization of treatments, resulting in balanced groups with respect to the confounding variables. We say that the relationship between treatment T and outcome O is confounded by a covariate C if C is associated with O and T but is not a consequence of T (i.e., not a mediator of the effect of T on O) [1].

A common strategy for evaluating the potential for confounding in such a study is to identify all covariates that may meet these criteria and evaluate their association with T. When treatment groups T are balanced on a variable C, that is, when T and C are probabilistically independent, then C cannot confound the estimation of the relationship between T and O.

A variety of techniques are typically employed to assess balance in observational samples, including estimation of simple univariate descriptive statistics, univariate tests of association, and estimation of standardized difference scores (defined as the difference in means between groups divided by a combined estimate of standard deviation). Depending on the situation, however, each of these three approaches may lead to erroneous conclusions. Univariate descriptive statistics may not adequately capture complex distributions (e.g., those with multiple modes) [2]. Tests of association are heavily dependent on sample size, and thus can be as indicative of sample size as they are of imbalance. And standardized difference scores—despite their popularity—are not sensitive to discrepancies in higher order moments (e.g., skewness, kurtosis) and/or multimodalities among continuous distributions.

In this article, we propose the use of an information-theoretic measure known as the Jensen–Shannon divergence (JSD) [3] to assess treatment group balance. The JSD offers several advantages over the aforementioned approaches. First, it is universally defined for binary, multilevel, and continuous distributions (although, in practice, computation for continuous distributions is facilitated by binning the variables into a number of discrete levels), for any number of treatment groups, and for multivariate distributions (i.e., vectorized covariate values $\vec{C}$) across treatment groups. Second, it allows for the identification of specific levels of C or T—and, moreover, specific combinations of C and T—that contribute most to imbalances across groups or treatments in relation to others. And third, it is sensitive to high order imbalances (e.g., differences in variability, skewness, bimodality, etc.) in addition to location shifts.

A brief introduction to information theory and the JSD is presented in the next section of this report (Section 2). Properties of the JSD as a measure of covariate imbalance are discussed in Section 3. Examples are presented in Section 4. We conclude with a brief summary (Section 5).

2. Information Theory and the JSD

The JSD is an information-theoretic measure of dissimilarity among two or more probability distributions [3]. It is derived from relative entropy (or Kullback–Leibler divergence) [4] and is therefore related to mutual information [5] (pp. 18–21). These measures are fundamentally tied to Shannon's entropy [6]. The goal of this section is to describe the JSD in intuitive terms, beginning with the definition of entropy.

2.1. Entropy

Let X be a discrete random variable which takes on values $x_i \in \{x_1, x_2, \ldots, x_M\}$. Let the probability distribution of X be denoted as $f(X)$. The entropy of X, denoted $H(X)$, is a measure of the uncertainty of the outcome of X and is defined as:

$$H(X) = E(-\log_2 f(X)) = -\sum_{i=1}^{M} f(x_i) \log_2 f(x_i). \tag{1}$$

The base of the logarithm is arbitrary. Log base two is often used, giving entropy units of bits (binary digits).

One approach to understanding the concept of entropy is to explore its relationship to the average number of bits (e.g., 0 s and 1 s) required to efficiently encode a sequence of outcomes of the random variable. Consider, for example, the case where the sample space is $\{A, B, C, D\}$ with corresponding probabilities $f(X) = \{0.25, 0.125, 0.5, 0.125\}$. With four possible outcomes, it may be tempting to encode a single outcome using two bits, e.g., $00 \rightarrow A$, $01 \rightarrow B$, $10 \rightarrow C$, and $11 \rightarrow D$. A more efficient mapping is $0 \rightarrow C$, $10 \rightarrow A$, $110 \rightarrow B$, and $111 \rightarrow D$. Since A, B, C, and D have probabilities of 0.25, 0.125, 0.5, and 0.125, respectively, and are encoded with 2, 3, 1, and 3 bits, respectively, the expected value of the number of bits required to transmit the outcome of X with this coding scheme is 0.25×2 bits + 0.125×3 bits + 0.5×1 bit + 0.125×3 bits = 1.75 bits.

Shannon demonstrated that $H(X)$ defines a limit beyond which codes cannot be made more efficient. Using either of the above coding schemes allows for the unambiguous encoding of a series of outcomes of X, but the second scheme is optimal in that the expected number of bits required to transmit the outcome of X is $H(X) = 1.75$ rather than two. To achieve (or to become arbitrarily close) to the efficiency specified by $H(X)$ may require a mapping that associates each code with a sequence of outcomes of X [5] (p. 104). For example, in the case of two possible outcomes A and B, with respective probabilities 2/3 and 1/3, a code that is more efficient than simply $0 \rightarrow A$ and $1 \rightarrow B$ is $0 \rightarrow AA$, $10 \rightarrow AB$, $110 \rightarrow BA$, and $111 \rightarrow BB$. The length of the inefficient code required to indicate the outcome is 1 bit, but the average length of the more efficient code, per outcome, is 0.9444 bits (compared to the ideal of $H(X) = 0.9183$ bits).

2.2. Joint and Conditional Entropy

The joint entropy $H(X,Y)$ of two random variables is a natural extension of the concept of entropy for a single random variable:

$$H(X,Y) = E(-\log_2 f(X,Y)) = -\sum_{i=1}^{M}\sum_{j=1}^{N} f(x_i, y_j) \log_2 f(x_i, y_j). \tag{2}$$

Similar to that described above for a single random varible, the joint entropy defines the lower limit of the average number of bits required to encode the observations from the joint distribution.

Conditional entropy, denoted $H(X|Y)$, is a measure of residual uncertainty in X, given the observation of some other random variable Y. It is defined as:

$$H(X|Y) = E(-\log_2 f(X|Y)) = -\sum_{i=1}^{M}\sum_{j=1}^{N} f(x_i, y_j) \log_2 f(x_i|y_j). \tag{3}$$

Conditional entropy is also equal to the difference between joint and marginal entropies, i.e., $H(X|Y) = H(X,Y) - H(Y)$. In this sense, conditional entropy represents the number of bits needed to encode X after the value of Y is observed. Joint and conditional entropy naturally extend to distributions that are defined across three or more random variables (we omit these equations for the purposes of this discussion).

2.3. Mutual Information

The mutual information between the random variables X and Y, denoted $I(X;Y)$, is the expected value of the amount of information that knowledge of the outcome of Y provides about the outcome of X. Mutual information is symmetric with respect to X and Y, and is a function of both the variables' marginal entropies and their joint entropy:

$$\begin{aligned} I(X;Y) \quad &= H(X) + H(Y) - H(X,Y) \\ &= H(X) - H(X|Y) \\ &= H(Y) - H(Y|X) \\ &= I(Y;X). \end{aligned} \tag{4}$$

2.4. Relative Entropy

Relative entropy is an information-theoretic measure expressing the divergence from a given probability distribution $f(X)$ to a reference (or target) distribution $g(X)$. It is defined as

$$D(g(X) \parallel f(X)) = E_{g(X)}\left[-\log_2 \frac{f(X)}{g(X)}\right] = \sum_{i=1}^{M} g(x_i) \log_2 \frac{g(x_i)}{f(x_i)}. \tag{5}$$

The relative entropy is interpreted as the number of bits required to "correct" the probabilities in the distribution f so that they match those of the reference distribution g (under an optimal coding scheme) [5] (p. 18).

Since the expectation in Equation (5) is taken with respect to the target distribution $g(X)$, the relative entropy function is asymmetric, i.e., it is not necessarily the case that $D(g(X) \parallel f(X)) = D(f(X) \parallel g(X))$. Given this asymmetry, it is not a suitable candidate for a measure of covariate balance among groups: the divergence between two groups would depend upon which group is taken to be the Reference group. Jeffrey's divergence (J) is a symmetric version of relative entropy, defined as $J(g(x); f(x)) = D(g(x) \| f(x)) + D(f(x) \| g(x))$ [7]. One reason why it is not a suitable candidate for the task of assessing covariate balance among groups is that there may be more than two groups.

2.5. Jensen–Shannon Divergence (JSD)

The JSD is a modified version of relative entropy, that addresses the asymmetry problem described above by expressing divergences with respect to a common distribution $\widetilde{f}(X)$. Assume that there are N distributions of X: $f_1(X)$, $f_2(X), \ldots, f_N(X)$. The common distribution is taken as the (unweighted) mean of the component densities:

$$\widetilde{f}(x) = \frac{1}{N} \sum_{k=1}^{N} f_k(x). \tag{6}$$

The JSD of the set of distributions $f_k(X)$ is defined as the average relative entropy from the common distribution $\widetilde{f}(X)$ to the specific distributions $f_k(X)$:

$$\mathrm{JSD} = \frac{1}{N} \sum_{k=1}^{N} D\big(f_k(X) \parallel \widetilde{f}(X)\big). \tag{7}$$

2.6. The JSD of Covariate Distributions Across Treatment Groups

Equations (6) and (7) can be modified to calculate the JSD for a set of N treatment groups. We replace the continuous random variable X with the discrete covariate random variable C. Similary, we replace the probability density function f with the probability mass function p. Assuming that C can assume M values, we have, for $i = 1, \cdots , M$:

$$\widetilde{p}(c_i) = \frac{1}{N} \sum_{k=1}^{N} p_k(c_i), \tag{8}$$

and

$$\mathrm{JSD} = \frac{1}{N} \sum_{k=1}^{N} D(p_k(C) \parallel \widetilde{p}(C)) = \frac{1}{N} \sum_{k=1}^{N} \sum_{i=1}^{M} p_k(c_i) \log_2\left(\frac{p_k(c_i)}{\frac{1}{N} \sum_{k=1}^{N} p_k(c_i)} \right). \tag{9}$$

3. Properties of the JSD

The JSD is non-negative and is equal to zero when the covariate distributions are identical for all treatment groups. It is interpreted as the average relative entropy from the common covariate distribution, $\widetilde{f}(C)$, to the group-specific distributions. As noted in the Introduction, the JSD can be applied to binary random variables, categorical random variables, or continuous random variables.

Being defined additively in terms of units of information, the JSD is decomposable. One may calculate the JSD across all the treatment groups or determine the contribution of a subset of groups to the overall JSD. Similarly, specific levels of the covariate(s) of interest may be examined to identify regions of the covariate space exhibiting the greatest degree of imbalance across groups. Furthermore, contributions of individual treatment/covariate combinations to the overall JSD can be studied and compared. The decomposability of the JSD is illustrated in Section 4.

As a function of the densities themselves (and not their moments), the JSD allows for the evaluation of balance in a manner that does not assume that continuous densities belong to any particular family of distributions. It is sensitive to shape discrepancies among groups. In contrast, the standardized difference score converges to zero (with increasing group sample sizes) whenever the means of the two samples are equal (see Figure 1).

In practice, computation of the JSD using observational data can be difficult for continuous densities, especially mixture distributions [2]. Our approach relies on the binning of continuous variables (as is done with histograms). When small numbers of categories are used, this simplification can mask subtle features of group-specific probability densities. A further limitation of the JSD is that density estimates for categorical variables are increasingly variable among small samples.

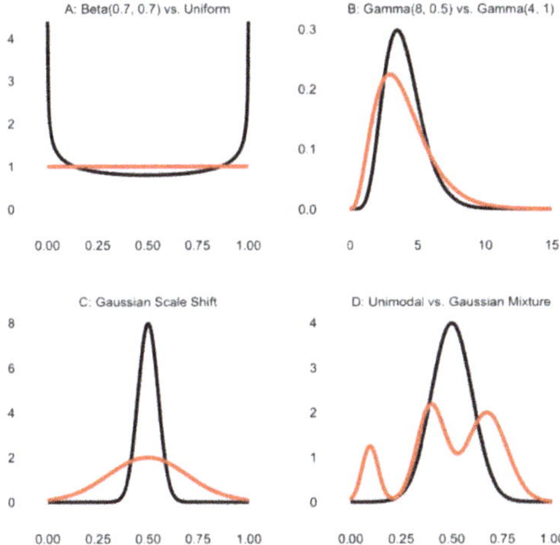

Figure 1. Four pairs of continuous distributions, each of which has a standardized difference score equal to zero.

4. Applications

Table 1 summarizes findings from 93,583 outpatients in the Cleveland Clinic Health System who had a lipid panel drawn between 2007 and 2010 (first visit meeting these criteria). The patients are partitioned into three treatment groups: Disadvantaged (age < 80 years and living in a census tract that is in the top 25% of all tracts in the United States with respect to the Area Deprivation Index [8]), Elderly (not living in a disadvantaged neighborhood per the above definition but aged 80 or older), and Reference (neither disadvantaged nor elderly). The covariate is baseline diabetes state defined by blood sugars < 109 mg/dL, 109–125 mg/dL, and > 125 mg/dL. A stand-alone R package for implementing the JSD computations illustrated in this section is provided at http://github.com/jarrod-dalton/jsd, and the code used for this section is given in the Appendix A.

Table 1. Number of individuals in three treatment groups (Disadvantaged, Elderly, Reference) and three covariate groups (defined by blood sugar ranges).

Glucose	Disadvantaged	Elderly	Reference
<109	7191	3637	64,265
109–125	1025	835	7298
>125	1715	685	6932

Table 2 presents the probability distributions of glucose levels within each treatment group. The average of these distributions, i.e., the common distribution, $\widetilde{f}(C)$, is shown in the final column.

Table 2. Probability distributions of glucose levels within each treatment group (Disadvantaged, Elderly, Reference). The common distribution, $\widetilde{f}(C)$, is shown in the final column.

Glucose	Disadvantaged	Elderly	Reference	$\widetilde{f}(C)$
<109	0.724	0.705	0.819	0.749
109–125	0.103	0.162	0.093	0.119
>125	0.173	0.133	0.088	0.131

Table 3 presents contributions of individual cells, the three treatment groups, and the three covariate groups to the overall JSD, which is 0.0144 bits. This is the average of the relative entropies from the common distribution to the treatment group-specific distributions. Given three treatment groups, the maximum possible JSD is $\log_2(3) = 1.5850$ bits.

Table 3. Contributions of individual cells, treatment groups, and levels of the covariate to the overall JSD (in units of bits).

Glucose	Disadvantaged	Elderly	Reference	Total
<109	−0.0119	−0.0206	0.0349	0.0023
109–125	−0.0072	0.0237	−0.0112	0.0053
>125	0.0228	0.0008	−0.0168	0.0067
Total	0.0036	0.0039	0.0068	0.0144 *

* Note: row/column sums do not equal 0.0144 due to rounding error.

The Reference group is the largest treatment group contributor to the JSD, and the Glucose > 125 category is the largest covariate group contributor to the JSD. Moreover, by considering the absolute values of the individual cell components, we conclude that the largest contributor to the JSD is from individuals in the Reference group with serum glucose values less than 109 mg/dL.

A problem with using any method to quantify covariate imbalance among treatment groups is that there is no obvious point that defines an acceptable amount of imbalance [9]. For the current example, the JSD value of 0.0144 bits is small relative to its maximal possible value of 1.5850 bits, but it is clear from Table 2 that individuals in the Reference group tend to have lower blood sugars than individuals in the other two treatment groups. An important factor in deciding what constitutes acceptable balance is the potential of the covariate to affect the outcome [10].

In order to further examine differences between the JSD and standardized difference scores, we consider the case in which there are two treatment groups with normally distributed covariates. Figure 2 plots the JSD as a function of the standardized difference score, when the standard deviation of one of the two distributions is one and the standard deviation of the other distribution is either one (plotted in black), two (plotted in blue), or three (plotted in red). Since there are two treatment groups, the JSD curves asymptote at one bit (since $\log_2(2) = 1$). The standardized difference score curves, on the other hand, are unbounded in the positive direction. As expected, both the JSDs and the standard difference scores increase as the two distributions diverge. The plot also illustrates the point made in Section 3 that the JSD, but not the standardized difference score, is sensitive to differences between the standard deviations of the two distributions when the means of two distributions are identical.

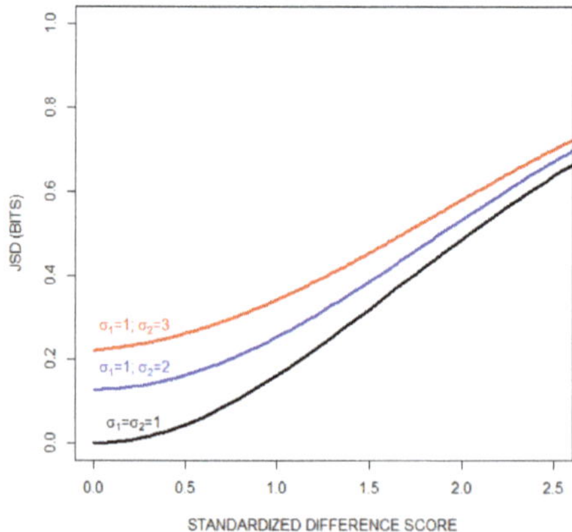

Figure 2. The JSD as a function of the standardized difference score when there are two treatment groups with normally distributed covariates. Three cases are shown: the standard deviation of one of the two distributions is set equal to one, while the standard deviation of the second distribution is set to equal either one (black curve), two (blue curve), or three (red curve).

5. Summary

We propose that the JSD be used to assess treatment group balance on known potential confounding variables in comparative clinical studies. This information-theoretic measure is equal to the average relative entropy between the covariate distributions for each treatment group and a common distribution, defined as the average of the individual distributions. Advantages of the JSD over alternative measures of treatment group balance include its sensitivity to the shape of distributions and its insensitivity to sample size. The JSD is applicable to both categorical and continuous random variables. Moreover, the JSD is decomposable, allowing for comparisons among specific levels of covariates of interest.

Author Contributions: Conceptualization, J.E.D. and W.A.B.; writing—original draft preparation, J.E.D. and W.A.B.; writing—review and editing, J.E.D. and W.A.B.; funding acquisition, J.E.D; software, J.E.D., N.I.K. and W.A.B. All authors have read and agreed to the published version of the manuscript.

Funding: Research reported in this publication was supported by The National Institute on Aging of the National Institutes of Health under award number R01AG055480. The content is solely the responsibility of the authors and does not necessarily represent the official views of the National Institutes of Health.

Acknowledgments: The authors are grateful to the reviewers for their helpful suggestions.

Conflicts of Interest: The funders had no role in the design of the study; in the collection, analyses, or interpretation of data; in the writing of the manuscript, or in the decision to publish the results.

Appendix A

The jsd R package can be found at https://github.com/jarrod-dalton/jsd. The package can be installed using the following command (run the first command if the remotes package has not already been installed):

```r
#install.packages("remotes")
remotes::install_github("jarrod-dalton/jsd")
```

The library is then loaded as follows:

```r
library(jsd)
```

The glucose dataset contains the data used for the example in Section 4. Note that the actual glucose values are simulated.

```r
data(glucose)
head(glucose)
##      cohort    glucose
## 1 Reference   79.16898
## 2   Elderly  124.66537
## 3   Elderly   95.61820
## 4 Reference   90.11065
## 5 Reference  105.50799
## 6 Reference   89.35060
```

There is a helper function in the package, called chop, which will convert numeric variables into categorical variables. See help(chop) for details. Here, we convert the glucose variable into a categorical variable with 3 levels:

```
glucose$glucose_cat <- chop(glucose$glucose, cuts = c(0, 109, 125, Inf))
```

The jsd_balance function is then used to compute the JSD measures. The output of the jsd_balance function contains the cell contributions to the JSD, marginal contributions of each treatment group to the JSD, marginal contributions of each covariate level to the JSD and the overall JSD value (see Table 3 for details). The first argument to the jsd_balance function is a formula in which the group variable is on the left hand side of the tilde and the covariate(s) is/are on the right hand side of the tilde (separated by '+' – see help(jsd_balance) for details and examples):

```
jsd_balance(cohort ~ glucose_cat, data = glucose)
## $glucose_cat
## $freqs
##                cohort
## glucose_cat Disadvantaged Elderly Reference
##   [  0,109)          7191    3637     64265
##   [109,125)          1025     835      7298
##   [125,Inf]          1715     685      6932
##
## $cell_contribs
##                cohort
## glucose_cat Disadvantaged         Elderly       Reference
##   [  0,109)    -0.0119399530 -0.0205710889   0.0348528681
##   [109,125)    -0.0072178122  0.0237387920  -0.0111726329
##   [125,Inf]     0.0227710815  0.0007506513  -0.0168364804
##
## $group_contribs
## Disadvantaged       Elderly     Reference
##    0.003613316   0.003918354   0.006843755
##
## $cov_contribs
##      [  0,109)    [109,125)      [125,Inf]
## 0.002341826   0.005348347   0.006685252
##
## $jsd
## [1] 0.01437543
##
## attr(,"class")
## [1] "jsd_balance"
```

Entropy **2020**, *22*, 218

References

1. Friedman, L.M.; Furberg, C.D.; DeMets, D.L.; Reboussin, D.M.; Granger, C.B. *Fundamentals of Clinical Trials*, 5th ed.; Springer: New York, NY, USA, 2010.
2. Contreras-Reyes, J.E.; Cortés, D.D. Bounds on Rényi and Shannon Entropies for Finite Mixtures of Multivariate Skew-Normal Distributions: Application to Swordfish (Xiphias gladius Linnaeus). *Entropy* **2016**, *18*, 382. [CrossRef]
3. Lin, J. Divergence measures based on the Shannon entropy. *IEEE Trans. Inform. Theory* **1991**, *37*, 145–151. [CrossRef]
4. Kullback, S.; Leibler, R.A. On information and sufficiency. *Ann. Math. Stat.* **1951**, *2*, 79–86. [CrossRef]
5. Cover, T.M.; Thomas, J.A. *Elements of Information Theory*; John Wiley & Sons, Inc.: Hoboken, NJ, USA, 2012.
6. Shannon, C.E. A mathematical theory of communication. *Bell Syst. Tech. J.* **1948**, *27*, 379–423. [CrossRef]
7. Nielsen, F. On the Jensen-Shannon symmetrization of distances relying on abstract means. *Entropy* **2019**, *21*, 485. [CrossRef]
8. Kind, A.J.H.; Buckingham, W.R. Making Neighborhood-Disadvantage Metrics Accessible—The Neighborhood Atlas. *N. Engl. J. Med.* **2018**, *378*, 2456–2458. [CrossRef] [PubMed]
9. Austin, P.C. Using the standardized difference to compare the prevalence of a binary variable between two groups in observational research. *Commun. Stat.-Simul. Comput.* **2009**, *38*, 1228–1234. [CrossRef]
10. Ho, D.E.; Imai, K.; King, G.; Stuart, E.A. Matching as nonparametric preprocessing for reducing model dependence in parametric causal inference. *Political Anal.* **2007**, *15*, 199–236. [CrossRef]

MDPI

St. Alban-Anlage 66

4052 Basel

Switzerland

Tel. +41 61 683 77 34

Fax +41 61 302 89 18

www.mdpi.com

Entropy Editorial Office

E-mail: entropy@mdpi.com

www.mdpi.com/journal/entropy